Introduction to SOLID-STATE Electronics

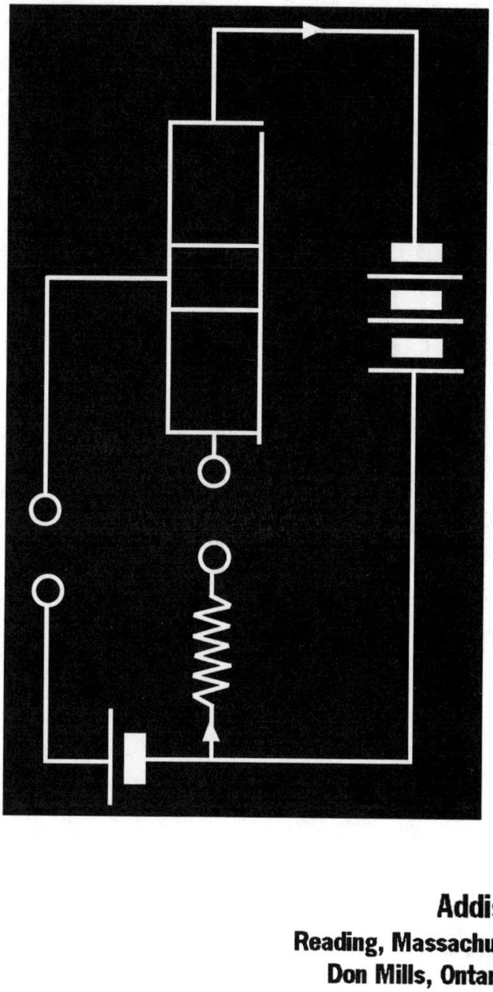

Ia Ipatova
Physical Technical Institute
Russian Academy of Sciences

and

Vladimir Mitin
Wayne State University
Detroit, Michigan

Addison-Wesley Publishing Company, Inc.
Reading, Massachusetts Menlo Park, California New York
Don Mills, Ontario Harlow, England Amsterdam Bonn
Sydney Singapore Tokyo Madrid San Juan
Paris Seoul Milan Mexico City Taipei

Jacket design by Suzanne Heiser
Text was typeset by the authors

ISBN 0-201-47962-1

1 2 3 4 5 6 7 8 9 10—MA—0099989796
First printing, September 1996

To Vera, Lioudmila, and Oleg — VM

To Peter, Konstantin, Alexei, Katya, and Yakov — II

Contents

Preface

The frontiers of all branches of physics are constantly shifting. What is a discovery today becomes common knowledge tomorrow. Therefore, there is a permanent need for new text books.

The development of semiconductor physics in the last decades brought us a clear understanding of the band structure of semiconductors on the basis of symmetry analysis. The concepts of nonparabolicity and effects of band degeneracy, became objects of common knowledge.

The study of effects of the electron-electron interaction resulted in a clear understanding of screening effects in semiconductors.

Disordered systems gave rise to the concepts of localized and delocalized states of electrons and phonons. Heavily doped semiconductors appear to be materials with metallic conductivity.

There are new levels of understanding of the physical properties of semiconductor surfaces, which display a strong tendency to reconstruct.

New types of semiconductor nanostructures with special low-dimensional properties have become the objects of very intensive study, which demonstrate the opportunities for the fabrication of new semiconductor devices.

The aim of this text book is to combine the modern presentation of semiconductor physics with a description of the principles of semiconductor devices. Since a crystal is an object of high symmetry, some simple techniques are used for the application of symmetry to the analysis of band structures which do not require knowledge of the mathematical theory of groups at professional level. Symmetry arguments are used everywhere where this is possible.

The special feature of our text book consists of careful simple derivations and detailed physical interpretations of equations, with special attention to the limits of applicability of these equations. We have tried to make the mathematical derivations both as simple as possible and complete in themselves without using arguments of the type "it can be easily shown that ...". This book will give an understanding of the main properties of semiconductors and their relations to device structures. We believe that new developments in semiconductor physics and in the design of new devices are possible only on the basis of such knowledge.

The structure of the book is the following.

Since the syllabi of many engineering departments do not include quantum mechanics, an introduction to quantum physics is given in Chapter 1. Those who are acquainted with quantum mechanics can start directly with Chapter 2.

Chapter 2 presents an introduction to the crystal structure of solids.

Chapter 3 introduces the concepts of crystal symmetry, and covers the application of symmetry considerations to the study of the behavior of an electron in a crystal.

Chapter 4 treats the band structure of typical semiconductors. The origin of many-valley band structures, nonparabolicity, band degeneracy, and spin-orbit-interaction effects is explained with simple symmetry arguments. Special attention is paid to the band structure of semiconductor nanostructures: quantum wells, quantum wires, quantum dots, and superlattices.

Chapter 5 covers the dynamics of the crystal lattice from the point of view of symmetry. The concept of the phonon is introduced without the use of second quantization.

Chapter 6 is devoted to the equilibrium properties of semiconductors. The electron density of states is obtained for an arbitrary dependence of the electron energy on quasimomentum. The concept of density of states is applied to quantum wells, quantum wires, and quantum dots.

Chapter 7 considers the electron-electron interaction in semiconductors. Effects of impurity atom screening, plasma oscillations, and the behavior of heavily doped semiconductors are discussed.

Chapters 8 gives a review of kinetic properties of semiconductors on the basis of the Boltzmann kinetic equation. Special attention is paid to anisotropic surface conductivity.

Chapters 9 and 10 treat the physics of excess carriers in nonequilibrium semiconductors, and the physical principles of p-n junctions, heterojunctions, metal-semiconductor junctions, bipolar transistors, and field-effect transistors.

Chapter 11 is devoted to optical properties of semiconductors. The concepts of exciton and selection rules defined by the band structure symmetry are also considered. This chapter ends with a discussion of semiconductor lasers.

Our textbook is intended for a one semester course of solid state electronics for graduate students training in engineering, and for research physicists working in this field. Being a text book, it contains almost no references to original papers. For information of this type, students should consult the monographs and reviews listed at the end of each chapter.

The course was developed and taught at Wayne State University. It unites a Russian teaching experience with the requirements of an introductory graduate course in Solid State Electronics for American engineering students. This material was distributed among the students in the form of handouts, and their feedback has been taken into account in the final version.

The authors are grateful to the Department of Electrical and Computer Engineering at Wayne State University, which provided the creative atmosphere which made this work possible, to C. Hughes, R. Gaska, S. Kersulis, and V. Korobov for their help in the preparation of the manuscript.

The authors are especially thankful to Professor M.I. Kaganov for valuable remarks concerning Chapters 1-4. We would like also to acknowledge Professor A. Maradudin of the Department of Physics and Astronomy of the University of California, Irvine, for editing our text and for many valuable remarks. We would also like to thank the Addison-Wesley Publishing Company for their help in completing this work. Last, but not least, we are grateful to our families and friends for their encouragement and help during this lengthy work, and we express our apologies for missing maybe too many cups of tea and coffee with them.

CHAPTER 1

Elements of Quantum Mechanics

1.1 EXPERIMENTAL BACKGROUND

As did all the branches of physics, quantum mechanics grew out of certain experimental observations. Probably the most important from today's point of view are the following three experiments.

A. The problem of Blackbody Radiation

All bodies emit radiation whose intensity depends on their temperature. The relation between the temperature of a body and the frequency spectrum of the emitted radiation reflects, generally speaking, the structure of the radiative body. But there is a class of hot bodies that emit spectra of universal form. These are bodies that absorb all the radiation incident on them. They are called ***blackbodies.*** Blackbodies don't reflect the light and hence look black.

An example of a blackbody used in laboratory experiments is a cavity which is connected to the outside world by a small hole, see Fig. 1.1.1. The radiation incident

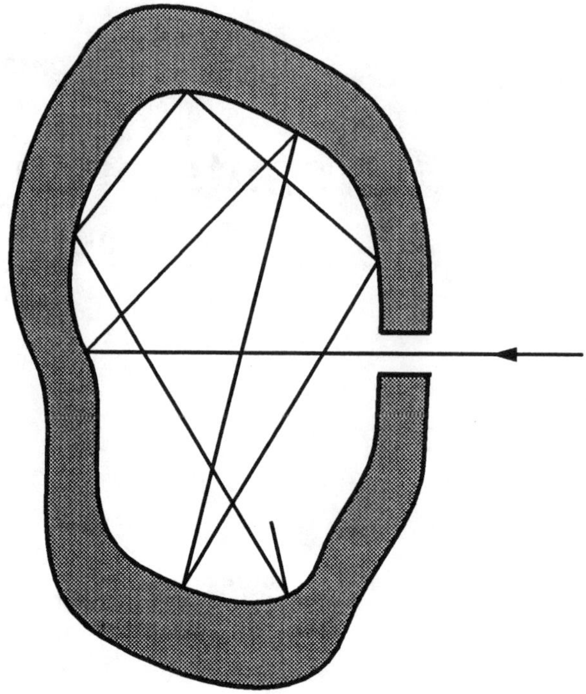

Figure 1.1.1 A cavity in a body connected by a small hole to the outside. Radiation incident on the hole is completely absorbed as a result of successive reflection from the inner surface of the cavity. The hole absorbs and emits like a blackbody.

through the hole undergoes multiple reflections from the inside walls of the cavity and, finally, is absorbed by the walls of the cavity. The hole itself has the properties of a blackbody. If the temperature of the cavity is T, the radiation out of the hole is typical of this temperature, since it is the only temperature in the system. Therefore, it is ***equilibrium radiation***.

The spectrum of the radiation emitted by the hole in the cavity is defined by the ***spectral radiancy*** $U(\omega, T)$, which is equal to the energy emitted per unit time in the frequency interval ω to $\omega + d\omega$ from a unit area of the surface at temperature T. A typical experimental curve of blackbody radiation is depicted in Fig.1.1.2 by the solid line. Rayleigh and Jeans calculated $U(\omega, T)$ by counting the number of standing electromagnetic waves in the cavity and using the fact that each vibrational degree of freedom carries the energy $k_B T$, where k_B is the Boltzmann constant. The corresponding result is depicted in Fig.1.1.2 by the dashed line. There is an obvious discrepancy with the solid experimental curve. They also calculated the total energy of the radiation and found that it is infinite

$$\int U(\omega, T) d\omega = \infty.$$

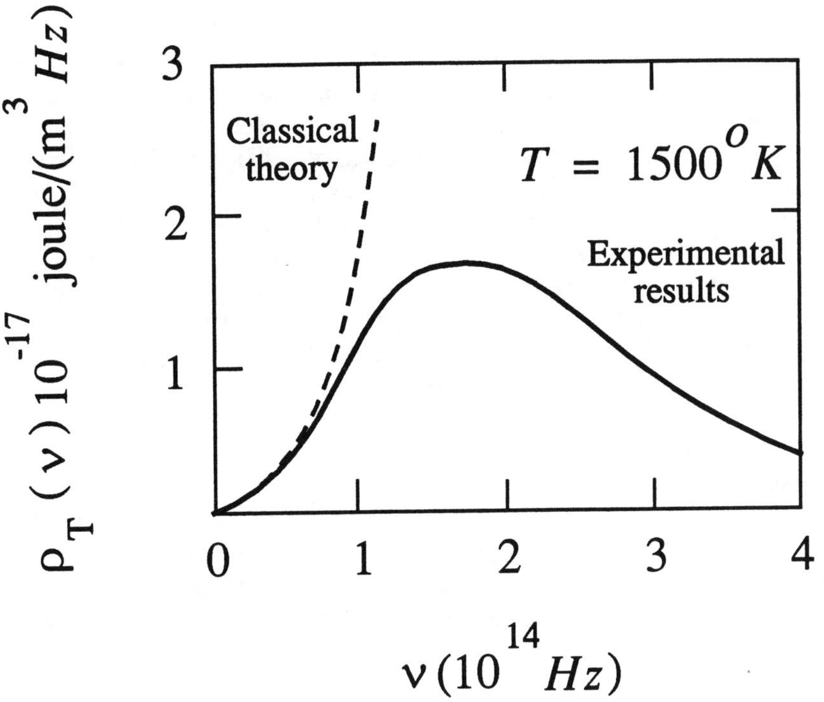

Figure 1.1.2 The Rayleigh-Jeans prediction (dashed line) compared with the experimental results (solid line) for the energy density of blackbody radiation. The serious discrepancy between these results is called the "ultraviolet catastrophe". (After R. Eisberg, R. Resnick. "Quantum Physics of Atoms, Molecules, Solids, Nuclei, and Particles." John Wiley & Sons, N. Y., 1985)

This result became known as the ***ultraviolet catastrophe!*** In 1910, Max Planck presented a talk to the German Physical Society which removed this "catastrophe" from physics. The main idea of the talk was the following. The energy in the cavity can increase or decrease only in portions (parcels) of magnitude

$$E = h\nu, \tag{1.1.1}$$

where ν is the natural frequency of the oscillator and h is the fundamental constant that is equal to $h = 6.625 \times 10^{-34}$ *joule s*. The fundamental constant has the name of the ***Planck constant.*** The physical dimension of it is the following: ***energy*** $\times$ ***time*** = ***length*** $\times$ ***momentum*** = ***angular momentum.*** A quantity of this dimension is called ***action.*** The Planck constant h is the smallest possible action (quantum of action), $h\nu$ is the quantum of electromagnetic radiation, correspondingly. Using this hypothesis, Planck obtained the spectral radiancy U that corresponds the solid experimental curve in Fig.1.1.2.

B. Photoelectric effect

The problem of the photoelectric effect was the second mystery of classical physics. It was shown by Lenard and then by Millikan that light facilitates the discharge between two electrodes due to the ejection of electrons from the surface of the cathode. The ejection of electrons from the surface of a metal by light incident on it is called the ***photoelectric effect***. It was expected from the view-point of classical physics that the kinetic energy of the ejected electrons would increase with an increase in the intensity of the light. But the experimental data showed that the increase of the intensity led only to an increase of the number of electrons ejected, and did not change their energies. Einstein presented a theory of this effect based on the assumption that the radiant energy E is related to its frequency by Eq.(1.1.1). It allowed a complete interpretation of the experimental data.

C. Stability of an Atom

In 1911, Rutherford concluded from experiments on the scattering of α-particles by atoms that the positive charge in an atom is concentrated in a small region, called the nucleus, at the center of the atom. On the basis of this knowledge Bohr explained the discrete structure of atomic energy levels. He completed Newton's equation of motion for the electron that moves in the Coulomb field of the nucleus

$$m_0 \frac{v^2}{a_0} = \frac{e^2}{a_0^2} \tag{1.1.2}$$

by the condition expressing the quantization of the angular momentum

$$m_0 v a_0 = \frac{h}{2\pi}. \tag{1.1.3}$$

Here m_0 is the mass of the electron, v is the velocity of the electron, e is the charge of the electron, and a_0 is the radius of the electron orbit in the atom. By solving for a_0 from Eqs. (1.1.2) and (1.1.3), one obtains

$$a_0 = \frac{h^2}{(2\pi)^2 m_0 e^2} = \frac{\hbar^2}{m_0 e^2}, \tag{1.1.4}$$

where $\hbar = h/2\pi$. The substitution of

$$m_0 = 0.9 \times 10^{-30}\ kg, \tag{1.1.5}$$

$$e = 1.6 \times 10^{-19}\quad Coulomb, \tag{1.1.6}$$

$$h = 6.62 \times 10^{-34}\quad joule\ s \tag{1.1.7}$$

into Eq.(1.1.4) leads to the value a_0=0.53 × 10^{-10} m. This value corresponds to the results of experimental measurements.

1.2 QUANTUM BEHAVIOR

Let us consider an electron as an example of the object that obeys quantum mechanics. An electron is one of the smallest particles. The mass of an electron is given in Eq.(1.1.5). The behavior of matter on such a small scale has been explained by Schrödinger, Heisenberg, and Bohr.

The complexity of understanding quantum mechanics follows from the fact that human experience is restricted to large objects only. Small scale objects behave differently. We may learn about them in a sort of abstract or imaginative "experiments" that are not connected with our direct experience. We shall introduce the imaginative set up and carry out a "thought" experiment.

A. Experiment with bullets

We start with a discussion of a classical object: a bullet. Our experimental set up consists of a machine gun that shoots bullets in the direction of the screen with two holes, see Fig.1.2.1a. Beyond the screen there is a wall that stops bullets. By counting the number of bullet traces on the wall, one can study the distribution of bullets over the wall. Each bullet can hit one of the holes, and by bouncing off the edge of the hole can go anywhere. We cannot say definitely where any particular bullet will go.

To describe the results of the experiment, we can introduce the ***probability*** that a bullet will arrive at the wall at the distance x from the center. Probability can be measured by counting the number of bullets $N(x)$ that arrive at x in a certain time and then taking the ratio of this number to the total number of bullets N which strike the wall:

$$P_{12}(x)=\frac{N(x)}{N}. \tag{1.2.1}$$

Since the bullets go either through hole 1 or through hole 2, we call this probability $P_{12}(x)$. The results of our experiments are plotted in Fig.1.2.1 b,c. The curve has a maximum at $x = 0$. To understand the origin of this behavior, we cover hole 2. Since bullets are now passing through hole 1 only, we can find the probability $P_1(x)$ that has maximum on the straight line that passes through the hole 1. If hole 1 is closed, we obtain the symmetric curve $P_2(x)$. Both probabilities are shown in Fig.1.2.1b. We see that the probability $P_{12}(x)$ plotted in Fig.1.2.1c is the sum of $P_1(x)$ and $P_2(x)$

$$P_{12}(x)=P_1(x)+P_2(x). \tag{1.2.2}$$

The result of both holes open is the sum of the curves corresponding to each hole open alone.

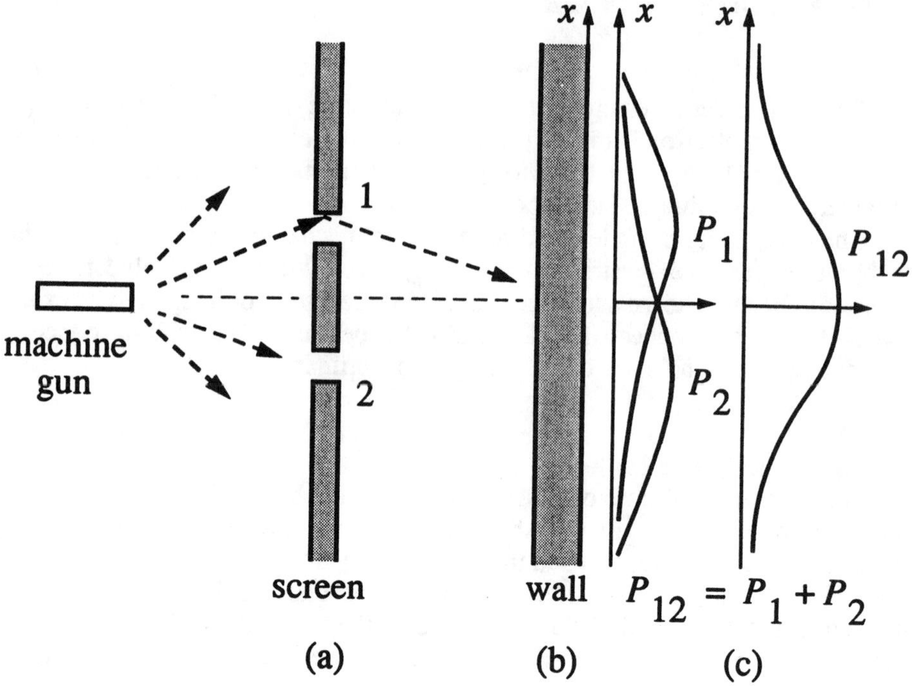

Figure 1.2.1 A machine gun is shooting bullets in the direction of a screen with two holes in it (a). The distribution of bullets on the wall in the case of that one of the holes is open, is shown in (b). When two holes are open simultaneously, the distribution of bullets takes the form (c). (After R. P. Feynman, R. B. Leighton, M. Sands. "The Feynman Lectures on Physics. Quantum Mechanics." Addison-Wesley Publishing Company, Menlo Park, CA, 1964)

B. Experiment with Electromagnetic Waves

Now we consider the propagation of an electromagnetic wave in the direction of the screen with two holes in it. To the right of the screen, we have another screen with a movable detector that registers the arrival of waves. The total set up is shown in Fig.1.2.2a. The incident wave is diffracted from the holes and circular waves propagate from each hole in the direction of the detector D. The movable detector measures the intensity of the light for various values of x.

If we cover hole 1 and measure the intensity for different values of x, we find the intensity $I_2(x)$ registered by the detector, see Fig.1.2.2b. If hole 2 is covered, we obtain the curve $I_1(x)$ that is shown in Fig.1.2.2b. When both holes are open, we find the intensity $I_{12}(x)$ which is shown in Fig.1.2.2c. The curve $I_{12}(x)$ is obviously not the sum of $I_1(x)$ and $I_2(x)$. We are observing a typical ***interference pattern*** from two coherent waves. This pattern has the following origin. The electromagnetic wave that comes from hole 1 has the form

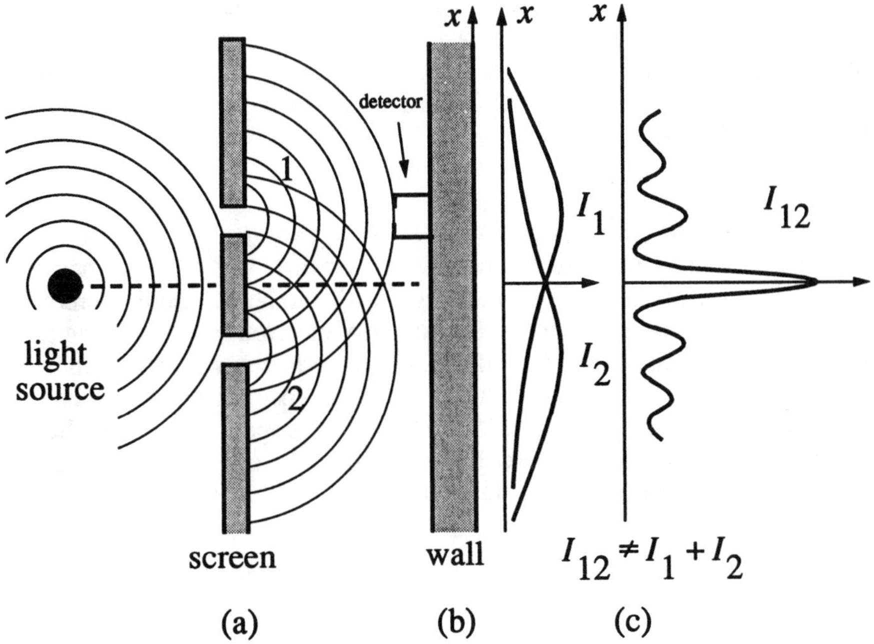

Figure 1.2.2 The propagation of an electromagnetic wave in the direction of a screen with two holes in it (a). If one of the holes is covered, the intensity registered by a detector is represented by curves I_1 or I_2 (b). When both holes are open, we find the interference pattern shown in (c). (After R. P. Feynman, R. B. Leighton, M. Sands. "The Feynman Lectures on Physics. Quantum Mechanics." Addison-Wesley Publishing Company, Menlo Park, CA, 1964)

$$\mathcal{E}_1(x,t) = \mathcal{E}_1(x)e^{-i\omega t}, \tag{1.2.3}$$

where $\mathcal{E}_1(x)$ is the amplitude of this wave. This amplitude is, in general, a complex number. The intensity is proportional to the squared absolute value of the amplitude:

$$I_1(x) = |\mathcal{E}_1(x)|^2 . \tag{1.2.4}$$

Similarly, the electromagnetic wave through hole 2 is given by

$$\mathcal{E}_2(x,t) = \mathcal{E}_2(x)e^{-i\omega t}, \tag{1.2.5}$$

and the corresponding intensity has the form

$$I_2(x) = |\mathcal{E}_2(x)|^2 . \tag{1.2.6}$$

It follows from the superposition principle of electromagnetic theory that when both holes are open, the electromagnetic field is equal to

$$\mathcal{E}_{12} = \mathcal{E}_1 + \mathcal{E}_2 . \tag{1.2.7}$$

The corresponding intensity is given by

$$I_{12}(x) = |\mathcal{E}_1 + \mathcal{E}_2|^2 = \mathcal{E}_1^2 + \mathcal{E}_2^2 + 2|\mathcal{E}_1| \cdot |\mathcal{E}_2| \cos\delta, \tag{1.2.8}$$

where δ is the phase difference between $\mathcal{E}_1$ and $\mathcal{E}_2$ at the point x. The interference pattern is defined by the phase difference δ. We observe constructive interference maxima at x-points where the distance from the detector to one hole is an integer multiple of the wavelength larger than the distance from the detector to the other hole. Here $I_{12} \neq I_1 + I_2$.

C. Experiment with Electrons

Let us imagine a similar experiment with electrons. The experimental set up consists of a source of electrons, a diaphragm with two holes, and the screen with a detector that can register electrons. The source of electrons could be a metal wire heated by an electric current and surrounded by a metal box with a hole in it. The voltage is applied between the wire and the box with positive bias on the box. The emitted electrons are moving towards the screen. After passing through the holes the electrons reach the screen where they are counted by a detector. The set up is shown in Fig.1.2.3a. Each time an electron reaches wall, the detector (counter of

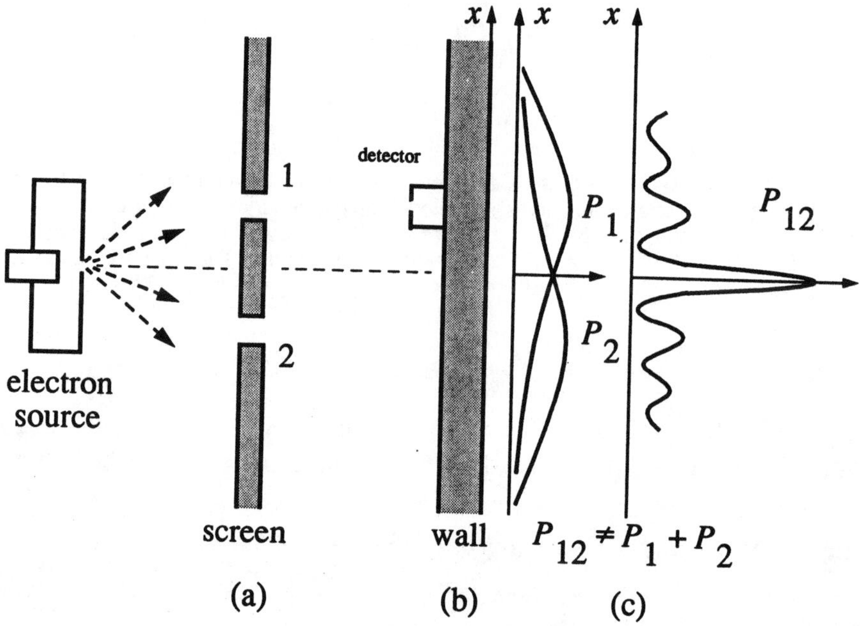

Figure 1.2.3. Electrons emitted by an electron source are moving towards a screen (a). Passing through the holes, they reach the wall that contains a detector. If one of the holes is closed, the electron probability distribution on the wall has the form shown in (b). If both holes are open, we find the interference pattern shown in (c). (After R. P. Feynman, R. B. Leighton, M. Sands. "The Feynman Lectures on Physics. Quantum Mechanics." Addison-Wesley Publishing Company, Menlo Park, CA, 1964)

electrons) shows a sharp signal. These signals are random in time.

When two detectors are put on the wall, they never give a signal at once. This means that electrons arrive as entire particles. The probability $P_1(x)$ for the group of electrons to arrive through hole 1 when hole 2 is covered is plotted in Fig.1.2.3b. This probability is proportional to the average rate of the signals at the point x. The probability $P_2(x)$ is obtained when hole 1 is covered. It is also plotted in Fig.1.2.3b. When both holes are open, we find the probability $P_{12}(x)$ that is shown in Fig.1.2.3c. We see that $P_{12}(x)$ is not the sum of $P_1(x)$ and $P_2(x)$

$$P_{12} \neq P_1 + P_2 . \tag{1.2.9}$$

Curve $P_{12}(x)$ at $x = 0$ is more than twice as large as $P_1(x)+P_2(x)$. The pattern is typical for an ***interference phenomenon***. In fact, $P_{12}(x)$ is very similar to I_{12} in Fig.1.2.2c.

In order to describe this new kind of interference, we will introduce two complex numbers, the probability wave amplitudes, $\phi_1(x)$ and $\phi_2(x)$. They have the physical sense of the amplitudes of some probability waves which represent the electrons. A certain wavelength is attributed to the electron. The probabilities $P_1(x)$ and $P_2(x)$ have the following forms in terms of ϕ_1 and ϕ_2

$$P_1(x) = |\,\phi_1(x)\,|^2, \quad P_2(x) = |\,\phi_2(x)\,|^2, \tag{1.2.10}$$

which are similar to $I_1(x)$ and $I_2(x)$ from Eqs.(1.2.4) and (1.2.6). The effect of two open holes is described by

$$P_{12}(x) = |\,\phi_1(x) + \phi_2(x)\,|^2, \tag{1.2.11}$$

which is an analog of Eq.(1.2.8).

Summing up we can say that, on the one hand, electrons arrive as particles, but on the other hand, the probability of arrival is distributed as in an interference pattern. This is typical behavior of an object with a quantum nature.

D. Observation of the Electrons

In order to clarify the physical meaning of probability waves, we complete our experimental set up by a light source placed between the holes and behind the screen. The new set up is shown in Fig.1.2.4a. The electric charge of the electron scatters the light from the light source. We can see a flash of light from the vicinity of the hole through which the electron goes. From the number of electrons which came to the point x through hole 1 (hole 2 is closed), we calculate the probability $P_1'(x)$ that is plotted in Fig.1.2.4b. The number of electrons which come to the point x of the screen through hole 2 (hole 1 is closed) gives the probability $P_2'(x)$ that is shown in Fig.1.2.4b, but the total probability $P_{12}'(x)$ with both holes open appears to be of the form

$$P'_{12} = P'_1 + P'_2 . \tag{1.2.12}$$

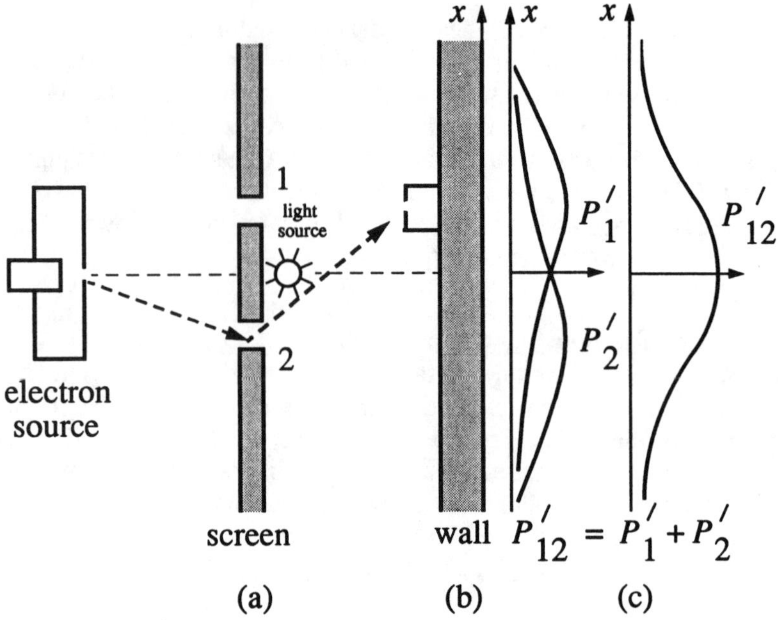

Figure 1.2.4 An experimental setup as in Fig.1.2.3 with a light source added (a). If one of the holes is closed, the electron probability distribution on the wall has the same form (b) as in Fig.1.2.3. If both holes are open, the light source disturbs the motion of an electron so much that the interference pattern is destroyed (c). (After R. P. Feynman, R. B. Leighton, M. Sands. "The Feynman Lectures on Physics. Quantum Mechanics." Addison-Wesley Publishing Company, Menlo Park, CA, 1964)

We have observed which hole the electron comes through, but have failed to obtain the interference curve P_{12}' plotted in Fig.1.2.3c. The reason for this disappointing result is that the light source disturbs the motion of the electron so much that the double slit interference pattern is destroyed.

By using light of longer wavelength, for example infrared (IR) light, we can try to reduce the effect of the light on the electron. But it is known from wave optics that there is a limitation on how close two holes can be in order to be seen separately. By increasing the wavelength we are reducing the possibility of determining through which hole the electron goes. We are losing information on the location of an electron.

Heisenberg suggested that due to the wave nature of the particle there is a limitation on our experimental capability: ***It is impossible to find an experimental set up which allows us to find which hole the electron goes through without disturbing the electron enough to destroy the interference pattern.*** This limitation is called the ***Uncertainty Principle.*** All knowledge in quantum mechanics is based on this principle.

The experiment which we are discussing has never been done in reality, since the equipment needed would be too small to see the effects. However, there are some other, real, experiments that demonstrate a similar wave-like quantum behavior of the electron. Experiments of Davisson and Germer (1927) demonstrated the diffraction of electrons by a crystal. The wave nature of the electron can be tested in the same way as the wave nature of X-rays. If a beam of electrons of appropriate energy falls on a crystal, the atoms of the crystal serve as a three-dimensional array of diffraction centers for the electron wave. Electrons are strongly scattered from the crystalline planes in certain directions only. The crystal plays the role of a diffraction grating. The diffraction occurs when the wave length of the electron probability wave λ is of the order of the lattice parameter a. Figure 1.2.5 shows diffraction patterns (a) of X-ray diffraction by polycrystalline zirconium and (b) of electron diffraction by gold. The figures are very similar.

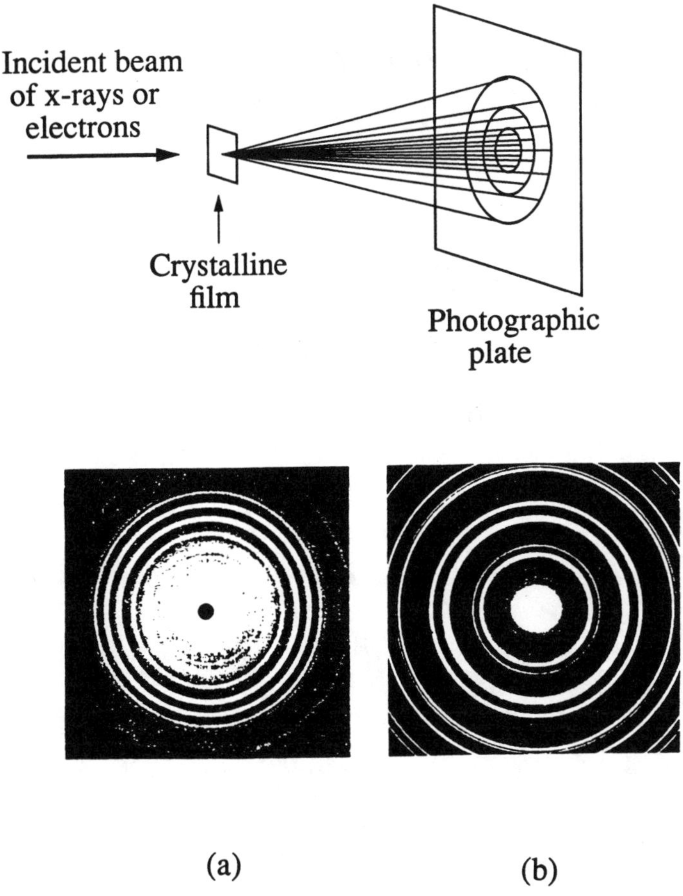

Figure 1.2.5 Top: The experimental setup for Debye-Scherer diffraction. (a) is the Debye-Scherer pattern of X-ray diffraction by zirconium oxide crystals. (b) Debye-Scherer pattern of electron diffraction by a gold crystal. (After R. Eisberg, R. Resnick. "Quantum Physics of Atoms, Molecules, Solids, Nuclei, and Particles." John Wiley & Sons, N. Y., 1985)

1.3 DE BROGLIE WAVES

On the basis of our "thought" experiments, we introduced the probability wave amplitude $\phi(x)$ for an electron. The probability of finding the electron at position x is proportional to the absolute square of the amplitude $|\phi(x)|^2$.

In case of a free electron, the probability wave amplitude varies in space and time like the usual traveling plane wave

$$\phi(x,t) = e^{-i(\omega t - kx)}, \tag{1.3.1}$$

where x is the position, ω is the frequency of the wave and k is the wave number of the wave that propagates in the x-direction. This sinusoidal plane wave Eq.(1.3.1) goes on forever, as is shown in Fig.1.3.1a. The corresponding probability of finding a free electron does not depend on the coordinate x, and is equal to

$$P = |\phi(x,t)|^2 = 1. \tag{1.3.2}$$

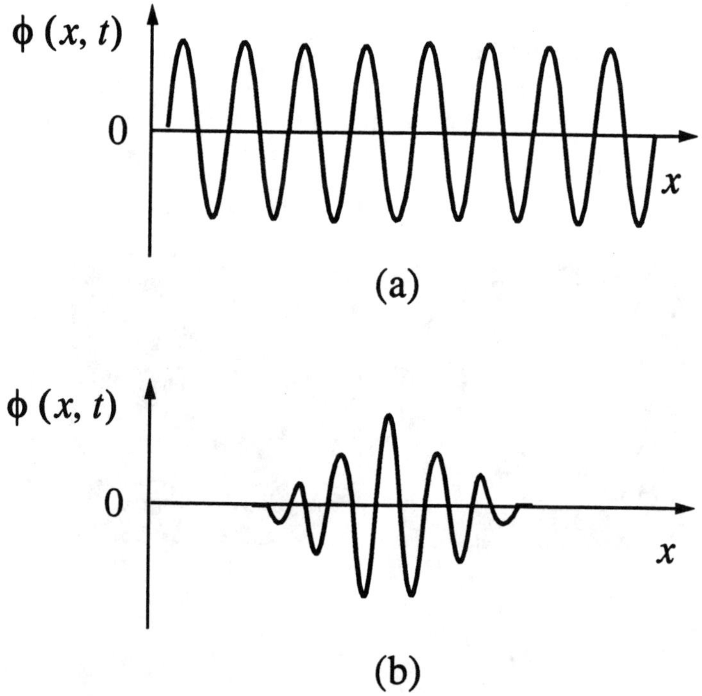

Figure 1.3.1 (a) Propagating wave. (b) Wave packet.

Equation (1.3.2) means that the electron has a great uncertainty in position. In correspondence with the Planck hypothesis, Eq.(1.1.1), the frequency of the probability wave ω is connected with the energy of a particle (an electron) by the relation

$$E = \hbar\omega. \tag{1.3.3}$$

Equation (1.3.3) differs from Eq.(1.1.1) since we have introduced an angular frequency ω instead of the natural frequency ν that we had in Eq.(1.1.1). The Planck constant h from Eq.(1.1.1) is replaced in Eq.(1.3.3) by $\hbar = h/2\pi$, correspondingly.

De Broglie made the next step in understanding the physical nature of the wave amplitude. He has suggested that all particles in nature show wave-like behavior under proper conditions, and that the wave vector $\vec{k}$ of the wave is directly connected with momentum $\vec{p}$ of the particle (electron) by the relation

$$\vec{p} = \hbar\vec{k}. \tag{1.3.4a}$$

Here the proportionality coefficient between $\vec{p}$ and $\vec{k}$ is the same Planck's constant $\hbar$ that enters Eq.(1.3.1). In fact, the de Broglie hypothesis concerns the wave length λ of the particle which is connected with the wave vector k by the relation $\lambda = 2\pi/k$. The original form of Eq.(1.3.4a) is

$$\lambda = \frac{2\pi}{k} = \frac{h}{p}. \tag{1.3.4b}$$

The relations (1.3.4) have very general nature. They associate with each particle a wave with the wavelength λ that depends on the momentum of the particle $\vec{p}$ and, therefore, on the mass of the particle. The wavelength λ can be attributed to any macroscopic body. But its mass is so large that features of the corresponding interference pattern are beyond experimental detection.

When the position x of the particle is known, the probability $P \neq 0$ of finding the particle should be confined to a certain region Δx in the vicinity of x. Outside this region the probability P should vanish. The probability wave amplitude ϕ should be zero outside Δx also. The electron is represented by a ***wave packet*** of width Δx in this case. The wave packet with the width Δx is shown schematically in Fig.1.3.1b in contrast to the propagating wave shown in Fig.1.3.1a. To obtain the wave packet with a length Δx, it is necessary to make a superposition of the plane waves from the range of wave vectors Δk in the vicinity of vector k,

$$\Delta k = \frac{2\pi}{\Delta x}. \tag{1.3.5}$$

By using the de Broglie relation Eq.(1.3.4) between the wave vector k and the momentum of the particle p and Eq.(1.3.5), we see that the range of momenta is restricted by

$$\Delta p = \hbar\Delta k \cong \frac{2\pi\hbar}{\Delta x}. \tag{1.3.6}$$

Equation (1.3.6) can be written in the form

$$\Delta p\Delta x \approx 2\pi\hbar. \tag{1.3.7}$$

Equation (1.3.7) shows that the product of the uncertainty of the momentum Δp in the x-direction and the uncertainty in position Δx of the particle is of the order of $2\pi\hbar$. Equation (1.3.7) is the mathematical expression of the ***uncertainty principle*** that was discussed qualitatively in Section 1.2D. Equation (1.3.7) allows estimating whether the object we consider is classical or quantum mechanical. If the action $p\,x$ is as small as $2\pi\hbar$, we deal with quantum behavior of the object. If the action is substantially larger than $2\pi\hbar$, we are in the region of classical physics.

It is important to understand that the same object behaves as a quantum object under certain conditions and manifests itself as a classical object under some other conditions. Consider as an example an electron. The uncertainty of its position in an atom is defined by the linear size of the atom: $\Delta r \approx 10^{-10}\ m$. The velocity of an electron in the atom is therefore known with the uncertainty

$$\Delta v = \frac{\hbar}{m_0 \Delta r} = 7\times 10^6\ m\ s^{-1}.$$

The velocity of an electron in an atom can be estimated from the energy $E \sim 13.5\ eV$ of the electron in an atom. It is equal to $v \approx 10^6\ m\ s^{-1}$. We see that the uncertainty of velocity is of the order of the velocity itself. This means that the classical approach does not work for an electron in an atom. The electron in an atom behaves as a quantum particle. The motion of an electron in a Wilson cloud chamber obeys classical physics, since the width of the electron track in the chamber, Δx, is of the order of $1\ mm$. The velocity of an electron with an energy of 1 MeV equals $v \approx 10^8\ m\ s^{-1}$, while the uncertainty in velocity is

$$\Delta v = \frac{\hbar}{m_0 \Delta r} \approx 0.07 m\ s^{-1}.$$

Therefore, we have $\Delta v \ll v$ and the electron moves along a classical trajectory.

The uncertainty principle is the fundamental restriction on the accuracy with which we can speak about the position and momentum of the particle. Since the action $\hbar$ is very small, these restrictions are not important in ***macrophysics***. It is on the ***submicroscopic*** level that quantum mechanics becomes important.

If the wave propagates in some medium, the frequency of the wave ω is a function of the wave vector k,

$$\omega = \omega(k). \tag{1.3.8}$$

This dependence of ω on k is called ***dispersion***. The simplest way to watch the propagation of the wave packet in a ***dispersive*** medium is to watch the motion of the maximum amplitude of the packet. The phase $\alpha = -\omega t + kx$ which corresponds to the maximum amplitude remains constant as the wave packet propagates

$$\alpha(x_{max}) = \omega(k)t - kx_{max} = Constant. \tag{1.3.9}$$

Here x_{max} is the position vector of the maximum amplitude. It follows form Eq.(1.3.9) that

$$\frac{d\alpha(x_{max})}{dk} = \left\{\frac{d\omega(k)}{dk}\right\}t - x_{max} = 0 \tag{1.3.10}$$

or

$$x_{max} = \left\{ \frac{d\omega(k)}{dk} \right\} t. \tag{1.3.11}$$

Equation (1.3.11) means that the maximum of the wave packet travels with the ***group velocity***

$$v_g = \frac{d\omega(k)}{dk}. \tag{1.3.12}$$

Another very important feature of the wave packet is that the wave packet spreads in time as it propagates. The width of the wave packet Δx is not constant in a dispersive medium. This follows from the dependence of the group velocity v_g on the wave vector k which has not been taken into consideration yet:

$$v_g = v_g(k).$$

The interval Δk of propagation vectors corresponds to a certain interval of group velocity values Δv_g

$$\Delta v_g = \frac{dv_g}{dk}\Delta k = \frac{d^2\omega(k)}{dk^2}\Delta k.$$

If at the moment $t = 0$, the width of the wave packet equals $(\Delta x)_0$, it follows from Eq.(1.3.11) that at the moment Δt, the width of the wave packet increases:

$$(\Delta x)_t = (\Delta x)_0 + \Delta v_g \Delta t.$$

The time of travel of the packet over the distance x increases with time. The wave packet spreads in time. There is no classical trajectory for the wave packet which represents the particle. The only thing that remains is the probability interpretation of the wave packet.

1.4 TIME-DEPENDENT SCHRÖDINGER EQUATION

We have presented experimental evidence that microscopic particles move according to the laws of wave propagation. In order to see how these laws govern the motion of particles, we should find a suitable wave equation. This equation was simply postulated by Schrödinger in 1926. The justification for it is that it predicts many results that are verified experimentally.

We give here the plausibility arguments that allow finding the correct form of this equation. This equation should be consistent with the de Broglie and Planck relations, Eqs.(1.3.3) and (1.3.4). It should also be consistent with the equation that connects the total energy of the particle E with the kinetic and potential energies

$$E = \frac{\vec{p}^{\,2}}{2m_0} + U. \tag{1.4.1}$$

The equation should be linear in $\phi(\vec{r},t)$, since the linearity ensures the validity of the superposition principle which allows us to sum wave functions in order to get patterns of constructive and destructive interference. We start from the special case of a free particle when the potential energy U reduces to the constant value $U_0 = 0$. In this case the traveling wave

$$\phi(\vec{r},t) = e^{-\frac{i}{\hbar}Et + \frac{i}{\hbar}(\vec{p}\cdot\vec{r})} \tag{1.4.2}$$

should be a solution of the equation. The free particle energy

$$E = \frac{\vec{p}^{\,2}}{2m_0} \tag{1.4.3}$$

enables us to construct an equation that satisfies our requirements. In order to find the equation, we calculate the first time derivative of $\phi(\vec{r},t)$ from Eq.(1.4.2)

$$\frac{\partial\phi(\vec{r},t)}{\partial t} = -\frac{i}{\hbar}E\phi(\vec{r},t) \tag{1.4.4}$$

and find the energy from this equation

$$E = \frac{1}{\phi}i\hbar\frac{\partial\phi}{\partial t}. \tag{1.4.5}$$

We also construct the second spatial derivatives of $\phi(\vec{r},t)$

$$\frac{\partial\phi}{\partial x} = \frac{i}{\hbar}p_x\phi; \qquad \frac{\partial^2\phi}{\partial x^2} = -\frac{p_x^2}{\hbar^2}\phi; \tag{1.4.6a}$$

$$\frac{\partial\phi}{\partial y} = \frac{i}{\hbar}p_y\phi; \qquad \frac{\partial^2\phi}{\partial y^2} = -\frac{p_y^2}{\hbar^2}\phi; \tag{1.4.6b}$$

$$\frac{\partial\phi}{\partial z} = \frac{i}{\hbar}p_z\phi; \qquad \frac{\partial^2\phi}{\partial z^2} = -\frac{p_z^2}{\hbar^2}\phi. \tag{1.4.6c}$$

It follows from Eqs.(1.4.6) that the momentum of the electron equals

$$p_x^2 = -\frac{1}{\phi}\hbar^2\frac{\partial^2\phi}{\partial x^2}; \qquad p_y^2 = -\frac{1}{\phi}\hbar^2\frac{\partial^2\phi}{\partial z^2}; \qquad p_z^2 = -\frac{1}{\phi}\hbar^2\frac{\partial^2\phi}{\partial z^2}. \tag{1.4.7}$$

The substitution of Eqs.(1.4.5) and (1.4.7) in Eq.(1.4.3) leads to

$$i\hbar\frac{\partial\phi}{\partial t} = -\frac{\hbar^2}{2m_0}\left\{\frac{\partial^2\phi}{\partial x^2} + \frac{\partial^2\phi}{\partial y^2} + \frac{\partial^2\phi}{\partial z^2}\right\}. \tag{1.4.8}$$

This is the wave equation for the probability amplitude $\phi(\vec{r},t)$ in case of a free electron. The right hand side of Eq.(1.4.8) is easily represented through the Laplacian operator

$$\nabla^2 \equiv \Delta \equiv \frac{\partial^2}{\partial x^2} + \frac{\partial^2}{\partial y^2} + \frac{\partial^2}{\partial z^2}. \tag{1.4.9}$$

With the help of Eq.(1.4.9), Eq.(1.4.8) takes the form

$$i\hbar\frac{\partial\phi}{\partial t} = -\frac{\hbar^2}{2m_0}\nabla^2\phi. \tag{1.4.10}$$

We now consider the particle in an external field represented by the potential energy $U(\vec{r})$. Since the total energy of the particle is the sum of the kinetic and potential energies Eq.(1.4.1), the kinetic energy equals $p^2/2m_0 = E - U$. By taking time and space derivatives from Eqs.(1.4.5) and (1.4.6 a,b,c) and substituting them into the kinetic energy we are led to the following wave equation

$$i\hbar\frac{\partial\phi(\vec{r},t)}{\partial t} = \left\{-\frac{\hbar^2}{2m_0}\nabla^2 + U(\vec{r})\right\}\phi(\vec{r},t). \tag{1.4.11}$$

Equation (1.4.11) is known as the ***time-dependent Schrödinger equation.*** It is a partial differential equation. Wave equations of classical physics, such as those for electromagnetic and sound waves, are of the second order in time. The amplitude of classic wave is real. Shrödinger equation is of the first order in time. As a result, the probability wave amplitude $\phi(\vec{r},t)$ is, in general, a complex quantity. It does not describe real waves propagating in a medium. The probability wave amplitude ϕ, is called a ***wave function.***

Since the motion of the particle is connected with the propagation of the associated wave $\phi(\vec{r},t)$, it is reasonable to suggest that the amplitude of the wave should have a considerable value at the place where the particle is located. The connection of the wave function with the behavior of the particle was established by Born in 1929, in terms of the probability density $P(\vec{r},t)$ from Eq.(1.2.10). Born stated that

$$P(\vec{r},t) = \phi^*(\vec{r},t)\phi(\vec{r},t) = |\phi(\vec{r},t)|^2, \tag{1.4.12}$$

where $\phi^*(r,t)$ is the complex conjugate of $\phi(\vec{r},t)$. Equation (1.4.12) means that the wave function $\phi(\vec{r},t)$ specifies the probability density for the particle (electron), which is the probability of finding a particle near point $\vec{r}$ (in the interval from $\vec{r}$ to $\vec{r}+d\vec{r}$)

$$P(\vec{r},t)d\vec{r} = \phi^*(\vec{r},t)\phi(\vec{r},t)d\vec{r}. \tag{1.4.13}$$

The total probability of finding the electron somewhere in space is necessarily equal to unity

$$\int |\phi(\vec{r},t)|^2 d\vec{r} = 1. \tag{1.4.14}$$

The integral in Eq.(1.4.14) is taken over the entire coordinate space. The procedure expressed by Eq.(1.4.14) is called the ***normalization of the wave function.*** Because of the linearity of the Schrödinger equation (1.4.11), the wave function is defined only to within an arbitrary factor. This factor is defined by the normalization condition (1.4.14).

The probability interpretation allows finding mean or the ***expectation value*** of the coordinate $\vec{r}$ at the time t in the form

$$<\vec{r}> = \int \vec{r}P(\vec{r},t)d\vec{r} = \int \phi^*\,\vec{r}\,\phi d\vec{r}. \tag{1.4.15}$$

The symbol <...> means an average. The integrand in Eq.(1.4.15) is the value of the coordinate weighted by the probability of observing a particle at the point $\vec{r}$. The averaging of observed values is carried out by integration.

A similar averaging procedure can be used for the evaluation of the expectation value of an arbitrary function of $\vec{r}$

$$<f(\vec{r})>=\int \phi^*(\vec{r},t)f(\vec{r})\phi(\vec{r},t)d\vec{r}. \tag{1.4.16}$$

But the calculation of an average value of the momentum $\vec{p}$ in the form of Eq.(1.4.16),

$$<\vec{p}>=\int \phi^*(\vec{r},t)\vec{p}\phi(\vec{r},t)d\vec{r}, \tag{1.4.17}$$

encounters the following problem. To evaluate this integral we should know $\vec{p}$ as a function of $\vec{r}$ and t. This dependence is well known in classical mechanics; but in quantum mechanics the uncertainty principle does not allow us to find this dependence, since $\vec{p}$ and $\vec{r}$ are not simultaneously known with the same accuracy. This difficulty can be circumvented with the help of the free particle wave function Eq.(1.4.2). On calculating the first spatial derivative of $\phi(\vec{r},t)$, one obtains

$$\frac{\partial\phi(\vec{r},t)}{\partial\vec{r}}=i\frac{\vec{p}}{\hbar}\phi(\vec{r},t), \tag{1.4.18a}$$

or

$$-i\hbar\frac{\partial}{\partial\vec{r}}\phi(\vec{r},t)=\vec{p}\phi(\vec{r},t). \tag{1.4.18b}$$

Equation (1.4.18b) means that the action of the operator $(-i\hbar)\partial/\partial\vec{r}$ on the wave function $\phi(\vec{r},t)$ results in the multiplication of $\phi(\vec{r},t)$ by the constant factor $\vec{p}$. We see that the Eq.(1.4.17) is the ***eigenvalue equation*** for the linear operator $(-i\hbar)\partial/\partial\vec{r}$. The conventional notation for the momentum operator has the form

$$\hat{\vec{p}}=-i\hbar\frac{\partial}{\partial\vec{r}}=-i\hbar\nabla, \tag{1.4.19}$$

where the operator has the special "hat"-symbol and ∇ is the del-operator. In components, Eq.(1.4.19) has the form

$$\hat{p}_x=-i\hbar\frac{\partial}{\partial x};\quad \hat{p}_y=-i\hbar\frac{\partial}{\partial y};\quad \hat{p}_z=-i\hbar\frac{\partial}{\partial z}. \tag{1.4.20}$$

It is convenient to write Eq.(1.4.17) in the complete form of an eigenvalue problem

$$\hat{\vec{p}}\phi_n=p_n\phi_n. \tag{1.4.21}$$

Here the index n labels the possible eigenvalues $\vec{p}$ of the operator $\hat{\vec{p}}$. Substituting the operator (1.4.19) into the expectation value Eq.(1.4.17) one finds

$$<\hat{\vec{p}}>=\int \phi^*(\vec{r},t)\left(-i\hbar\frac{\partial}{\partial\vec{r}}\right)\phi(\vec{r},t)d\vec{r}. \tag{1.4.22}$$

When $\phi(\vec{r},t)$ is known, the integral in Eq.(1.4.22) can be easily evaluated and the expectation value obtained.

This is the first time in our consideration when we meet the necessity of representing physical quantity (the momentum $\hat{\vec{p}}$) by an operator.

1.5 TIME-INDEPENDENT SCHRÖDINGER EQUATION

The time-dependent Schrödinger equation (1.4.11) allows the separation of variables $\vec{r}$ and t, if the operator U does not depend on t. This procedure for solving the partial differential equation (1.4.11) reduces the equation to a time dependent ordinary differential equation and time independent partial differential equation. The procedure consists of representing the wave function in the form of the product of two functions, one of which is a function of the spatial coordinate $\vec{r}$ and the second is a function of time t:

$$\phi(\vec{r},t)=\psi(\vec{r})T(t). \tag{1.5.1}$$

Substitution of Eq.(1.5.1) in Eq.(1.4.11) and division by $\phi(\vec{r},t)$ results in

$$\frac{1}{\phi(\vec{r})}\left\{-\frac{\hbar^2}{2m_0}\nabla^2+U(\vec{r})\right\}\phi(\vec{r})=i\hbar\frac{1}{T(t)}\frac{\partial T(t)}{\partial t}=E. \tag{1.5.2}$$

The left-hand side of Eq.(1.5.2) is a function of the coordinate $\vec{r}$ alone. The middle term in Eq.(1.5.2) depends on t only. Since the variables $\vec{r}=(x,y,z)$ and t are independent, the common value in Eq.(1.5.2) must be equal to the separation constant which we designated by E. The time-dependent part of Eq.(1.5.2) has the form

$$i\hbar\frac{1}{T(t)}\frac{dT(t)}{dt}=E. \tag{1.5.3}$$

This is a first-order differential equation. The standard solution of it has the form

$$T(t)=e^{-\frac{i}{\hbar}Et}. \tag{1.5.4}$$

A comparison of this solution with Eq.(1.4.2) shows that the separation constant has the meaning of the energy of the particle E and that it is constant. The substitution of Eq.(1.5.4) in Eq.(1.5.1) gives the total wave function

$$\phi(\vec{r},t)=\psi(\vec{r})e^{-\frac{i}{\hbar}Et}. \tag{1.5.5}$$

Since the squared module of this function is time independent, this function is called ***the wave function of the stationary state***.

The space-dependent part of Eq.(1.5.2) becomes

$$\left\{-\frac{\hbar^2}{2m_0}\nabla^2+U(\vec{r})\right\}\psi(\vec{r})=E\psi(\vec{r}). \tag{1.5.6}$$

For the probability amplitude to have a physical meaning, the wave function $\psi(\vec{r})$ should be a ***single-valued*** function of position $\vec{r}$ and must be ***finite*** everywhere. Since the Schrödinger equation (1.5.6) is of the second order in the spatial variables, both $\psi(\vec{r})$ and its first derivative $\psi'(\vec{r})$ should be ***continuous functions***.

The Schrödinger equations (1.4.11) and (1.5.6) are linear equations. Therefore a linear combination of two solutions ψ_1 and ψ_2,

$$\psi=C_1\psi_1+C_2\psi_2, \tag{1.5.7}$$

is also a solution of these equations. The squared moduli of the coefficients $|C_1|^2$ and $|C_2|^2$ are equal to the probabilities of realizing state 1 and state 2, respectively. For example, if ψ_1 describes the electron which goes through hole 1 in our imaginative experiments from Section 1.2, and ψ_2 corresponds to the electron which goes through hole 2, the quantities $|C_1|^2$ and $|C_2|^2$ give the probabilities for the electron to go either through hole 1 or through hole 2, respectively.

The general problem connected with the time-independent Schrödinger equation (1.5.6) is the determination of the stationary states of the particle, which correspond to certain values of the energy E. It is well known in mathematics that the eigenvalue problem for a linear operator has solutions only for special discrete eigenvalues. Therefore, it is natural to relate the problem of the quantization of the energy states of an electron with the eigenvalue problem for a linear operator. The time independent Schrödinger Equation (1.5.6) is already written in operator form. The quantity in the square brackets is a linear differential operator corresponding to the energy E. It is called the ***Hamiltonian*** $\hat{H}$:

$$\hat{H} = -\frac{\hbar^2}{2m_0}\nabla^2 + U(\vec{r}). \tag{1.5.8}$$

Taking into account the definition Eq.(1.4.19) of the momentum operator, one can see that the first term in Eq.(1.5.6) is the kinetic energy operator $\hat{\vec{p}}^2/2m_0$, and the second term is the potential energy operator $U(\vec{r})$. Equation (1.5.6) in operator form is

$$\hat{H}\psi(\vec{r}) = E\psi(\vec{r}). \tag{1.5.9}$$

Equation (1.5.9) represents a ***eigenvalue problem***, with E the ***eigenvalue*** of $\hat{H}$. The function $\psi(\vec{r})$ is called the ***eigenfunction*** of the operator $\hat{H}$. We shall see below that Eq.(1.5.9) has solutions only for discrete values of E for an electron which is confined to a finite region of space by the potential $U(\vec{r})$. These discrete values are the allowed energies of the quantum particle. This is the second time that we speak about operators in quantum mechanics.

The time-dependent Schrödinger equation (1.4.11) in terms of $\hat{H}$ takes the form

$$-i\hbar\frac{\partial\phi(\vec{r},t)}{\partial t} = \hat{H}\,\phi(\vec{r},t). \tag{1.5.10}$$

1.6 OPERATORS

Linear operators are associated in quantum mechanics not only with momentum and energy, but with all known physical quantities. The linear differential operator $\hat{Q}$ satisfies the following condition

$$\hat{Q}(\psi_1+\psi_2)=\hat{Q}\,\psi_1+\hat{Q}\,\psi_2\,. \tag{1.6.1}$$

Equation (1.6.1) means that the addition of linear operators is similar to the addition of ordinary numbers.

The multiplication of operators differs from the multiplication of ordinary numbers, since the permutation of factors is often not allowed:

$$\hat{Q}_1\hat{Q}_2\neq\hat{Q}_2\hat{Q}_1\,. \tag{1.6.2}$$

For example, if $\hat{Q}_1=x$ and $\hat{Q}_2=(-i\hbar)\partial/\partial x$, the product of these quantities depends on the order of the factors:

$$\hat{Q}_1\hat{Q}_2\psi(x)=-i\hbar x\frac{\partial}{\partial x}\psi(x)=-i\hbar x\frac{\partial\psi}{\partial x},$$

$$\hat{Q}_2\hat{Q}_1\psi(x)=-i\hbar\frac{\partial}{\partial x}x\psi(x)=-i\hbar x\frac{\partial\psi(x)}{\partial x}-i\hbar\psi(x).$$

We see that $\hat{Q}_1\hat{Q}_2$ is not equal to $\hat{Q}_2\hat{Q}_1$.

The eigenvalue equation for any linear operator $\hat{Q}$ has the following form: the $\hat{Q}$-operator acting on a function reproduces the same function multiplied by a constant factor,

$$\hat{Q}\chi_n(\vec{r})=q_n\chi_n(\vec{r}). \tag{1.6.3}$$

Here q_n is the eigenvalue of the $\hat{Q}$-operator, corresponding to the eigenfunction $\chi_n(\vec{r})$. The eigenfunctions of a linear operator satisfy the conditions of orthogonality and normalization

$$\int\chi_n^*(\vec{r})\chi_{n'}(\vec{r})d\vec{r}=\delta_{nn'}, \tag{1.6.4}$$

where $\delta_{nn'}$ is the Kronecker delta

$$\delta_{nn'}=\begin{Bmatrix}1 & \text{if} \quad n=n\\ 0 & \text{if} \quad n\neq n'\end{Bmatrix}. \tag{1.6.5}$$

The expectation value of the $\hat{Q}$-operator in the state n is defined by

$$<\hat{Q}>=\int\chi_n^*(\vec{r})\hat{Q}\chi_n(\vec{r})d\vec{r}. \tag{1.6.6}$$

Substitution of Eq.(1.6.3) into Eq.(1.6.6) leads to the following result:

$$<\hat{Q}>=\int \chi_n^*(\vec{r})q_n\chi_n(\vec{r})d\vec{r}=q_n\int|\chi_n|^2 d\vec{r}=q_n. \qquad (1.6.7)$$

We see from Eq.(1.6.7) that the expectation value of the $\hat{Q}$-operator in the state n is equal to its eigenvalue q_n. Since the results of experimental measurements of the expectation values of any particular quantity should be real numbers, certain limitations are imposed on operators representing physical quantities. The requirement that expectation values be real means that:

$$\int \chi^*(\vec{r})\hat{Q}\chi(\vec{r})d\vec{r}=\left\{\int \chi^*(\vec{r})\hat{Q}\chi(\vec{r})d\vec{r}\right\}^*=\int \chi(\vec{r})\left\{\hat{Q}\chi(\vec{r})\right\}^* d\vec{r}. \qquad (1.6.8)$$

Usually we place ψ^* to the left of ψ. Following this rule in Eq.(1.6.8), one obtains

$$\int \chi^*\hat{Q}\chi d\vec{r}=\int\left\{\hat{Q}\chi(r)\right\}^*\chi(\vec{r})d\vec{r}. \qquad (1.6.9)$$

Operators satisfying Eq.(1.6.8) are called ***hermitian***. In fact, a more general relation is used in quantum mechanics to define a hermitian operator:

$$\int \chi_1^*(\vec{r})\hat{Q}\chi_2(\vec{r})d\vec{r}=\int\left\{\hat{Q}\chi_1(\vec{r})\right\}^*\chi_2(\vec{r})d\vec{r}. \qquad (1.6.10)$$

1.7 MEASUREMENT IN QUANTUM PHYSICS

The central point of quantum physics is that the process of the measurement affects the measured quantity, and this influence can not be reduced to zero.

Due to the strong interaction of a quantum particle with an experimental set up, results of measurements have a probabilistic nature. We shall illustrate this statement with two thought experiments which we describe with the help of the quantum mechanical approach discussed in Sections 1.2 and 1.6.

We complete the experimental set up of Section 1.2.D now for measurements of the electron momentum, see Fig.1.7.1. The set up consists of a source of electrons that are incident on a diaphragm with two holes. This time the diaphragm is able to move up and down when electrons bounce off the edges of the holes. By measuring the recoil of the diaphragm, one can find the expectation value of the momentum $<\hat{\vec{p}}>$. Since our set up is built with the intention of measuring the momentum $\hat{\vec{p}}$, the electron is described by the momentum eigenfunction from Eq.(1.4.21). Using this equation and the normalization condition, one can calculate the expectation (measured) value of this operator

$$<\hat{\vec{p}}>=\int \phi_n^*(\vec{r},t)(-ih\nabla)\phi_n(\vec{r},t)dr=\vec{p}_n\int \phi_n^*\phi_n d\vec{r}=\vec{p}_n. \qquad (1.7.1)$$

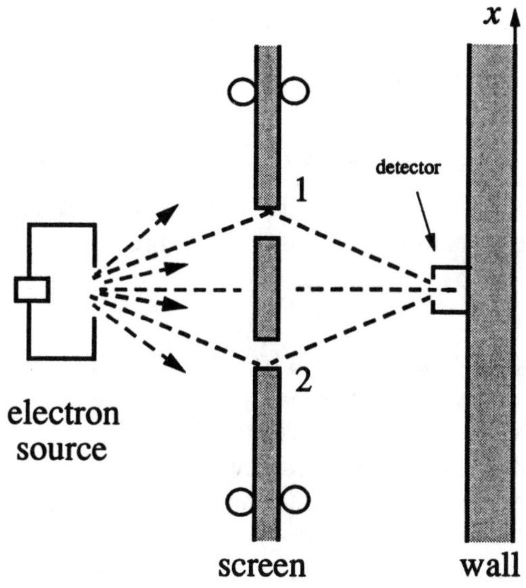

Figure 1.7.1 An experiment in which the recoil of the wall is measured. (After R.P. Feynman, R.B. Leighton, M. Sands. "The Feynman Lectures on Physics. Quantum Mechanics." Addison-Wesley Publishing Company, Menlo Park, CA, 1964).

We see that with this special apparatus, the exact value of the momentum can be measured. It is equal to its eigenvalue $p_{\vec{n}}$. If we make an attempt to measure by this experimental set up the coordinate of the electron, we need to determine the quantity

$$<\vec{r}>= \int \phi_n^*(\vec{r})\vec{r}\phi_n(\vec{r})d\vec{r}. \tag{1.7.2}$$

The wave function $\phi_n(\vec{r})$ here is not an eigenfunction of the $\vec{r}$-operator. The operator $\vec{r}$ has its own eigenfunction ψ_i. For simplicity, we suggest that $\vec{r}$ has a discrete set of eigenvalues $\vec{r}_i$. Then the eigenvalue equation for $\vec{r}$ takes the form

$$\hat{\vec{r}}\psi_i = \vec{r}_i\psi_i. \tag{1.7.3}$$

In order to calculate the expectation value from Eq.(1.7.2), we take advantage of the superposition principle of Eq.(1.5.7) and represent $\phi(\vec{r})$ in the form of a linear combination of ψ_i-functions with some coefficients C_i

$$\phi_p(\vec{r}) = \sum_i C_i\psi_i(\vec{r}). \tag{1.7.4}$$

The substitution of Eq.(1.7.4) into Eq.(1.7.2) results in

$$<\hat{\vec{r}}>= \int\left\{\sum_i C_i\psi_i\right\}^*\hat{\vec{r}}\left\{\sum_k C_k\psi_k\right\}d\vec{r}.$$

Interchanging the integral and summations, one finds

$$<\hat{\vec{r}}> = \sum_i C_i^* \sum_k C_k \int \psi_i^*(\vec{r})\hat{\vec{r}}\psi_k(r)d\vec{r}.$$

Inserting $\vec{r}\psi_k$ from Eq.(1.7.3), we find

$$<\hat{\vec{r}}> = \sum_i C_i^* \sum_k C_k \vec{r}_k \int \psi_i^*(\vec{r})\psi_k(\vec{r})d\vec{r}.$$

By making use of orthogonality conditions of the type of Eq.(1.6.4), one obtains the final result for the expectation value of the coordinate $\vec{r}$,

$$<\hat{\vec{r}}> = \sum_{ik} C_i^* C_k \vec{r}_k \delta_{ik} = \sum_i |C_i|^2 \vec{r}_i. \tag{1.7.5}$$

A wide spectrum of possible values of $\vec{r}$ is found with the help of the experimental set up made for the measurement of the momentum $\vec{p}$. We can measure any of the possible values $\vec{r}_i$, each one occurring with probability $|C_i|^2$.

It is possible to build another experimental arrangement that is adjusted for measurements of the coordinate $\vec{r}$. This arrangement was already used in Section 1.2, Fig.1.2.3. Two holes in the screen are used for measuring the coordinate of the electron $\vec{r}$. The electron in this experiment is described by the coordinate eigenfunction $\psi_i(\vec{r})$ from Eq.(1.7.3). The calculation of the expectation value of the coordinate results in

$$<\hat{\vec{r}}> = \int \psi_i^*(\vec{r})\hat{\vec{r}}\psi_i(\vec{r})d\vec{r} = \vec{r}_i \int |\psi_i|^2 d\vec{r} = \vec{r}_i. \tag{1.7.6}$$

This means that the set up allows measuring exact eigenvalue of the coordinate $\vec{r}_i$.

An attempt to measure the momentum by this set up leads to the expectation value

$$<\hat{\vec{p}}> = \int \psi_i^*(r)\hat{\vec{p}}\psi_i(r)d\vec{r}, \tag{1.7.7}$$

where the $\psi_i(\vec{r})$-functions are not eigenfunctions of the operator $\hat{\vec{p}}$. Expanding $\psi_i(\vec{r})$ in terms of momentum eigenfunction $\phi_n(\vec{r})$ from Eq.(1.4.21),

$$\psi_i(\vec{r}) = \sum_n C_n \phi_n(\vec{r}),$$

we find for the expectation value of the momentum

$$<\hat{\vec{p}}> = \sum_{nn'} C_n^* C_{n'} \vec{p} \int \phi_n^*(\vec{r})\phi_{n'}(\vec{r}) = \sum_n |C_n|^2 \vec{p}. \tag{1.7.8}$$

By measuring the momentum with the set up prepared for the measurement of the coordinate, we obtain a wide spectrum of $\hat{\vec{p}}$-values with probabilities $|C_n|^2$ each.

The reason for the statistical nature of the results in Eq.(1.7.5) and Eq.(1.7.8) is the inevitable destruction of the electron state in the process of improper measurements. These examples show that in measuring one quantity, we disturb the measurement of the other. ***The quantum nature of interactions on the atomic scale introduces uncontrollable disturbances of the system being measured, leading to a real uncertainty in the results.***

All variables in quantum mechanics are divided into two groups. One group consists of quantities which do not disturb each other in the process of measurement. We are able to measure these quantities simultaneously. The second group includes the quantities which are disturbed by the measurement of a quantity of the first group. This impossibility to find all the quantities simultaneously in the same experiment is of fundamental importance in quantum mechanics. This behaviour reflects the uncertainty principle.

The greatest accuracy that exists in quantum physics is obtained in the simultaneous exact measurements of quantities from the same group. ***The quantum state of a particle is given exactly, if all the simultaneously measured quantities are known.*** A microscopic physical object never shows all of its features. Depending on which effect the electron participates in, it shows itself in a different way.

1.8 THE LAW OF MOMENTUM CONSERVATION

The conservation laws of energy E, momentum $\vec{p}$ and angular momentum $\vec{L}=\vec{r}\times\vec{p}$ are of great importance in classical mechanics. One of the possible representations of these conservation laws is the following

$$\frac{dE}{dt}=0 \quad ; \quad \frac{d\vec{p}}{dt}=0 \quad ; \quad \frac{d\vec{L}}{dt}=0.$$

These equations mean that E, $\vec{p}$, and $\vec{L}$ do not change in time, so that these quantities are conserved. These conservation laws of classical mechanics originate from general symmetry: homogeneity of time, homogeneity of space, and isotropy of space, respectively.

These symmetries hold in quantum mechanics, and similar conservation laws hold in quantum mechanics. The calculation of the time derivative in quantum mechanics is not a simple generalization of classical equations. The time derivative in quantum physics should take into account the spread in time of the wave packet that represents the particle. Since the Schrödinger equation describes this spread, the proper procedure consists of using this equation for calculating the time derivative.

The time derivative of the momentum $\vec{p}$ is defined as the quantity whose expectation value $< d\hat{\vec{p}}/dt >$ equals the time derivative of the expectation value of $\hat{\vec{p}}$:

$$<\frac{d\hat{\vec{p}}}{dt}>=\frac{d}{dt}<\hat{\vec{p}}>. \tag{1.8.1}$$

Let us calculate the right hand side of Eq.(1.8.1)

$$<\frac{d\hat{\vec{p}}}{dt}> = \int \phi^* \frac{d\hat{\vec{p}}}{dt} \phi d\vec{r} = \frac{d}{dt} \int \phi^* \hat{\vec{p}} \phi d\vec{r} =$$

$$\int \phi^* \frac{\partial \hat{\vec{p}}}{\partial t} \phi d\vec{r} + \int \frac{\partial \phi^*}{\partial t} \hat{\vec{p}} \phi d\vec{r} + \int \phi^* \hat{\vec{p}} \frac{\partial \phi}{\partial t} d\vec{r}. \tag{1.8.2}$$

Taking the values of $\partial\phi/\partial t$ and $\partial\phi^*/\partial t$ from the Schrödinger equation Eq.(1.5.10), one can find

$$<\frac{d\hat{\vec{p}}}{dt}> = \int \phi^* \frac{\partial \hat{\vec{p}}}{\partial t} \phi d\vec{r} + \frac{i}{\hbar} \int \left(\hat{H}\phi^*\right) \hat{\vec{p}} \phi d\vec{r} - \frac{i}{\hbar} \int \phi^* \hat{\vec{p}} \left(\hat{H}\phi\right) d\vec{r}. \tag{1.8.3}$$

Since the Hamiltonian $\hat{H}$ is a hermitian operator, it satisfies the condition Eq.(1.6.10),

$$\int \left(\hat{H}\phi^*\right)\left(\hat{\vec{p}}\phi\right) d\vec{r} = \int \phi^* \hat{H} \left(\hat{\vec{p}}\phi\right) d\vec{r}. \tag{1.8.4}$$

The substitution of Eq.(1.8.4) into Eq.(1.8.3) leads to

$$<\frac{d\hat{\vec{p}}}{dt}> = \int \phi^* \left\{ \frac{\partial \hat{\vec{p}}}{\partial t} + \frac{i}{\hbar} \hat{H} \hat{\vec{p}} - \frac{i}{\hbar} \hat{\vec{p}} \hat{H} \right\} \phi d\vec{r}. \tag{1.8.5}$$

Comparing Eq.(1.8.5) with Eq.(1.8.1), one finds the relation in operator form

$$\frac{d\hat{\vec{p}}}{dt} = \frac{\partial \hat{\vec{p}}}{\partial t} + \frac{i}{\hbar} \left\{ \hat{H} \hat{\vec{p}} - \hat{\vec{p}} \hat{H} \right\}.$$

If $\hat{\vec{p}}$ does not depend on time explicitly, $\partial\hat{\vec{p}}/\partial t = 0$ and

$$\frac{d\hat{\vec{p}}}{dt} = \frac{i}{\hbar} \left(\hat{H} \hat{\vec{p}} - \hat{\vec{p}} \hat{H} \right) = \frac{i}{\hbar} \left[\hat{H}, \hat{\vec{p}} \right], \tag{1.8.6}$$

where the square brackets are used to shorten the notation. Equation (1.8.6) has the same meaning as Eq.(1.8.5): the wave functions are not written in Eq.(1.8.6) in an explicit form. But one should keep in mind that they are, in fact, present in Eq.(1.8.6).

The quantity on the right hand side of Eq.(1.8.6) is called the ***commutator*** of two operators. Commutators are indicated by square brackets. In Eq.(1.8.6) both operators are differential ones. The result of their action on the wave function ϕ depends, generally speaking, on the order of their successive action. If $\hat{H}\hat{\vec{p}} - \hat{\vec{p}}\hat{H} = 0$, it is said that operators $\hat{H}$ and $\hat{\vec{p}}$ ***commute***. The time derivative of momentum vanishes,

$$\frac{d\hat{\vec{p}}}{dt} = 0,$$

and the derivative of the expectation value is zero, correspondingly,

$$\frac{d}{dt} < \hat{\vec{p}} > = 0,$$

The average value of a conserved quantity remains constant in time. The momentum of the particle is conserved if the momentum operator $\hat{\vec{p}} = -i\hbar\nabla$ commutes with the Hamiltonian $\hat{H}$. The law of momentum conservation has the following form in quantum mechanics

$$\left[\hat{H}, \hat{\vec{p}}\right] = 0. \tag{1.8.7}$$

We shall now prove that commuting operators have the same wave functions. By considering the time-independent Schrödinger equation

$$\hat{H}\psi = E\psi, \tag{1.8.8}$$

and acting from the left by the momentum operator $\hat{\vec{p}}$ one finds,

$$\hat{\vec{p}}\hat{H}\psi = \hat{\vec{p}}E\psi = E\hat{\vec{p}}\psi.$$

Taking into account the commutator Eq.(1.8.6) results in

$$\hat{H}\left(\hat{\vec{p}}\psi\right) = E\hat{\vec{p}}\psi. \tag{1.8.9}$$

The linear equations (1.8.8) and (1.8.9) have the same form. Therefore, ψ and $\hat{\vec{p}}\psi$ can differ only by a constant factor that is usually defined by the normalization condition. The equality of eigenfunctions of commuting operators means that these operators are measurable simultaneously with the same accuracy. By checking the commutators of different operators, we can select those that commute and, therefore, can be measured simultaneously. ***The complete set of commuting observables gives a total description of the system in quantum mechanics.***

For example, consider the very important commutator of the momentum operator $\hat{p}_x = -i\hbar\partial/\partial x$ with the corresponding coordinate x,

$$\left[\hat{p}_x, x\right]\psi = \left(\hat{p}_x x - x\hat{p}_x\right)\psi = -i\hbar\left(\frac{\partial}{\partial x}x - x\frac{\partial}{\partial x}\right)\psi =$$

$$-i\hbar\left(\psi + x\frac{\partial\psi}{\partial x} - x\frac{\partial\psi}{\partial x}\right) = -i\hbar\psi \tag{1.8.10a}$$

or, in operator form,

$$\left[\hat{p}_x, x\right] = -i\hbar. \tag{1.8.10b}$$

This famous equation was written for the first time by Born and Jordan. By generalizing it to any components of $\hat{\vec{p}}$ and $\vec{r}$, one gets

$$\left[\hat{p}_i, x_k\right] = -i\hbar\delta_{ik}. \tag{1.8.11}$$

Here $i,k = x,y,z$, and δ_{ik} is the Kronecker delta. We see that the operators $\hat{\vec{p}}$ and $\vec{r}$ don't commute. Consequently, they don't have common wave functions. They can never be measured simultaneously with the same accuracy. We see that the commutator Eq.(1.8.11) expresses the ***Uncertainty Principle***.

1.9 MOTION OF A FREE PARTICLE

To illustrate the ideas of quantum physics, we solve the eigenvalue problem for a free electron. The time-independent Schrödinger equation in this case, has the form

$$\hat{H}\psi = E\psi, \tag{1.9.1}$$

where

$$\hat{H} = -\frac{\hbar^2}{2m_0}\nabla^2. \tag{1.9.2}$$

We seek the eigenvalues E and the eigenfunctions ψ in the case of free electrons. The conservation law of momentum is very helpful in solving this problem. The momentum operator $\hat{\vec{p}} = -i\hbar\nabla$ commutes with the Hamiltonian in Eq.(1.9.2), because

$$[\nabla^2, \nabla] = 0.$$

This means that $\hat{\vec{p}}$ and $\hat{H}$ have common wave functions. One can use the eigenvalue equation for the momentum $\hat{\vec{p}}$ for finding these common wave functions

$$\hat{\vec{p}}\psi \equiv -i\hbar\nabla\psi = \vec{p}\psi, \tag{1.9.3}$$

where $\vec{p}$ is the eigenvalue of the operator $\hat{\vec{p}}$. Equation (1.9.3) is a first order differential equation that has an exponential solution

$$\psi = Ae^{\frac{i}{\hbar}\vec{p}\cdot\vec{r}}. \tag{1.9.4}$$

The eigenfunction (1.9.4) is also an eigenfunction of the Schrödinger equation (1.9.1). By substituting ψ from Eq.(1.9.4) in Eq.(1.9.1), one finds

$$\left(-\frac{\hbar^2}{2m_0}\nabla^2\right)e^{\frac{i}{\hbar}\vec{p}\cdot\vec{r}} = Ee^{\frac{i}{\hbar}\vec{p}\cdot\vec{r}}$$

or, calculating the second order derivative,

$$\frac{\vec{p}^2}{2m_0}\psi = E\psi,$$

where E is the eigenvalue of $\hat{H}$. Finally we have for the energy and the space dependent part of the wave function,

$$E_{\vec{p}} = \frac{\vec{p}^{\,2}}{2m_0}, \quad \psi_{\vec{p}} = A e^{\frac{i}{\hbar}\vec{p}\cdot\vec{r}}. \tag{1.9.5}$$

The conserved quantity $\vec{p} = \langle \hat{\vec{p}} \rangle$ is the quantum number labeling the free electron energy and wave function Eq.(1.9.5).

1.10 REFLECTION AND TRANSMISSION OF A PARTICLE BY A POTENTIAL STEP

The wave-like quantum mechanical behavior of a particle is clearly demonstrated by the phenomenon of particle scattering by simple potentials.

We now discuss the motion of a free particle that moves in the x-direction and meets the potential step shown in Fig.1.10.1. Real potentials do not change abruptly. Nevertheless, the potential in Fig.1.10.1 approximates some important features of physical systems with sharp changes of physical properties.

Generally speaking, the motion of the particle is described by the time dependent Schrödinger equation (1.4.10) with a potential energy that depends on the x – coordinate only. The wave function of a particle is a product of amplitude $\psi(\vec{r})$ and oscillating in time function according to $e^{-\frac{i}{\hbar}Et}$ (see Eq.(1.5.5)). The space amplitude $\psi(\vec{r})$ satisfies the time-independent Schrödinger Eq.(1.5.6), where the

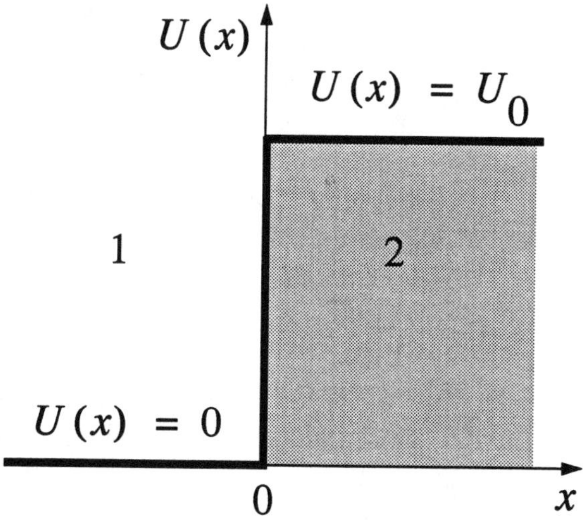

Figure 1.10.1 Potential step

potential U is a function of the x-coordinate. The three-dimensional Eq.(1.5.6) allows the separation of variables and the wave function is represented in the form

$$\psi(\vec{r}) = \chi(y,z)\psi(x)\,, \tag{1.10.1}$$

where $\chi(y,z)$ describes the motion of the particle in the direction perpendicular to the x-axis and $\psi(x)$ corresponds to one-dimensional motion of the particle in the potential $U(x)$. Substituting Eq.(1.10.1) into Eq.(1.5.6) and separating variables, we find that $\chi(y,z)$ satisfies the two-dimensional free particle Schrödinger equation

$$-\frac{\hbar^2}{2m_0}\left\{\frac{d^2}{dy^2}+\frac{d^2}{dz^2}\right\}\chi = E\chi. \tag{1.10.2}$$

The solution of this equation has a form similar to the solutions given by Eqs.(1.9.5),

$$E = \frac{(p_y^2 + p_z^2)}{2m_0}, \quad \chi(y,z) = Ae^{\frac{i}{\hbar}(p_y y + p_z z)}, \tag{1.10.3}$$

where A is the normalization constant. The wave function (1.10.3) means that the particle moves in the y and z directions as a free particle.

The equation for $\psi(x)$ has the form

$$\left\{-\frac{\hbar^2}{2m_0}\frac{d^2}{dx^2}+U(x)\right\}\psi(x) = E\psi(x) \tag{1.10.4}$$

where the potential energy corresponding to Fig.1.10.1 is given by

$$U(x) = \left\{\begin{matrix} U_0 & \text{for} \quad x \geq 0 \\ 0 & \text{for} \quad x \leq 0 \end{matrix}\right\}. \tag{1.10.5}$$

It is seen from Eq.(1.10.4) that the motion of the particle is different for $x < 0$ (region 1) and for $x > 0$ (region 2). In region 1, the potential energy vanishes, and we have a Schrödinger equation of the form

$$-\frac{\hbar^2}{2m_0}\frac{d^2\psi_1(x)}{dx^2} = E\psi_1(x). \tag{1.10.6}$$

This is the equation for the free particle that is incident from the left on the potential step Eq.(1.10.5). In region 2, we have the following Schrödinger equation

$$\left\{-\frac{\hbar^2}{2m_0}\frac{d^2}{dx^2}+U_0\right\}\psi_2(x) = E\psi_2(x). \tag{1.10.7}$$

This equation describes the particle that is scattered by the potential $U(x)$. Eqs.(1.10.6) and (1.10.7) can be rewritten in the form of typical wave equations

$$\frac{d^2\psi_1(x)}{dx^2} = -k_1^2\psi_1(x), \tag{1.10.8a}$$

where

$$k_1 = \left(\frac{2m_0E}{\hbar^2}\right)^{1/2}, \tag{1.10.8b}$$

and

$$\frac{d^2\psi_2(x)}{dx^2} = -k_2^2\psi_2(x), \tag{1.10.9a}$$

where

$$k_2 = \left(\frac{2m_0(E-U_0)}{\hbar^2}\right)^{1/2}. \tag{1.10.9b}$$

Since the wave equations (1.10.8) and (1.10.9) are second order differential equations, the general solution is an appropriate combination of exponentials $e^{\pm ik_1x}$ and $e^{\pm ik_2x}$. The plane de Broglie wave that is approaching the potential step from the left should behave at $x = 0$ similarly to an electromagnetic wave at the boundary of two regions with different refractive indices. This means that there is a finite probability for the particle to be transmitted into the region 2 or to be reflected back into the region 1. Particle can go in either direction.

In region 1 ($x < 0$), there are incident and reflected waves that travel in opposite directions. The corresponding wave function has the form

$$\psi_1(x) = a_1e^{ik_1x} + b_1e^{-ik_1x}, \tag{1.10.10}$$

where a_1 and b_1 are the amplitudes of the incident and reflected waves, respectively. The incident wave travels to the right in the positive direction of the x-axis. The reflected wave travels to the left. For simplicity, we take the amplitude of the incident wave to be equal to unity

$$a_1 = 1.$$

In region 2 ($x > 0$), we have only the wave transmitted through the potential step and traveling to the right

$$\psi_2(x) = a_2e^{ik_2x}. \tag{1.10.11}$$

Since the Schrödinger equation is a differential equation of the second order, the wave functions and their first derivatives should be continuous at the boundary

$$\psi_1(x)\,|_{x=0} = \psi_2(x)\,|_{x=0}, \tag{1.10.12a}$$

$$\psi'_1(x)\,|_{x=0} = \psi'_2(x)\,|_{x=0}\,. \tag{1.10.12b}$$

a. We begin with the case when the energy of the particle E is greater than the step height U_0, see Fig.1.10.2,

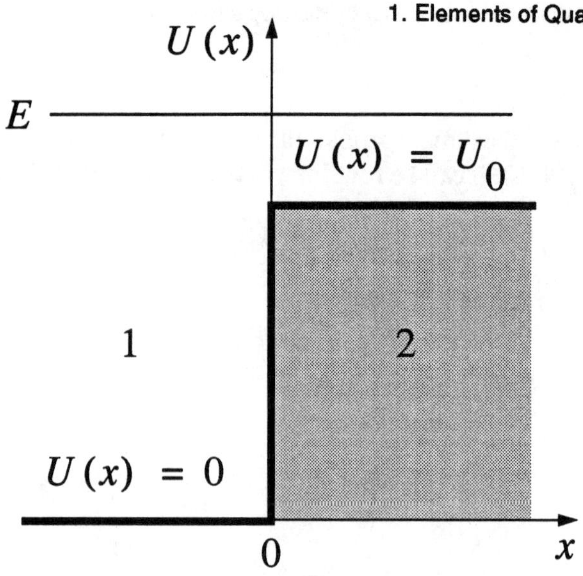

Figure 1.10.2 The relation between total and potential energies for particle incident upon a potential step with total energy larger than the height of the step: $E > U(x)$.

$$E > U_0. \tag{1.10.13}$$

By substituting the wave functions (1.10.10) and (1.10.11) in the boundary conditions (1.10.12), we obtain linear equations for the determination of the coefficients b_1 and a_2, keeping in mind that $a_1 = 1$,

$$1 + b_1 = a_2, \tag{1.10.14a}$$

$$1 - b_1 = \frac{k_2}{k_1} a_2. \tag{1.10.14b}$$

Solving this system for b_1 and a_2, we find

$$b_1 = \frac{k_1 - k_2}{k_1 + k_2}; \quad a_2 = \frac{2k_1}{k_1 + k_2}. \tag{1.10.15}$$

The substitution of Eqs.(1.10.15) into Eqs.(1.10.10) and (1.10.11) leads to

$$\psi_1(x) = e^{ik_1x} + \frac{k_1 - k_2}{k_1 + k_2} e^{-ik_1x}, \tag{1.10.16a}$$

$$\psi_2(x) = \frac{2k_1}{k_1 + k_2} e^{ik_2x}. \tag{1.10.16b}$$

The reflection and transmission coefficients of the particle are defined in terms of ratios of probability fluxes. The probability flux is the probability per second that a particle is passing a certain point in a particular direction.

The incident probability flux is the probability per second of finding a particle passing the point x in region 1 in the positive direction of the x-axis. The reflected probability flux R is the probability per second of finding a particle passing the point x at $x < 0$ in the direction of decreasing x. Because the probability per second is proportional to the distance the particle travels per second, the probability flux is proportional not only to the intensity of the wave $|b_1|^2$ but also to the velocity of the particle. According to Eq.(1.10.15) the reflection coefficient equals

$$R = \frac{v_1 |b_1|^2}{v_1 |a_1|^2} = |b_1|^2 = \left\{\frac{k_1 - k_2}{k_1 + k_2}\right\}^2, \tag{1.10.17}$$

where

$$v_1 = \frac{p_1}{m_0} = \frac{\hbar k_1}{m_0}.$$

This is the reflection coefficient for the particle that moves above the potential step. The classical particle would travel easily from region 1 into region 2. It slows down at $x = 0$ due to the reduction of its kinetic energy from the value E in region 1 down to the value $(E - U_0)$ in region 2. The quantum particle, being a de Broglie wave, behaves differently. It has a finite probability of being reflected at $x = 0$ even having an energy E above the height of the potential step.

The second important characteristic of the particle is the transmitted probability flux T that equals the probability per second that a particle will cross some point in the direction of increasing x. Taking a_2 from Eq.(1.10.15), we find

$$T = \frac{v_2 |a_2|^2}{v_1 |a_1|^2} = \frac{v_2}{v_1}\left\{\frac{2k_1}{k_1 + k_2}\right\}^2. \tag{1.10.18}$$

Taking into account that $v_1 = p_1/m_0 = \hbar k_1/m_0$ and $v_2 = p_2/m_0 = \hbar k_2/m_0$, we find

$$T = \frac{4k_1 k_2}{(k_1 + k_2)^2}. \tag{1.10.19}$$

The coefficients R and T satisfy the relation

$$R + T = 1.$$

This result has the clear probabilistic meaning that at the boundary $x = 0$ the particle is either reflected or transmitted. Substituting k_1 and k_2 from Eqs.(1.10.8b) and (1.10.9b) into Eqs.(1.10.17) and (1.10.18), one obtains

$$R = \frac{1 - \sqrt{1 - U_0/E}}{1 + \sqrt{1 - U_0/E}}, \tag{1.10.20}$$

$$T = 1 - R = \frac{2\sqrt{1 - U_0/E}}{1 + \sqrt{1 - U_0/E}}. \tag{1.10.21}$$

The numerical estimation of R and T for a particle with $E = 2\,U_0$ gives

$$R = 3\% \quad \text{and} \quad T = 97\%.$$

It is seen from Eq.(1.10.20) that for $E \to \infty$ the reflection coefficient $R \to 0$ and the reflected flux vanishes. Figure 1.10.3 illustrates the probability density $|\psi(x)|^2$. In region 2, $x > 0$, the wave function $\psi(x)$ is a plane wave traveling to the right. The corresponding probability density is constant. In region 1 $(x < 0)$, the

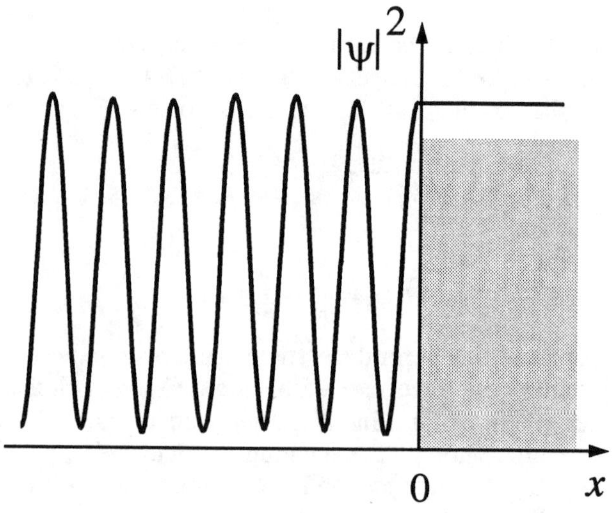

Figure 1.10.3 The probability density for the eigenfunction Eq.(1.10.16) in the case $E > U(x)$.

wave function is the superposition of incident and reflected waves. This superposition oscillates and has a minimum value greater than zero.

b. We next consider the case when the total energy is less than the step height U_0, see Fig.1.10.4,

$$E < U_0. \tag{1.10.22}$$

In classical physics there is no transition from region 1 into region 2, because the energy conservation law requires E to be always larger than or equal to the potential energy U_0. In quantum physics, when the particle is represented by a de Broglie wave, the situation is different. In region 1, $x < 0$, we have a superposition of the incident and reflected waves. The wave function $\psi_1(x)$ has the form Eq.(1.10.10). In region 2, $x > 0$, the wave vector k_2 under condition (1.10.22) becomes an imaginary quantity

$$k_2 = i\left\{\frac{2m_0(U_0 - E)}{\hbar^2}\right\}^{1/2} = ik^*. \tag{1.10.23}$$

The wave function ψ_2 from Eq.(1.10.11) takes the form

$$\psi_2(x) = a_2 e^{-k^* x}. \tag{1.10.24}$$

The wave function (1.10.24) decays exponentially in region 2. The probability density $|\psi_2(x)|^2$ defines the finite probability of finding an electron inside the potential step. A microscopic particle can penetrate into a region forbidden to a macroscopic classical particle.

The penetration into the classical forbidden region 2 is an impressive result of quantum physics. This penetration does not mean that the particle is stored inside the potential step. In dealing with waves, we always have an uncertainty in the coordinate of the order of the electron wavelength $\lambda = 2\pi/k^*$. The penetration depth is just of the order of λ.

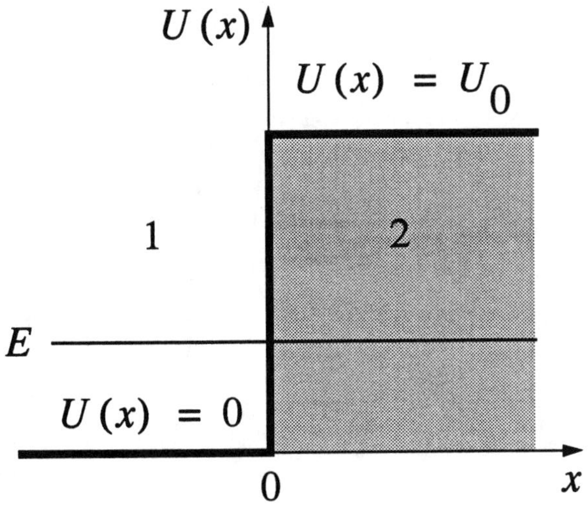

Figure 1.10.4 The relation between total and potential energies for a particle incident upon a potential step with total energy lower than the height of the step: $E < U(x)$.

Substituting the wave functions (1.10.10) and (1.10.11) into the boundary conditions (1.10.12), one obtains the equations for the coefficients b_1 and a_2; a_1 is taken to be unity

$$1 + b_1 = a_2,$$

$$1 - b_1 = \frac{ik^*}{k_1} a_2.$$

The solution of these equations has the form

$$b_1 = \frac{k_1 - ik^*}{k_1 + ik^*}; \quad a_2 = \frac{2k_1}{k_1 + ik^*}.$$

The reflection coefficient R equals unity

$$R = | b_1 |^2 = | \frac{k_1 - ik^*}{k_1 + ik^*} |^2 = 1. \qquad (1.10.25)$$

This means that the reflection is total. The transmission coefficient therefore vanishes,

$$T \equiv 1 - R = 0.$$

The results do not contradict the statement made above that there is a finite probability of finding the particle inside the step. The reflection of the particle does not occur at the boundary itself. The particle, being a wave, penetrates the barrier to a depth of the order of the wavelength λ and then returns to region 1. The corresponding probability density is shown in Fig.1.10.5. The penetration depth is equal to $\Delta x \approx 1/k^*$. The squared modulus of the wave function, $| \psi_2(x) |^2 \sim e^{-2k^* x}$, goes to zero for $x \gg \Delta x$. Since $k^* = [2m_0(U_0 - E)]^{1/2}/\hbar$, the penetration depth is

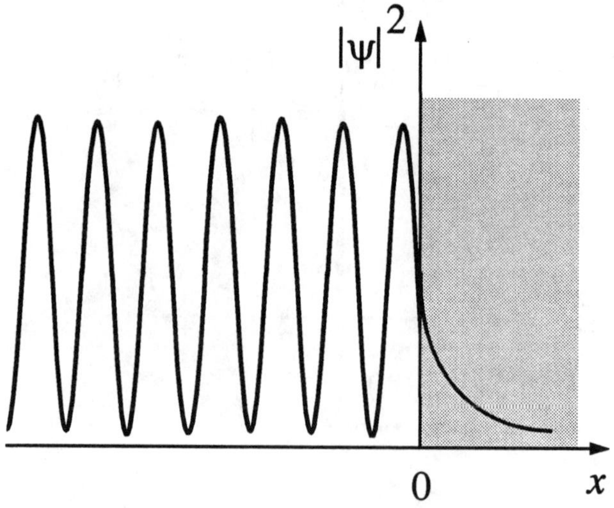

Figure 1.10.5 The probability density $|\psi|^2$ in the case when $E < U_0$.

$$\Delta x \geq \left(\frac{\hbar^2}{2m_0(U_0-E)}\right)^{1/2}.$$

If the experiment localizes the particle over a distance Δx the corresponding uncertainty in the particle momentum reaches (according to uncertainty principle) the value

$$\Delta p > \frac{\hbar}{\Delta x} = (2m_0(U_0-E))^{1/2}.$$

The corresponding uncertainty in the energy of the particle is

$$\Delta E = \frac{(\Delta p)^2}{2m_0} > (U_0-E). \tag{1.10.26}$$

The result means that in the process of measurement, the particle receives an energy ΔE that is larger than the "deficit" of kinetic energy in the step region. That is why the particle is able to enter region 2. There is no violation of the energy conservation law, since it holds with an accuracy given by Eq.(1.10.26). Making a numerical estimate of the particle penetration depth, we find that for $(U_0 - E) \approx 1eV$ the penetration depth $\Delta x \approx 1Å$ and the probability density

$$|\psi_2|^2 \approx 30\%.$$

If we take $\Delta x \approx 5Å$, the probability density decreases rapidly

$$|\psi_2|^2 \approx 0.5\%.$$

The probability density for $\Delta x \approx 10Å$ becomes negligible

$$|\psi_2|^2 \approx 5\times10^{-8}\%.$$

When the height of the potential step is taken to be infinite, $U_0 \rightarrow \infty$, it follows that

$$\lim_{U_0\to\infty} b_1 = \lim_{U_0\to\infty} \frac{1-\sqrt{1-U_0/E}}{1+\sqrt{1-U_0/E}} = -1 . \qquad (1.10.27)$$

The substitution of $b_1 = -1$ in the wave function Eq.(1.10.10), with a_1 equal to unity, leads to

$$\psi(x) = e^{ik_1x} - e^{-ik_1x} = 2i \sin k_1 x.$$

At the boundary $x = 0$ we have

$$\psi(x)\,|_{x=0} = 0. \qquad (1.10.28)$$

The wave function vanishes at the boundary, which corresponds to a potential step of infinite height. The behavior of the probability density $|\psi(x)|^2$ is shown for this case in Fig.1.10.6. Actually, this vanishing of the wave function holds at any boundary where the potential energy U_0 is infinite.

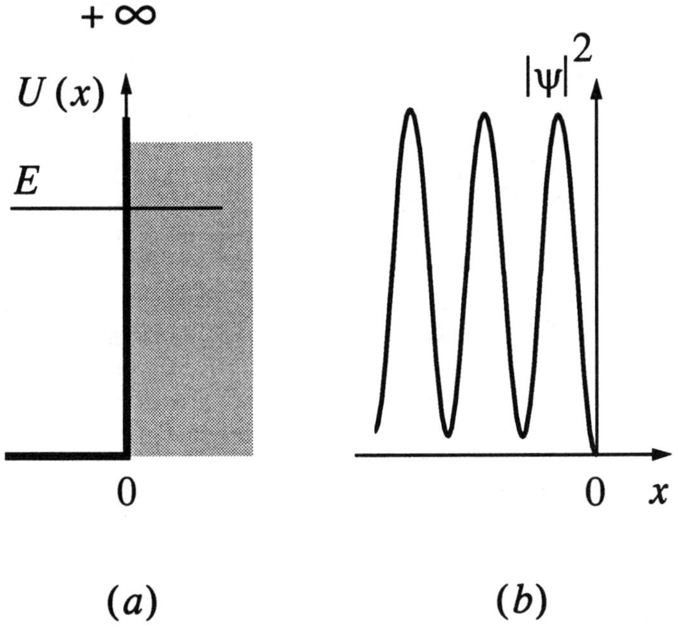

Figure 1.10.6 Vanishing of the probability density at a potential step of infinite height: (a) infinite potential wall; (b) probability density.

1.11 TUNNELING

Now we consider the potential barrier defined by the potential energy

$$U(x) = \begin{cases} U_0 & \text{for } 0 \le x \le a \\ 0 & \text{for } x \le 0, x \ge a \end{cases} \qquad (1.11.1)$$

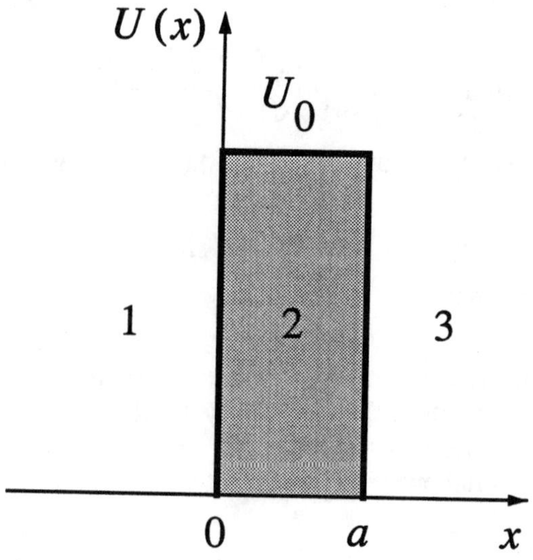

Figure 1.11.1 A potential barrier.

which is shown in Fig.1.11.1. A classical particle incident on the barrier Eq.(1.11.1) from the left in the direction of increasing x is reflected by the barrier, if the energy of the particle E is less than the potential barrier height U_0 ($E < U_0$), or it is easily transmitted when $E > U_0$.

The discussion of the step potential in Section 1.10 has shown that there appears a finite probability of finding a quantum particle inside the step. In the case of a barrier that is thin enough, there is a finite probability of finding the particle behind the barrier in the region $x > a$. This process of penetration through the barrier results from the wave-like behavior of the quantum particle.

In case of the potential barrier, there are three different regions of the particle motion. In the region 1 on the left side from the barrier we have incident and scattered waves given by Eq.(1.10.10) with $a_1 = 1$:

$$\psi_1(x) = e^{ik_1x} + b_1 e^{-ik_1x}. \tag{1.11.2}$$

Region 2 corresponds to $0 < x < a$. If the total energy of the particle is less than the height of the potential, $E < U_0$, we have an exponential dependence of the wave function on x that results from the imaginary value of $k_2 = ik^*$

$$\psi_2(x) = a_2 e^{-k^*x} + b_2 e^{+k^*x} \tag{1.11.3}$$

where

$$k^* = \left(\frac{2m_0(U_0 - E)}{\hbar^2}\right)^{1/2}.$$

The wave transmitted into region 3 has the form of a wave traveling to the right.

$$\psi_3(x) = b_3 e^{ik_1x}, \tag{1.11.4}$$

where $k_3 = k_1$.

Boundary conditions of the type of Eqs.(1.10.12a,b) hold at both $x = 0$ and $x = a$. The substitution of the wave functions Eqs.(1.11.2), (1.11.3), and (1.11.4) allows finding the coefficients b_1, a_2, b_2, and b_3 in terms of k_1, k^*, and $k_3 = k_1$.

When the "tail" of the wave function inside the barrier reaches the second boundary at $x = a$, it is partly transmitted into region 3 and partly reflected back into region 2. The wave function ψ_2 from Eq.(1.11.3) describes multiple back and forward scattering of the wave inside the barrier. The solution is simplified in the case of a steep barrier whose width a is larger than the de Broglie wavelength of the particle $\lambda = 1/k^*$

$$a >> \lambda = \frac{1}{k^*}. \tag{1.11.5}$$

In this case the "tail" of the wave function ψ_2 decreases exponentially and becomes very small at the boundary $x = a$:

$$\frac{\psi_2(x=a)}{\psi_2(x=0)} \approx e^{-k^*a} \ll 1. \tag{1.11.6}$$

The small value of $\psi_2(a)$ defines the amplitude of the wave in the region 3 . The magnitude of the coefficient b_3 in Eq.(1.11.4) is also small

$$b_3 = \frac{\psi_2(a)}{\psi_2(0)} \approx e^{-k^*a} \ll 1. \tag{1.11.7}$$

The probability density that corresponds to the wave functions Eqs.(1.11.2), (1.11.3), and (1.11.6) is given in Fig.1.11.2. In region 1 ($x < 0$), the wave function

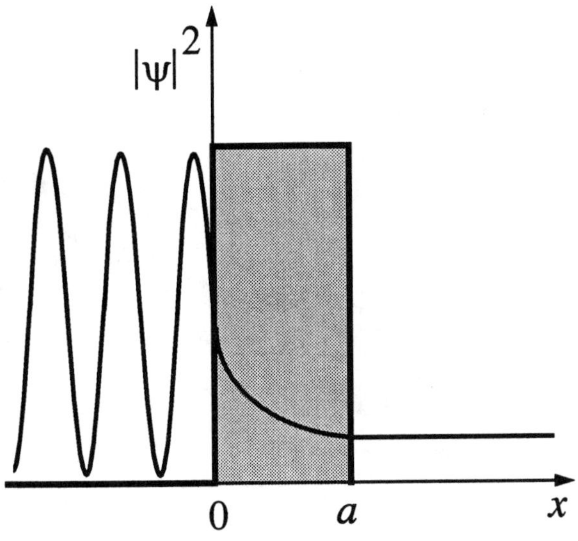

Figure 1.11.2 The probability density $|\psi|^2$ for a typical barrier penetration situation.

is a combination of incident and reflected waves, Eq.(1.11.2). The corresponding probability density is an oscillating function. In the region 3 ($x > a$), we have a traveling wave, and the corresponding probability density is constant.

The transmission coefficient T in this case equals

$$T = \frac{v_3 \mid b_3 \mid^2}{v_1 \mid a_1 \mid^2} = \mid b_3 \mid^2 = \frac{16E}{U_0}\left(1 - \frac{E}{U_0}\right) e^{-2a\left(\frac{2m_0(U_0 - E)}{\hbar^2}\right)^{1/2}} . \qquad (1.11.8)$$

This result means that a particle of mass m_0 and total energy E incident on the potential barrier $U_0 > E$ has a finite probability T of penetrating the barrier. The effect is called ***tunneling***.

There exists an optical analog of tunneling that is called ***frustrated internal reflection.*** Total internal reflection occurs when a light beam in some medium impinges on an interface with a medium of lower refractive index. When the angle of incidence is greater than a certain critical value total internal reflection occurs. Frustrated internal reflection occurs when the light wave propagates through an air gap that separates two media and is thin enough. The scheme of the optical experiment is shown in Fig.1.11.3.

Very impressive examples of barrier penetration by electrons exist in solid state electronics. An example is the tunnel diode that is used in fast electronic circuits. Barrier penetration is used for rapid switching of the current on and off. These switches operate at frequencies up to 10 GHz.

We finish by considering a particle tunneling through a potential barrier of arbitrary form shown in Fig.1.11.4. The steep barrier is approximated in Fig.1.11.4 by a set of step potentials. According to the rule of multiplication of probabilities, the total transmission coefficient is equal to the product of transmission coefficients of each step

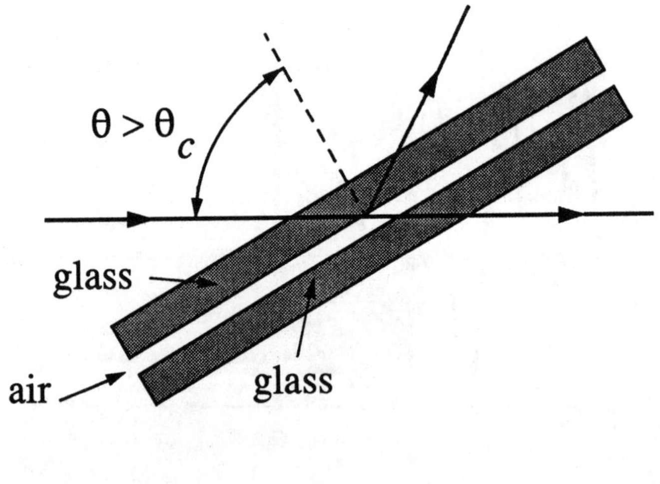

Figure 1.11.3 Frustrated total internal reflection. Some of the light is transmitted through the air gap if the gap is sufficiently narrow.

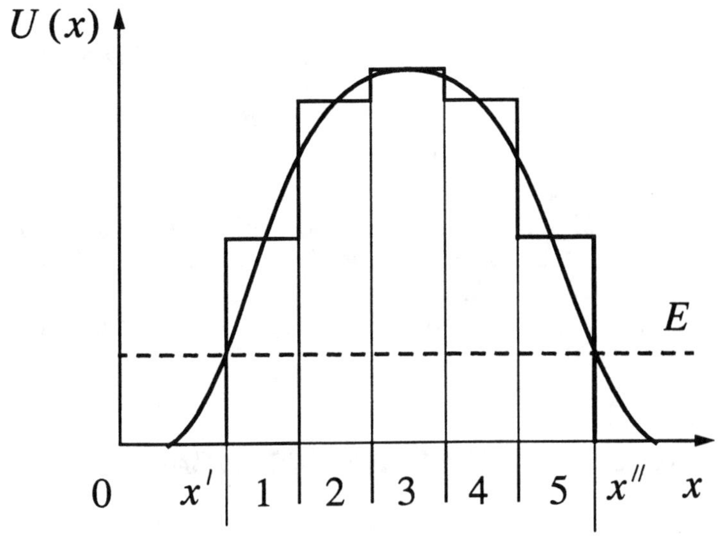

Figure 1.11.4 To illustrate the derivation of an approximate expression for the transmission coefficient of the step barrier. The transmission coefficient is the product of the transmission coefficients for all the rectangular step-barriers, x' and x'' are turning points which are defined by the energy conservation law $E < U(x)$.

$$T = T_1 \times T_2 \times T_3 \times \ldots\ldots \times T_n, \tag{1.11.9}$$

or

$$\ln T = \sum_i \ln T_i, \tag{1.11.10}$$

where

$$\ln T_i = -2\left(\frac{2m_0(U(x_i) - E)}{\hbar^2}\right)^{1/2} dx_i. \tag{1.11.11}$$

Here dx_i is the width of the i-th step. The summation over i in Eq.(11.10) can be replaced by integration over x

$$\ln T = -2\int_{x'}^{x''} dx\left(\frac{2m_0(U(x) - E)}{\hbar^2}\right)^{1/2}. \tag{1.11.12}$$

Here the limits of integration are x' and x''. They define the region of the space $x' < x < x''$ in which the particle is allowed to move by the energy conservation law, which states that the total energy of the particle

$$E = \frac{m_0 v^2}{2} + U(x) = Constant \tag{1.11.13}$$

has a constant value. Since the kinetic energy $m_0 v^2/2$ is a positive quantity, the total energy E obeys the condition

$$E \geq U(x). \tag{1.11.14}$$

If the potential energy has a maximum, see Fig.1.11.4, the allowed regions of classical particle motion are $x < x'$ and $x > x''$. The points x' and x'' are defined by the condition

$$E = U(x). \tag{1.11.15}$$

They are called ***turning points***.

It follows from Eq.(1.11.12) that T decreases with the mass of the particle. The transmission coefficient has a specific dependence on the energy of the particle E: T increases with increasing E. This happens for two reasons: (1) the decrease of the integral in Eq.(1.11.12) and (2) the decrease of the region of integration defined by the turning points x' and x''.

1.12 EIGENVALUE PROBLEM FOR A POTENTIAL WELL

The next impressive results in quantum physics are obtained for particles bound in a potential well. The discrete quantization of their energy occurs in these systems.

The simplest problem of the particle localization is the problem of a particle in a square potential well with infinitely high walls, see Fig.1.12.1. The potential energy has the form

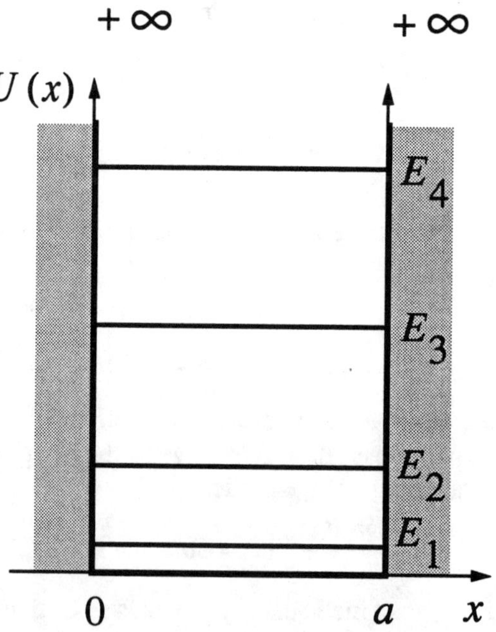

Figure 1.12.1 A potential well with infinite walls. The first few eigenvalues are shown.

$$U(x)=\begin{cases}\infty & \text{for} \quad x>a \\ 0 & \text{for} \quad 0\le x\le a \\ \infty & \text{for} \quad x<0\end{cases}. \tag{1.12.1}$$

Here a is the width of the potential well.

We are interested in the allowed values of the energy of the particle. A classical particle with an arbitrary energy E would oscillate back and forth between the walls of the potential well. The behavior of a quantum particle is different. It obeys the time-independent Schrödinger equation (1.5.6). Since $U(x)$ depends on x only, the problem is one-dimensional. The solution of the second order differential Schrödinger equation is the proper combination of $e^{\pm ikx}$.

The wave function $\psi(x)$ should satisfy boundary conditions. Since the height of the potential well walls is infinite, the wave function should vanish at their positions according to Eq.(1.10.28), that is

$$\psi(x)\,|_{x=0}=0, \tag{1.12.2}$$

$$\psi(x)\,|_{x=a}=0. \tag{1.12.3}$$

The linear combination of exponentials which satisfies the boundary conditions Eqs.(1.12.2) and (1.12.3) has the form of a standing sinusoidal wave

$$\psi(x)=\sin kx. \tag{1.12.4a}$$

This wave function vanishes at the boundary $x=a$ if

$$ak=n\pi, \quad n=1,2,3.... \tag{1.12.4b}$$

By combining Eq.(1.12.4b) with de Broglie relation Eq.(1.3.4) one can find the allowed values of the electron momentum

$$\frac{p_n a}{\hbar}=n\pi, \quad n=1,2,3.... \tag{1.12.5}$$

It is seen from Eq.(1.12.5) that the values of $\vec{p}$ are quantized.

The corresponding allowed values of the energy follow from the relation $E = p^2/2m_0 = \hbar^2k^2/2m_0$, where p is substituted from Eq.(1.12.5)

$$E_n=\frac{\pi^2\hbar^2}{2m_0a^2}n^2. \tag{1.12.6}$$

Equation (1.12.6) means that only certain discrete values of the energy are allowed. It is said that the ***energy*** of the particle in the potential well ***is quantized***. The smallest eigenvalue, corresponding to $n=1$, equals

$$E_1=\frac{\pi^2\hbar^2}{2m_0a^2}\times 1. \tag{1.12.7}$$

We see that the energy of the quantum particle can not vanish. The energy E_1 defined by Eq.(1.12.7) is called the ***zero point energy***. The result is the direct consequence of the uncertainty principle. If the particle is found inside a potential well of the width a, its coordinate is known with uncertainty

$$\Delta x\approx a.$$

According to the uncertainty principle Eq.(1.3.7), the particle momentum equals

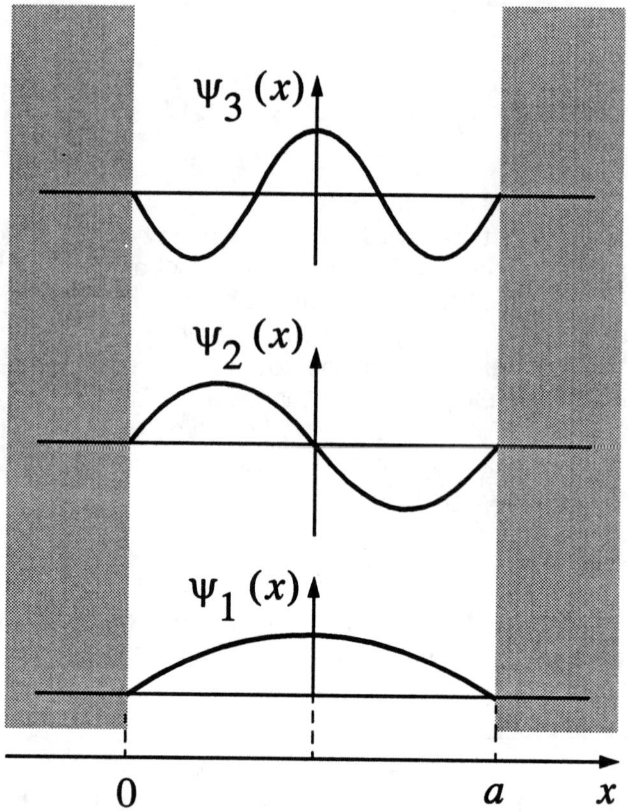

Figure 1.12.2 The first few eigenfunctions of an infinite square well.

$$\Delta p = \frac{\hbar}{\Delta x} = \frac{\hbar}{a}.$$

This value of the momentum leads to the zero point energy

$$E_1 = \frac{(\Delta p)^2}{2m_0} \cong \frac{\hbar^2}{m_0 a^2} \qquad (1.12.8)$$

which is of the order of magnitude of E_1, according to Eq.(1.12.7).

The second energy level from Eq.(1.12.6) with $n = 2$ is

$$E_2 = \frac{\pi^2 \hbar^2}{2m_0 a^2} \times 4 = 4E_1\,,$$

the third is

$$E_3 = \frac{\pi^2 \hbar^2}{2m_0 a^2} \times 9 = 9E_1\,,$$

and so on. The set of allowed energies is shown in Fig.1.12.1. Fig.1.12.2 shows the corresponding wave functions. We see that the number of nodes of each wave function is equal to $n + 1$, where n is the quantum number of the wave function.

A more realistic model of the potential well is one with walls of finite height U_0, see Fig.1.12.3. Potential well of this type occurs near the interface of two

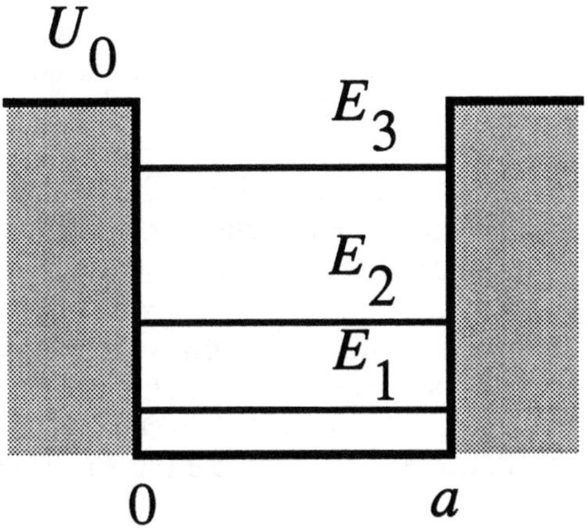

Figure 1.12.3 A square potential well with finite walls and its three bound eigenvalues.

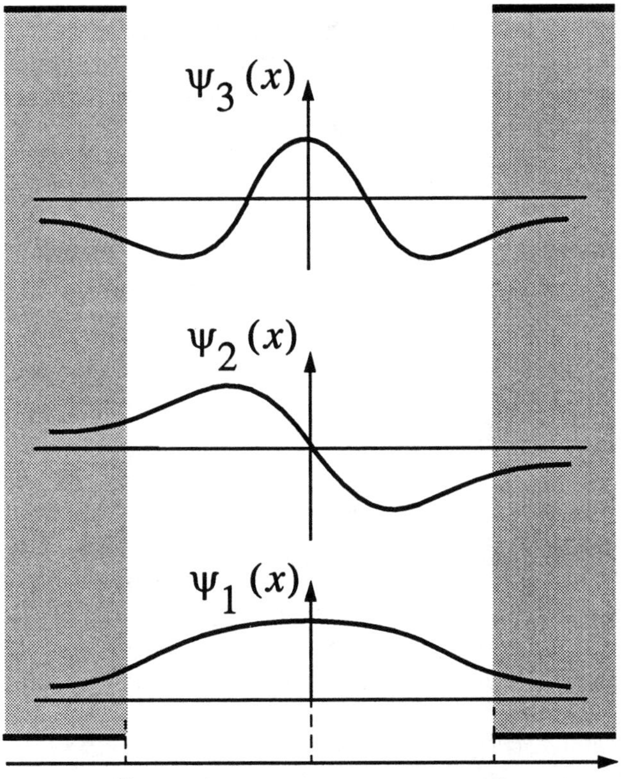

Fiqure 1.12.4 Three bound eigenfunctions for a square potential well of finite depth.

different materials in semiconductor devices. The finite height of the barrier results in tunneling into the barriers at $x < 0$ and $x > a$, since the boundary condition (1.10.28) is no longer valid. The oscillating wave functions from inside the barrier go smoothly into evanescent wave functions for $x < 0$ and $x > a$. The boundary conditions of the type given by Eq.(1.10.12) result in this case in transcendental equations for the definition of the allowed values of the wave vector k. The qualitative behavior of the wave functions in the well with finite walls is shown in Fig.1.12.4.

In any one-dimensional potential well there is always at least one energy level. This is specific to the one-dimensional case. In three-dimensions, an energy level occurs when the potential energy of the particle U is equal to or larger than the kinetic energy $p^2/2m_0 \cong \hbar^2/m_0a^2$, $|U| \gg \hbar^2/m_0a^2$, where a is the linear dimension of the well.

1.13 HYDROGEN ATOM

The bound state of an electron in a hydrogen atom is the next problem of importance in quantum physics. The problem is more complicated than before since the atom consists of the negative electron and the positive nucleus that are bound by the three dimensional Coulomb potential

$$U(r) = -\frac{e^2}{4\pi\varepsilon_0 r}.$$

Here e is the charge of the electron, ε_0 is the permittivity of vacuum, and r is the distance from the charge e to the positively charged nucleus. It is convenient to introduce the notation

$$e'^2 = \frac{e^2}{4\pi\varepsilon_0}, \tag{1.13.1}$$

and to reduce the interaction Eq.(1.13.1) to the simpler form

$$U(r) = -\frac{e'^2}{r}. \tag{1.13.2}$$

The Coulomb potential energy Eq.(1.13.2) is plotted in Fig.1.13.1. Since the mass of the nucleus is 1000 times heavier then the mass of electron, we consider the nucleus to be at rest. The energy levels and wave functions for an electron in a hydrogen atom are found from the solution of the time-independent Schrödinger equation (1.5.6) with the kinetic energy of the electron only and the potential energy given by Eq.(1.13.2),

$$\left\{-\frac{\hbar^2}{2m_0}\nabla^2 - \frac{e'^2}{r}\right\}\psi(\vec{r}) = E\psi(\vec{r}). \tag{1.13.3}$$

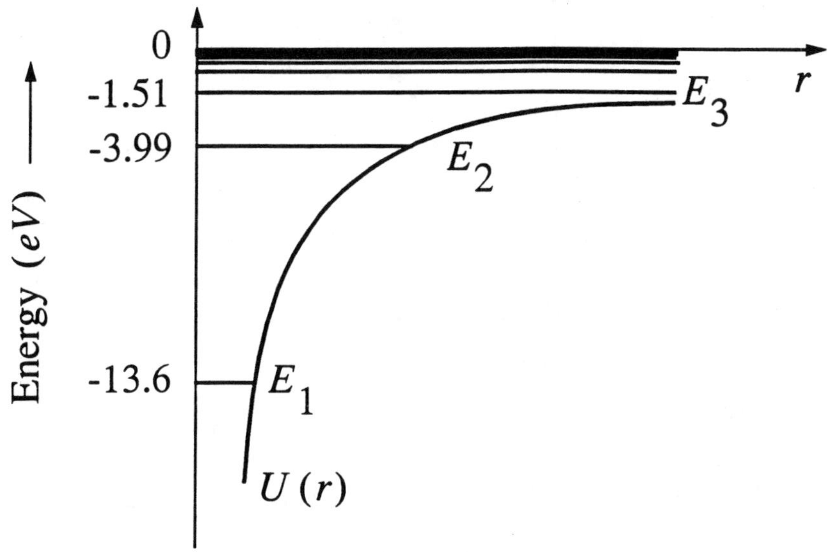

Figure 1.13.1 The Coulomb potential $U(r)$ and its eigenvalues E_n. (After R. Eisberg, R. Resnick. "Quantum Physics of Atoms, Molecules, Solids, Nuclei, and Particles." John Wiley & Sons, N. Y., 1985)

The potential energy in Eq.(1.13.3) depends on r only and has spherical symmetry. It is said that the electron moves in the ***centro-symmetric field*** of the nucleus. Taking this symmetry into account, we shall solve Eq.(1.13.3) by the technique of separation of variables. This procedure is possible in the spherical polar coordinates shown in Fig.1.13.2

$$x = r\sin\theta\cos\phi, \tag{1.13.4a}$$

$$y = r\sin\theta\sin\phi, \tag{1.13.4b}$$

$$z = r\cos\theta. \tag{1.13.4c}$$

In terms of these spherical coordinates the Laplacian operator in Eq.(1.13.3) becomes

$$\nabla^2 = \frac{1}{r^2}\frac{\partial}{\partial r}\left(r^2\frac{\partial}{\partial r}\right) + \frac{1}{r^2}\left\{\frac{1}{\sin\theta}\frac{\partial}{\partial\theta}\left(\sin\theta\frac{\partial}{\partial\theta}\right) + \frac{1}{\sin^2\theta}\frac{\partial^2}{\partial\phi^2}\right\}. \tag{1.13.5}$$

The Schrödinger equation (1.13.3) in spherical coordinates then takes the form

$$-\frac{\hbar^2}{2m_0}\left\{\frac{1}{r^2}\frac{\partial}{\partial r}\left(r^2\frac{\partial\psi}{\partial r}\right) + \frac{1}{r^2}\left[\frac{1}{\sin\theta}\frac{\partial}{\partial\theta}\left(\sin\theta\frac{\partial\psi}{\partial\theta}\right) + \frac{1}{\sin^2\theta}\frac{\partial^2\psi}{\partial\phi^2}\right]\right\} + U(r)\psi = E\psi. \tag{1.13.6}$$

The multiplication of Eq.(1.13.6) by $r^2 2m_0/\hbar^2$ results in

$$-\frac{\partial}{\partial r}\left(r^2\frac{\partial\psi}{\partial r}\right) + \left\{\frac{1}{\sin\theta}\frac{\partial}{\partial\theta}\left(\sin\theta\frac{\partial\psi}{\partial\theta}\right) + \frac{1}{\sin^2\theta}\frac{\partial^2\psi}{\partial\phi^2}\right\} + r^2\frac{2m_0}{\hbar^2}U(r) = \frac{2m_0}{\hbar^2}r^2E\psi. \tag{1.13.7}$$

This equation allows the separation of variables in the form

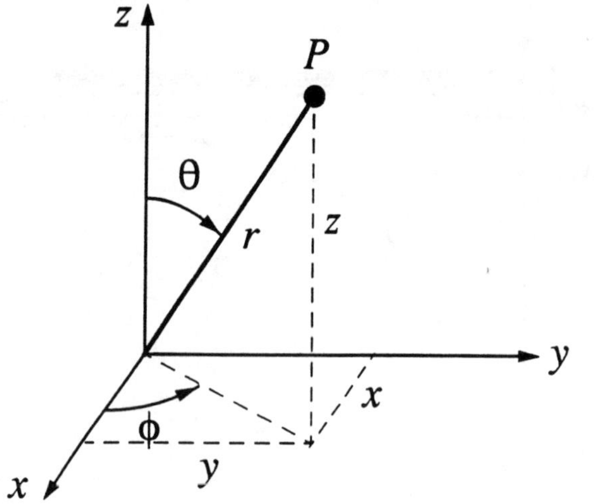

Figure 1.13.2 The spherical coordinates r, θ, ϕ of a point P, and its rectangular coordinates.

$$\psi(r,\theta,\phi) = R(r)Y(\theta,\phi). \tag{1.13.8}$$

Substituting the wave function Eq.(1.13.8) in Eq.(1.13.7) and dividing by ψ, one obtains

$$\frac{1}{R(r)}\left\{-\frac{\partial}{\partial r}\left(r^2\frac{\partial R(r)}{\partial r}\right)+r^2\frac{2m_0}{\hbar^2}U(r)R(r)-\frac{2m_0}{\hbar^2}r^2ER(r)\right\} =$$
$$-\frac{1}{Y(\theta,\phi)}\left\{\frac{1}{\sin\theta}\frac{\partial}{\partial\theta}\left(\sin\theta\frac{\partial Y(\theta,\phi)}{\partial\theta}\right)+\frac{1}{\sin^2\theta}\frac{\partial^2 Y(\theta,\phi)}{\partial\phi^2}\right\} = \lambda. \tag{1.13.9}$$

The left hand side of this equation is a function of r alone. The central part depends on θ and ϕ only. Since r, θ, and ϕ are independent variables, the equality holds, if both parts are equal to the separation constant λ on the right-hand side of Eq.(1.13.9).

The angle-dependent part of Eq.(1.13.9) is

$$\left\{\frac{1}{\sin\theta}\frac{\partial}{\partial\theta}\left(\sin\theta\frac{\partial}{\partial\theta}\right)+\frac{1}{\sin^2\theta}\frac{\partial^2}{\partial\phi^2}\right\}Y(\theta,\phi) = -\lambda Y(\theta,\phi). \tag{1.13.10}$$

The differential operator in square brackets of Eq.(1.13.10) multiplied by $-\hbar^2$ is known in quantum mechanics to be the operator of the square of the electron's angular momentum $\hat{L}^2$ in spherical coordinates. The angular momentum in classical mechanics plays a very important role in understanding the motion of a particle in a centro-symmetric field, because it is one of the conserved quantities. A similar conservation law for the angular momentum holds in quantum mechanics as well.

The electron angular momentum operator $\hat{\vec{L}}$ is defined in quantum mechanics as in classical physics as the vector product of the coordinate $\vec{r}$ with the electron momentum $\hat{\vec{p}}$:

$$\hat{\vec{L}} = \vec{r} \times \hat{\vec{p}} = -i\hbar(\vec{r} \times \nabla). \tag{1.13.11}$$

The components of $\hat{\vec{L}}$ in Cartesian coordinates have the forms

$$\hat{L}_x = y\hat{p}_z - z\hat{p}_y = -i\hbar\left(y\frac{\partial}{\partial z} - z\frac{\partial}{\partial y}\right), \tag{1.13.12a}$$

$$\hat{L}_y = z\hat{p}_x - x\hat{p}_z = -i\hbar\left(z\frac{\partial}{\partial x} - x\frac{\partial}{\partial z}\right), \tag{1.13.12b}$$

$$\hat{L}_z = x\hat{p}_y - y\hat{p}_x = -i\hbar\left(x\frac{\partial}{\partial y} - y\frac{\partial}{\partial x}\right). \tag{1.13.12c}$$

It is possible to show by the direct substitution of Eq.(1.13.12a-c) that the Cartesian components of $\hat{\vec{L}}$ do not commute

$$\left[\hat{L}_x, \hat{L}_y\right] = i\hbar\hat{L}_z; \quad \left[\hat{L}_y, \hat{L}_z\right] = i\hbar\hat{L}_x; \quad \left[\hat{L}_z, \hat{L}_x\right] = i\hbar\hat{L}_y. \tag{1.13.13}$$

It has been shown in Section 1.8 that noncommuting operators do not have common wave functions, which means that there is no way to measure them simultaneously. The squared angular momentum operator $\hat{L}^2$ equals $\hat{L}^2 = \hat{L}_x^{\,2} + \hat{L}_y^{\,2} + \hat{L}_z^{\,2}$. It commutes with the components of $\hat{\vec{L}}$

$$\left[\hat{L}^2, \hat{L}_x\right] = 0; \quad \left[\hat{L}^2, \hat{L}_y\right] = 0; \quad \left[\hat{L}^2, \hat{L}_z\right] = 0. \tag{1.13.14}$$

These equations mean that $\hat{L}^2$ and one of the components of $\hat{\vec{L}}$, say, $\hat{L}_z$, can be measured simultaneously. It is shown in quantum mechanics that the operator $\hat{L}^2$ has the following form

$$\hat{L}^2 = -\hbar^2\left\{\frac{1}{\sin\theta}\frac{\partial}{\partial\theta}\left(\sin\theta\frac{\partial}{\partial\theta}\right) + \frac{1}{\sin^2\theta}\frac{\partial^2}{\partial\phi^2}\right\} \tag{1.13.15}$$

in the spherical coordinates (1.13.4).

A comparison of $\hat{L}^2$ from Eq.(1.13.15) with the differential operator in the brackets of the Schrödinger equation (1.13.10) shows that they differ only by the constant factor $-\hbar^2$. Substituting Eq.(1.13.15) into Eq.(1.13.10), one obtains

$$\hat{L}^2 Y(\theta,\phi) = \hbar^2\lambda Y(\theta,\phi). \tag{1.13.16}$$

The problem of solving the angle-dependent part of the hydrogen atom Schrödinger equation (1.13.10) is reduced now to the eigenvalue problem for $\hat{L}^2$, Eq.(1.13.16), here $\hbar^2\lambda$ is the eigenvalue of the operator $\hat{L}^2$. Direct substitution shows that the operator $\hat{L}^2$ commutes with the Hamiltonian operator of the Schrödinger equation (1.13.6) for the hydrogen atom:

$$\left[\hat{H},\hat{L}^2\right]=0. \tag{1.13.17}$$

The commutator in Eq.(1.13.17) means that $\hat{L}^2$ is conserved in the motion of a particle in a centro-symmetric field. This means that the electron in the hydrogen atom has stationary states in which the square of the angular momentum has a definite value $\hbar^2\lambda$.

Another conserved quantity in this problem is the z-component of the angular momentum operator $\hat{L}_z$. In spherical coordinates Eq.(1.13.12c) for $\hat{L}_z$ has the very simple form

$$\hat{L}_z=-i\hbar\frac{\partial}{\partial\phi}. \tag{1.13.18}$$

The operator $\hat{L}_z$ commutes with the Hamiltonian of the hydrogen atom

$$\left[\hat{H},\hat{L}_z\right]=0. \tag{1.13.19}$$

The commutators (1.13.14), (1.13.17), and (1.13.19) show that the operators $\hat{H}, \hat{L}^2$, and $\hat{L}_z$ have the same eigenfunctions. The stationary states of the electron have definite values of $\hat{L}^2$ and $\hat{L}_z$ as quantum numbers.

We start with the simplest eigenvalue equation, namely the one for the operator $\hat{L}_z$, in order to find the eigenvalues of $\hat{L}_z$ and the ϕ-dependence of the $Y(\theta,\phi)$-eigenfunction

$$\hat{L}_z Y(\theta,\phi)\equiv -i\hbar\frac{\partial}{\partial\phi}Y(\theta,\phi)=L_z Y(\theta,\phi). \tag{1.13.20}$$

Here L_z is the eigenvalue of $\hat{L}_z$. Equation (1.13.20) is a first order linear differential equation in ϕ which has an exponential solution

$$Y(\theta,\phi)=f(\theta)e^{\frac{i}{\hbar}L_z\phi}, \tag{1.13.21}$$

where the amplitude $f(\theta)$ is still an arbitrary function of θ. The wave function should be a finite single-valued function of ϕ. This means that $Y(\theta,\phi)$ is a periodic function of ϕ with the period 2π, i.e. $e^{\frac{i}{\hbar}L_z(\phi+2\pi)}=e^{\frac{i}{\hbar}L_z\phi}$ or $e^{\frac{i}{\hbar}L_z 2\pi}=1$. It follows from this that the eigenvalue L_z takes the discrete set of values

$$L_z=\hbar m, \quad m=0,\pm1,\pm2,\pm3,\ldots \tag{1.13.22}$$

By combining Eq.(1.13.22) with Eq.(1.13.21), one obtains

$$Y(\theta,\phi) = f_m(\theta)e^{im\phi}, \qquad (1.13.23)$$

where m is the quantum number that results from the conservation of L_z, and is called the ***magnetic quantum number***. Taking into account Eq.(1.13.15) and substituting Eq.(1.13.23) into Eq.(1.13.16) leads to the equation for $f_m(\theta)$

$$\hat{L}^2 f_m(\theta) = -\hbar^2\left\{\frac{1}{\sin\theta}\frac{\partial}{\partial\theta}\left(\sin\theta\frac{\partial}{\partial\theta}\right) - \frac{m^2}{\sin^2\theta}\right\} f_m(\theta) = \hbar^2\lambda f_m(\theta). \qquad (1.13.24)$$

where $-m^2$ is substituted for $\partial^2/\partial\phi^2$ according to Eqs.(1.13.18) and (1.13.23). The eigenvalues and the eigenfunctions of $\hat{L}^2$ are known from quantum mechanics. The acceptable finite solutions of Eq.(1.13.24) exist only for certain values of λ which are shown in quantum mechanics to be

$$\lambda = l(l+1), \qquad (1.13.25)$$

where l has the integer values

$$l = 0, 1, 2\ldots.. \qquad (1.13.26)$$

related to the magnetic quantum number m in the following way:

$$m = 0, \pm 1, \pm 2 \ldots . \pm l. \qquad (1.13.27)$$

The quantum number l is called the ***orbital quantum number***. The eigenfunctions of Eq.(1.13.24), $f_m(\theta)$, are also known from quantum mechanics, and are expressed in terms of ***associated Legendre polynomials***. The total angle-dependent wave functions $Y_{lm}(\theta,\phi)$ are given by ***spherical harmonics***

$$Y_{lm}(\theta,\phi) = P_l^m(\cos\theta)e^{im\phi}, \qquad (1.13.28)$$

where $P_l^m(\cos\theta)$ is the associated Legendre polynomial. Spherical harmonics are normalized and orthogonal

$$\int_0^{2\pi}\int_0^{\pi} Y^*_{l'm'}(\theta,\phi)Y_{lm}(\theta,\phi)\sin\theta d\theta d\phi = \delta_{ll'}\delta_{mm'},$$

where $\delta_{ll'}$ and $\delta_{mm'}$ are Kronecker deltas. This normalization condition results in some numerical coefficients, dependent on m and l, in the definitions of the spherical harmonics.

Some examples of spherical harmonics with the smallest values of l and m are

$$Y_{00} = \frac{1}{\sqrt{4\pi}}, \qquad l = 0, m = 0; \qquad (1.13.29a)$$

$$Y_{10} = \left(\frac{3}{8\pi}\right)^{1/2}\cos\theta, \qquad l = 1, m = 0; \qquad (1.13.29b)$$

$$Y_{11} = -\left(\frac{3}{8\pi}\right)^{1/2}\sin\theta e^{i\phi}, \qquad l = 1, m = 1; \qquad (1.13.29c)$$

$$Y_{1-1} = \left(\frac{3}{8\pi}\right)^{1/2}\sin\theta e^{-i\phi}, \qquad l = -1, m = -1. \qquad (1.13.29d)$$

There is another accepted notation for the orbital quantum number l. Instead of using $l = 0,1,2,3...$ the notations $l = s,p,d,f...$ which are inherited from the field of atomic spectroscopy, are also used in quantum mechanics.

Since the angular momentum involves only the angular variables, the form of the radial wave function $R(r)$ from Eq.(1.13.9) still remains undetermined. The substitution of λ from Eq.(1.13.25) and the division of the radial part of Eq.(1.13.9) by $2m_0r^2/\hbar^2$ results in the radial equation

$$-\frac{\hbar^2}{2m_0}\left\{\frac{1}{r^2}\frac{\partial}{\partial r}\left(r^2\frac{\partial}{\partial r}\right)\right\}R(r)+\frac{\hbar^2 l(l+1)}{2m_0 r^2}R(r)+U(r)R(r)=ER(r). \quad (1.13.30a)$$

Equation (1.13.30) is a one-dimensional equation in the variable r. The term $\hbar^2 l(l+1)/2m_0r^2$ contributes to the potential energy. It is called the ***centrifugal potential energy***. The radial wave functions are known from Mathematical Physics: they are closely related with the ***associated Laguerre polynomials.*** With the given value of l in Eq.(1.13.30a), the equation has finite solutions only for certain values of total energy E which are numerated by the ***principle quantum number*** n.

The lowest energy of an electron in the hydrogen atom corresponds to $l = 0$, and $m = 0$. There is no centrifugal potential energy in Eq.(1.13.30) in this case. To represent the localization of an electron near the nucleus, the wave function $R(r)$ should decrease rapidly with distance from the nucleus. This means that at infinity we have the boundary condition

$$R(r)\,|_{r\to\infty}= 0. \quad (1.13.30b)$$

The simplest radially symmetric wave function that is finite at the origin and falls off rapidly with distance from the nucleus has the exponential form

$$R_{100}(r) \;=\; e^{-\alpha r}. \quad (1.13.31)$$

Substituting the radial wave function Eq.(1.13.31) in Eq.(1.13.30) and equating coefficients of equal powers of r shows that a solution exists when the parameters α and E are equal to

$$\alpha=\frac{m_0 e'^2}{\hbar^2} \quad \text{and} \quad E_1=-\frac{\hbar^2\alpha^2}{2m_0}, \quad (1.13.32)$$

respectively. Substituting the quantity e^2 from Eq.(1.13.1), yields

$$E_1=-\frac{1}{2}\frac{m_0 e'^4}{\hbar^2}=-\frac{1}{2}\frac{m_0 e^4}{(4\pi\varepsilon_0\hbar)^2}. \quad (1.13.33)$$

A numerical evaluation of E_1 leads to

$$E_1=-13.5eV=2.18\times 10^{-18}\,j. \quad (1.13.34)$$

The negative value of the energy E_1 means that an electron is attracted to the nucleus, and has a bound state in the potential of the nucleus. The energy of the lowest state E_1 is called the ***ground state energy*** and the corresponding wave function Eq.(1.13.31) has spherical symmetry.

The quantity α from Eq.(1.13.32) has the dimension of an inverse length, and α^{-1} is, in fact, the radius of the Bohr orbit in the atom. To show this, we calculate

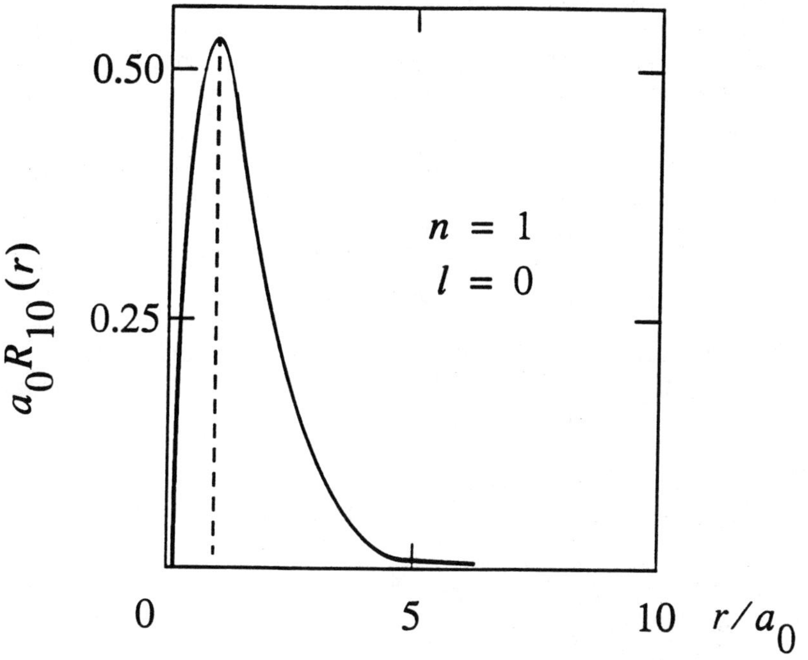

Figure 1.13.3 The radial probability density function for the hydrogen atom in the lowest energy state. (After R. Eisberg, R. Resnick. "Quantum Physics of Atoms, Molecules, Solids, Nuclei, and Particles." John Wiley & Sons, N. Y., 1985)

the probability of finding the electron in the interval of radii from r to $r+dr$. The radial probability equals $dP = |R_{100}|^2 4\pi r^2 dr$. The function $f(r) = |R|^2 r^2$ has a maximum, see Fig. 1.13.3, at value of r which is defined by the equation

$$\frac{df}{dr} = \frac{d}{dr}(r^2 e^{-2\alpha r}) = (1 - r\alpha)2re^{-2\alpha r} = 0.$$

Therefore, the radius of the first Bohr orbit a_0 equals to α^{-1}:

$$a_0 = r_{\max} = \frac{1}{\alpha} = \frac{\hbar^2}{m_0 e'^2} = \frac{4\pi\varepsilon_0\hbar^2}{m_0 e^2}. \tag{1.13.35}$$

The lowest electron energy in terms of a_0 has the form

$$E_1 = \frac{e^2}{2a_0}.$$

The other wave functions of the radial equation (1.13.30) belong to the excited states of the hydrogen atom. They correspond to $l \neq 0$ and are not spherically symmetric. We have shown in Section 1.12 that the excited states of a particle in a potential well with infinite walls are functions with nodes. We now take the wave function of the first excited state $n = 2$, $l = 1$, and $m = 0$ with the node at the origin

$$R_{210}(r) = re^{-\beta r}. \tag{1.13.36}$$

The wave function (1.3.36) obeys the boundary condition (1.13.30b) that it vanish at infinity. There is also a second boundary condition which states that the wave function has to be finite at the origin $r = 0$. This requirement results in a limitation imposed on the possible values of the orbital quantum number l given by Eq.(1.13.26).

$$l = 1,2,3,\ldots.n-1 \tag{1.13.37}$$

The substitution of the wave function Eq.(1.13.36) in the radial equation (1.13.30) shows that a solution exists when

$$\beta = \frac{m_0 e'^2}{2\hbar^2} = \frac{1}{2a_0}, \quad E_2 = -\frac{\hbar^2\beta^2}{2m_0} = -\frac{1}{4}E_1\,. \tag{1.13.38}$$

The total wave function of an electron in a hydrogen atom from Eq.(1.13.8) is the product of the radial wave functions Eq.(1.13.31) or Eq.(1.13.36) and the angular wave function Eq.(1.13.27). Several normalized and orthogonal total wave functions are given below

$$\psi_{100} = \frac{1}{\sqrt{\pi}}\frac{1}{a_0^{1/2}}e^{-\frac{r}{a_0}}; \quad n=1, l=0, m=0; \tag{1.13.39a}$$

$$\psi_{200} = \frac{1}{4\sqrt{2}\pi}\frac{1}{a_0^{3/2}}\left(2-\frac{r}{a_0}\right)e^{-\frac{r}{2a_0}}; \quad n=2, l=0, m=0; \tag{1.13.39b}$$

$$\psi_{210} = \frac{1}{4\sqrt{2\pi}}\frac{1}{a_0^{3/2}}\frac{r}{a_0}e^{-\frac{r}{2a_0}}\cos\theta; \quad n=2, l=1, m=0; \tag{1.13.39c}$$

$$\psi_{21\pm1} = \frac{1}{8\sqrt{\pi}}\frac{1}{a_0^{3/2}}\frac{r}{a_0}e^{-\frac{r}{2a_0}}\sin\theta e^{\pm i\phi}; \quad n=2, l=1, m=\pm1. \tag{1.13.39d}$$

The general expression for the energy levels of the hydrogen atom in terms of the Bohr radius can be obtained by a generalization of Eqs.(1.13.32) and (1.13.38)

$$E_n = -\frac{1}{n^2}\frac{e'^2}{2a_0} = -\frac{1}{n^2}\frac{m_0 e^4}{(4\pi\varepsilon_0)^2 2\hbar^2}. \tag{1.13.40}$$

We see that the energy depends only on the principal quantum number n. Therefore the wave functions with different values of l and m belong to the same energy. When several different wave functions belong to the same energy level, it is said that energy level is ***degenerate.*** One can see from Eqs.(1.13.29a-b) that in case of the hydrogen atom the ***first excited state with*** $n=2$ ***is four-fold degenerate.*** This high degree of degeneracy is a direct consequence of the spherical symmetry of the atom. The energy level diagram for the hydrogen atom is shown in Fig.1.13.3. We see that as the principal quantum number n increases the separation of the energy levels decreases, and the levels approach the limit that corresponds to a free electron with binding energy $E \to 0$. The quantity $|E_{n=1}| = 0 - E_{n=1}$ is called the ***ionization energy*** of the hydrogen atom.

1.14 PAULI EXCLUSION PRINCIPLE

We have so far considered the orbital angular momentum $\hat{\vec{L}}$ of an electron that allows a classification of atomic energy levels and wave functions to be made. This classification was verified by Stern and Gerlach in 1922, in the following experiments.

It is known from classical physics that a particle with angular momentum $\vec{L}$ can have a magnetic dipole moment $\vec{M}$ that is proportional to $\vec{L}$

$$\vec{M} = \gamma\vec{L}, \tag{1.14.1}$$

where $\gamma = -e/2m_0c$ is called the ***gyromagnetic ratio***. The potential energy of the magnetic dipole moment $\vec{M}$ in an external magnetic field $\vec{B} = (0,0,B)$ equals

$$U_M = -\vec{B}\cdot\vec{M} = -BM_z = -\gamma BL_z. \tag{1.14.2}$$

Because in quantum mechanics L_z is quantized according to Eq.(1.13.22), $L_z = m\hbar$, the z-component of the magnetic dipole moment M_z is quantized as well,

$$M_z = \gamma m\hbar. \tag{1.14.3}$$

When a nonuniform magnetic field is applied to an atom with magnetic moment M_z, the force F_z acting on the magnetic dipole moment in the z-direction in nonuniform magnetic field $B(z)$ equals

$$F_z = -\frac{\partial U_M(z)}{\partial z} = \gamma\frac{\partial B(z)}{\partial z}M_z. \tag{1.14.4}$$

Since the force is proportional to M_z, each atom is deflected by an amount that is proportional to M_z. As a result, the deflected atomic beam is split into components corresponding to different values of M_z. Finally, the deflected atoms reach the screen where they condense in visible traces. If atoms are in an excited state with $l = 1$ and $m = 0, \pm 1$, three visible traces should appear on the screen, see Fig.1.14.1.

Stern and Gerlach have studied the deflection for silver atoms and observed only two visible traces. Special experiments were done with hydrogen atoms in the ground state with $l = 0, m = 0$. Since $M_z = 0$ in this case, it was expected that the beam would not be deflected by the inhomogeneous magnetic field $B(z)$. But the experiments showed that the beam is split into ***two symmetrically deflected components,*** see Fig.1.14.2. It was direct evidence of the existence of some magnetic dipole moment which has not been taken into account yet.

Uhlenbeck and Goudsmit, 1925, assumed that the electron has an intrinsic angular momentum which is responsible for the results of the experiment. In order to obtain two symmetrically detected components, Pauli, in 1927, introduced the intrinsic angular momentum $\vec{S}$ of an electron called ***spin***, whose z-component takes the half integral values $+\frac{\hbar}{2}, -\frac{\hbar}{2}$ only. The spin gyromagnetic ratio for intrinsic angular momentum is twice as large as that for the orbital motion

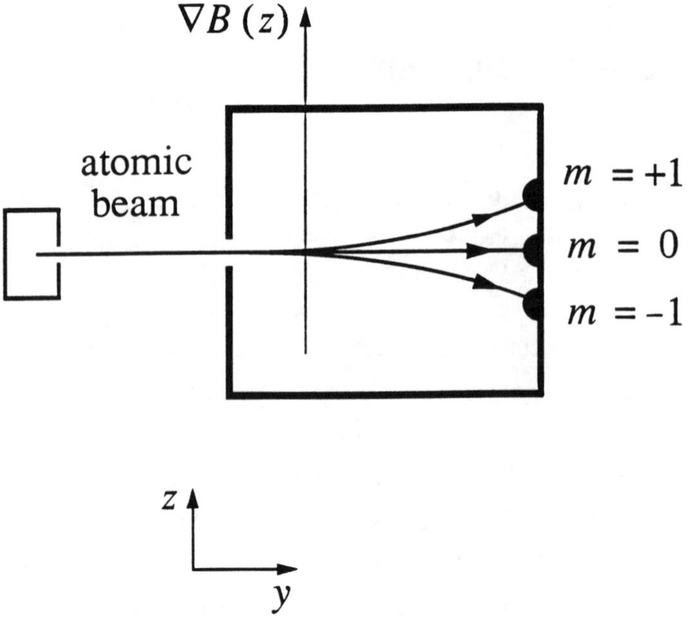

Figure 1.14.1 In a Stern-Gerlach experiment a beam of atoms with $l = 1$ is split into three beams in an external inhomogeneous magnetic field.

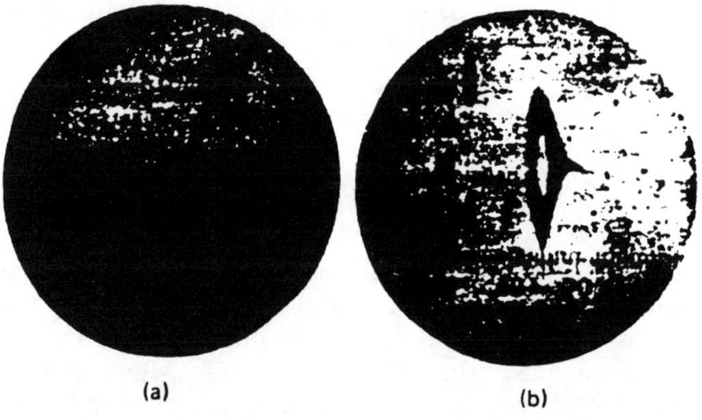

Figure 1.14.2 The result of Stern-Gerlach experiment for beam of atoms with $l = 0$. **(a)** The beam deposit in the absence of a magnetic field. (b) The beam deposit in the presence of inhomogeneous magnetic field. (After Garrison, Sposito. "An Introduction to Quantum Physics". John Wiley & Sons, N.Y., 1970)

$$\gamma_s = -\frac{e}{m_0 c}. \tag{1.14.5}$$

The orbital motion of the electron in an atom is classified by the angular momentum $\hat{\vec{L}}$ quantum numbers. Spin is not connected with a real rotation of the particle in space. It is an internal constant characteristic of the particle. Spin has no classical interpretation.

Being an angular momentum, spin has the same commutation rules and eigenvalue equations as the orbital angular momentum $\hat{\vec{L}}$

$$\hat{S}^2\chi = \hbar^2 s(s+1)\chi, \quad \hat{S}_z\chi = \hbar s_z\chi. \tag{1.14.6}$$

Here χ is the spin wave function that is defined by the internal spin variable. In the case of electrons, the eigenvalues of $\hat{S}^2$ and $\hat{S}_z$ are given by

$$s = \frac{1}{2}, \quad s_z = \pm\frac{1}{2}. \tag{1.14.7}$$

The total angular momentum of the particle $\hat{\vec{J}}$ is composed of its orbital angular momentum $\hat{\vec{L}}$ and its spin $\hat{\vec{S}}$

$$\hat{\vec{J}} = \hat{\vec{L}} + \hat{\vec{S}}. \tag{1.14.8}$$

The operators $\hat{\vec{L}}$ and $\hat{\vec{S}}$ act on different variables; therefore they commute. The square of the total angular momentum equals

$$\hat{J}^2 = \hat{L}^2 + \hat{S}^2 + 2\hat{\vec{L}}\cdot\hat{\vec{S}}. \tag{1.14.9}$$

The magnitude of the vector $\hat{\vec{J}}$ varies from $\left(\hat{\vec{L}} - \hat{\vec{S}}\right)$ to $\left(\hat{\vec{L}} + \hat{\vec{S}}\right)$. The possible eigenvalues j of the total angular momentum $\hat{\vec{J}}$ are

$$l + s \geq j \geq |\, l - s \,|. \tag{1.14.10}$$

It follows from Eq.(1.14.10) that there are $(2j+1)$ states of the particle corresponding to the $(2j+1)$ possible orientations of the total angular momentum $\vec{J}$. In the case of the electron with angular momentum $l = 1$ and spin that takes the value $s = 1/2$, the total angular momentum has two values, which are equal to

$$j = 1 + 1/2 = 3/2 \quad \text{and} \quad j = 1 - 1/2 = 1/2. \tag{1.14.11}$$

Diagrams that illustrate the "***law of addition***" of $\hat{\vec{L}}$ and $\hat{\vec{S}}$ are shown in Fig.1.14.3. Vector $\hat{\vec{J}}$ has a definite magnitude and definite z-component. Both x- and y-com-

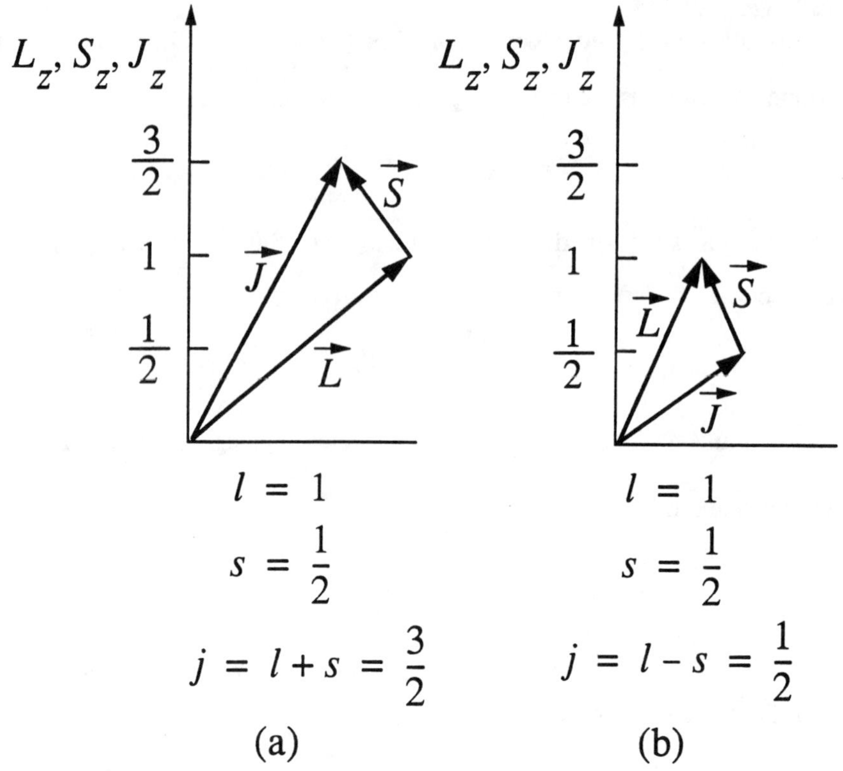

Figure 1.14.3 The law of addition of angular momentum $\hat{\vec{L}}$ and spin $\hat{\vec{S}}$. (a) $l = 1; s = 1/2; j = l + s = 3/2$, (b) $l = 1; s = 1/2; j = l - s = 1/2$.

ponents of $\vec{J}$ are not definite.

For particles with spin, the wave functions should define not only the probability of different positions in space but the probability of possible spin orientations also. The wave function depends now not only on the spatial coordinate $\vec{r}$ but on the discrete spin variable s_z as well:

$$\psi(x, y, z; s_z), \tag{1.14.12}$$

where $s_z = +1/2, -1/2$. If the electron is in the state with $s_z = +1/2$, the wave function is very often labeled by the symbol "arrow up $\uparrow$". If it is in the state with $s_z = -1/2$, it is labeled by "arrow down $\downarrow$",

$$\psi(\vec{r}; \uparrow) \quad \text{and} \quad \psi(\vec{r}; \downarrow).$$

The spread of the wave packets representing electrons leads to the fact that identical particles lose their individuality completely in the sense that each electron does not have a definite trajectory. There is only a probabilistic description of the electron by the squared modulus of the wave function $|\psi|^2$ in quantum mechanics. In a system of many particles there is no physical way to measure some effects which result from the interchange of two identical particles when both their position and spin coordinates are interchanged.

The symmetry under an interchange of particles is called in quantum mechanics the ***principle of particle identity.*** We demonstrate this principle using as an example a system of two noninteracting electrons. It follows from the identity of electrons that the states of the system obtained by simply interchanging the two particles must be completely equivalent physically to the initial one. As a result of the electron interchange, the wave function of the system can change only by a physically unimportant phase factor:

$$\psi(r_1 s_{1z}; r_2 s_{2z}) = e^{i\alpha}\psi(r_2 s_{2z}; r_1 s_{1z}), \tag{1.14.13}$$

where α is some real constant. The squared modulus of the wave function does not change because of the interchange of particles. By repeating the interchange twice, we return to the initial state of the system, while the wave function is multiplied by the factor $e^{i2\alpha}$. Therefore,

$$e^{i2\alpha} = 1 \quad \text{and} \quad e^{i\alpha} = \pm 1. \tag{1.14.14}$$

Combining Eq.(1.14.14) with Eq.(1.14.13), we obtain

$$\psi(\vec{r}_1 s_{1z}; \vec{r}_2 s_{sz}) = \pm\psi(r_2 s_{2z}; r_1 s_{1z}). \tag{1.14.15}$$

We see that the wave function must be either unchanged when particles are interchanged or it must change its sign when the interchange occurs. Wave functions of the first type are called ***symmetric***. Functions of the second type are called ***antisymmetric***. This result can be generalized to a system consisting of any number of identical particles.

Whether the particles are described by symmetric or antisymmetric wave function depends on their nature. It is known so far that all particles described by ***antisymmetric*** wave functions obey ***Fermi-Dirac statistics*** (they are called ***fermions***). Those particles that are described by ***symmetric*** functions obey ***Bose-Einstein statistics*** (they are ***bosons***).

It is shown in relativistic quantum mechanics that the statistics of particles is uniquely related with their spin: particles with ***half-integral spin*** (e.g. ***electrons***) are ***fermions***, and particles with ***integral spin*** are ***bosons*** (***e.g. photons***).

There is no spin in the Schrödinger equations we have considered so far. The Hamiltonian of a system contains only the electrical interaction of the particles and does not depend on spin. However, the type of statistics affects the motion of a particle significantly. The wave function of a system consisting of two noninteracting particles $\psi(\vec{r}_1 s_{1z}, \vec{r}_2 s_{2z}) = \psi_{p_1}(\vec{r}_1 s_{1z})\psi_{p_2}(\vec{r}_2 s_{2z})$ should be either even or odd according to Eq.(1.14.15), p_1, p_2 being quantum numbers of the separate particles. It is always possible to construct linear combinations which are even

$$\psi_{even}(\vec{r}_1 s_1; \vec{r}_2 s_2) = \frac{1}{\sqrt{2}}\{\psi_{p_1}(\vec{r}_1 s_1)\psi_{p_2}(\vec{r}_2 s_2) + \psi_{p_1}(\vec{r}_2 s_2)\psi_{p_2}(\vec{r}_1 s_1)\} \tag{1.14.16}$$

or odd

$$\psi_{odd}(\vec{r}_1 s_1; \vec{r}_2 s_2) = \frac{1}{\sqrt{2}}\{\psi_{p_1}(\vec{r}_1 s_1)\psi_{p_2}(\vec{r}_2 s_2) - \psi_{p_1}(\vec{r}_2 s_2)\psi_{p_2}(\vec{r}_1 s_1)\}. \tag{1.14.17}$$

The requirement of being even or odd implies strong correlation in the possible motions of particles. If two electrons are in the same state $p_1 = p_2$ it follows from Eq.(1.14.17) that

$$\psi_{odd}(\vec{r}_1 s_1; \vec{r}_2 s_2) = 0. \tag{1.14.18}$$

Equation (1.14.18) means that probability density of finding ***two electrons*** in the same state is equal to zero. Electrons with equal quantum numbers, including the spin quantum number, eliminate each other. The result is easily generalized to any number of electrons. We see that ***in a system consisting of identical electrons no two (or more) electrons can be in the same state at the same time***. This result was established by Pauli in 1925 and is called ***Pauli's exclusion principle.***

The potential energy of electrons in nonrelativistic consideration has an electrostatic origin. The Hamiltonian does not depend on spin of electron. It allows to use the separation of variables and to represent the wave function of the system $\Psi(\vec{r}_1 s_1; \vec{r}_2 s_2)$ as the product of two functions:

$$\Psi(\vec{r}_1 s_1; \vec{r}_2 s_2) = \psi(\vec{r}_1, \vec{r}_2)\chi(s_1, s_2).$$

Here $\psi(\vec{r}_1, \vec{r}_2)$ is the coordinate wave function that is defined by Schrödinger equation, and $\chi(s_1, s_2)$ is the spin wave function which remains arbitrary. Schrödinger equation does not impose any limitations on $\chi(s_1, s_2)$.

Nevertheless, when the principle of the particle identity is taken into account there appears the specific dependence of the energy of the system on the total spin.

In case of electrons the total wave function must be antisymmetric with respect to the interchange of two particles. If the coordinate wave function $\psi(\vec{r}_1, \vec{r}_2)$ is symmetric, the spin wave function $\chi(s_1, s_2)$ should be antisymmetric to provide the total wave function $\Psi(\vec{r}_1 s_1; \vec{r}_2 s_2)$ to be antisymmetric. It is shown in quantum mechanics that antisymmetric spin wave function corresponds to zero value of the total spin. The total spin is zero when two electrons have antiparallel spins.

If Schrödinger equation gives antisymmetric coordinate wave function, the spin wave function $\chi(s_1, s_2)$ should be symmetric. Symmetric χ corresponds to the total spin equal to the unity. Spins of two electrons should be parallel.

We see that there is the specific interaction of electrons called the ***exchange interaction*** which results in the dependence of the total energy on the total spin.

It should also be noted that the exchange energy has a large value in 1D-, and 2D-semiconductor structures (quantum wells and wires) with localized electrons.

1.15 PERTURBATION APPROXIMATION

We have discussed so far the simplest physical situations where it is possible to find an exact solution of the Schrödinger equation; but for the majority of systems an exact solution does not exist. Nevertheless, there are approximate methods of solution which are based on the fact that quantities of different orders of magnitude enter the conditions of the problem. It is assumed that the Hamiltonian in the Schrödinger equation has the form

$$\hat{H} = \hat{H}^{(0)} + \hat{H}^{(1)}, \tag{1.15.1}$$

where $\hat{H}^{(0)}$ corresponds to a system that is simple enough that the corresponding Schrödinger equation can be solved. The second term $\hat{H}^{(1)}$ is small and it is considered as a small perturbation on $\hat{H}^{(0)}$.

When $\hat{H}^{(1)}$ is time-independent, we have ***stationary perturbation theory*** which enables us to obtain the changes in the discrete energy levels and wave functions when a small perturbation $\hat{H}^{(1)}$ is applied to the system described by the Hamiltonian $\hat{H}^{(0)}$. It is convenient to write the Hamiltonian Eq.(1.15.1) in the form

$$\hat{H} = \hat{H}^{(0)} + \lambda \hat{H}^{(1)}, \tag{1.15.2}$$

where $\lambda < 1$ is a parameter that represents the smallness of the perturbation $\hat{H}^{(1)}$. The eigenfunctions and eigenvalues of the total Hamiltonian Eq.(1.15.1) are expanded in powers of λ

$$\psi = \psi^{(0)} + \lambda\psi^{(1)} + \lambda^2\psi^{(2)} + \ldots, \tag{1.15.3a}$$

$$E = E^{(0)} + \lambda E^{(1)} + \lambda^2 E^{(2)} + \ldots. \tag{1.15.3b}$$

Substituting Eqs.(1.15.2) and (1.15.3a-b) in the time-independent Schrödinger equation with the Hamiltonian Eq.(1.15.1) one obtains

$$\left(\hat{H}^{(0)} + \lambda\hat{H}^{(1)}\right)(\psi^{(0)} + \lambda\psi^{(1)} + \lambda^2\psi^{(2)}) =$$
$$(E^{(0)} + \lambda E^{(1)} + \lambda^2 E^{(2)} + \ldots)(\psi^{(0)} + \lambda\psi^{(1)} + \lambda^2\psi^{(2)} + \ldots). \tag{1.15.4}$$

When $\lambda = 0$, Eq.(1.15.4) reduces to the zero approximation

$$\hat{H}^{(0)}\psi_k^{(0)} = E_k^{(0)}\psi_k^{(0)}. \tag{1.15.5a}$$

The zero order equation is assumed to be solved, and the energies $E_k^{(0)}$ and eigenfunctions $\psi_k^{(0)}$ known. The zero order wave functions are orthogonal

$$\int \psi_k^{*(0)}\psi_l^{(0)} d\vec{r} = \delta_{kl}. \tag{1.15.6}$$

If λ is nonzero, we get the eigenvalue problem for the total system. Equating terms of the same order of magnitude in Eq.(1.15.4) results in the system of equations

$$\hat{H}^{(1)}\psi^{(0)} + \hat{H}^{(0)}\psi^{(1)} = E^{(1)}\psi^{(0)} + E^{(0)}\psi^{(1)}, \qquad (1.15.5b)$$

$$\hat{H}^{(0)}\psi^{(2)} + \hat{H}^{(1)}\psi^{(1)} = E^{(2)}\psi^{(0)} + E^{(1)}\psi^{(1)} + E^{(0)}\psi^{(2)}, \qquad (1.15.5c)$$

....

Equation (1.15.5b) is the first order equation. To solve it, we expand $\psi^{(1)}$ in terms of the unperturbed functions $\psi_n^{(0)}$

$$\psi^{(1)} = \sum_n C_n^{(1)}\psi_n^{(0)}, \qquad (1.15.7)$$

where $C_n^{(1)}$ are unknown coefficients. Substitution of Eq.(1.15.7) into Eq.(1.15.5b) gives

$$\hat{H}^{(1)}\psi_k^{(0)} + \hat{H}^{(0)}\sum_n C_n^{(1)}\psi_n^{(0)} = E^{(1)}\psi_k^{(0)} + E_k^{(0)}\sum_n C_n^{(1)}\psi_n^{(0)}. \qquad (1.15.8)$$

Multiplying Eq.(1.15.8) from the left by $\psi_j^{*(0)}$ and integrating over all space, we obtain

$$\int \psi_j^{*(0)}\hat{H}^{(1)}\psi_k^{(0)}d\vec{r} + \sum_n C_n^{(1)}E_n^{(0)}\int \psi_j^{*(0)}\psi_n^{(0)}d\vec{r} =$$

$$E^{(1)}\int \psi_j^{*(0)}\psi_k^{(0)}d\vec{r} + E_k^{(0)}\sum_n C_n^{(1)}\int \psi_j^{*(0)}\psi_n^{(0)}d\vec{r}. \qquad (1.15.9)$$

Since the wave functions of the zeroth approximation are assumed to be orthogonal, Eq.(1.15.6), we find

$$\int \psi_j^{*(0)}\hat{H}^{(1)}\psi_k^{(0)}d\vec{r} + C_j^{(1)}E_j^{(0)} = E^{(1)}\delta_{ik} + E_k^{(0)}C_j^{(1)}. \qquad (1.15.10)$$

When $j = k$, Eq.(1.15.10) becomes

$$E_k^{(1)} = \int \psi_j^{*(0)}\hat{H}^{(1)}\psi_j^{(0)}d\vec{r} = <\psi_j^{(0)} \mid \hat{H}^{(1)} \mid \psi_j^{(0)}>. \qquad (1.15.11)$$

We see from Eq.(1.15.11) that the first-order correction $E_k^{(1)}$ to the energy $E_k^{(0)}$ is given by the matrix element that is denoted by $<\psi_j^{(0)} \mid \hat{H}^{(1)} \mid \psi_j^{(0)}>$.

When $j \neq k$ Eq.(1.15.10) gives the expansion coefficients which enter Eq.(1.15.7)

$$C_j^{(1)} = \frac{<\psi_j^{(0)} \mid \hat{H}^{(1)} \mid \psi_k^{(0)}>}{E_k^{(0)} - E_j^{(0)}}. \qquad (1.15.12)$$

The value of $C_k^{(1)}$ remains arbitrary. Since the system is assumed to be initially in the state k, the zero approximation energy $E_k^{(0)}$ and the wave function $\psi_k^{(0)}$ of this state are known, $C_k^{(1)}$ must be taken equal to zero

$$C_k^{(1)} = 0. \qquad (1.15.13)$$

Substituting Eqs.(1.15.12) and (1.15.13) into Eq.(1.15.7), one finds the first-order correction to the wave function

$$\psi_j^{(1)} = \sum_k{}' \frac{<\psi_j^{(0)} | \hat{H}^{(1)} | \psi_k^{(0)}>}{E_k^{(0)} - E_j^{(0)}} \psi_k^{(0)}. \tag{1.15.14}$$

The prime on the sum means that the term with $k = j$ is excluded. Second order corrections can be obtained in a similar way, by using the first order solutions in Eq.(1.15.5c). The second order correction to the energy is

$$E_k^{(2)} = \sum_{n \neq k} \frac{|<\psi_k^{(0)} | \hat{H}^{(1)} | \psi_n^{(0)}>|^2}{E_k^{(0)} - E_n^{(0)}}. \tag{1.15.15}$$

If $E_k^{(0)}$ corresponds to the lowest value of the energy, all terms in the sum Eq.(1.15.15) are negative and the correction is negative. Equations (1.15.11) and (1.15.15) reveal the condition of applicability of the perturbation approximation: the matrix element of the perturbation should be small compared to the separation of the corresponding unperturbed energy levels.

We consider, as an example, the application of the stationary perturbation approximation to the determination of the splitting of the first excited energy level of the hydrogen atom ($n = 2; l = 1; m = 0, \pm 1$), E_2, from Eq.(1.13.40), in the uniform external magnetic field $\vec{B} = (0,0,B)$. The magnetic dipole moment $\vec{M}$ from Eq.(1.14.1) interacts with the external magnetic field $\vec{B}$ through the interaction presented in Eq.(1.14.2). Considering this interaction as a small, time-independent perturbation, one can calculate the first order correction Eq.(1.15.11) to the energy of the first excited state of the hydrogen atom. It was shown in Section 1.13 that eigenfunctions of the hydrogen atom are eigenfunctions of $\hat{L}_z$ with eigenvalues $\hbar m$, where m is the magnetic quantum number. Calculating the matrix element, one obtains in the first approximation the triplet of energy levels

$$E = <\psi_{21m} | \hat{H}^{(1)} | \psi_{21m}> = \left\{ \begin{matrix} -\dfrac{e}{2m_0} B\hbar & \text{for} & m = -1 \\ 0 & \text{for} & m = 0 \\ +\dfrac{e}{2m_0} B\hbar & \text{for} & m = +1 \end{matrix} \right\}. \tag{1.15.16}$$

The $(2l + 1)$-fold degeneracy of states with a given n and l is removed in an external magnetic field. These triplets are well seen in optical spectra of hydrogen in a magnetic field. This phenomenon is called the ***Zeeman effect***.

1.16 TIME-DEPENDENT PERTURBATION

There are many phenomena in which electrons are subjected to time-dependent external perturbations. We can not speak of corrections to the energy levels and to the stationary wave functions in such cases, since the Hamiltonian is time-dependent,

$$\hat{H} = \hat{H}^{(0)} + \hat{H}^{(1)}(t), \tag{1.16.1}$$

and the energy of the particle is not conserved. There are no stationary states in this case. The system makes transitions from an initial stationary state to another state under the action of the time-dependent perturbation $H^{(1)}(t)$. The problem consists of approximate calculations of the wave function in the final state through the wave functions of the stationary states of the unperturbed system.

The wave function of the stationary state of the unperturbed systems satisfies the time dependent equation

$$i\hbar\frac{\partial\phi^{(0)}}{\partial t} = \hat{H}^{(0)}\phi^{(0)}, \tag{1.16.2}$$

and has the following dependence on time, see Eq.(1.5.6),

$$\phi_k^{(0)}(\vec{r},t) = e^{-\frac{i}{\hbar}E_k^{(0)}t}\,\psi(\vec{r}). \tag{1.16.3}$$

We seek the solutions of the time-dependent Schrödinger equation

$$i\hbar\frac{\partial\phi}{\partial t} = \left\{\hat{H}^{(0)} + \hat{H}^{(1)}(t)\right\}\phi. \tag{1.16.4}$$

The solution of Eq.(1.16.4) is represented in the form of expansion

$$\phi(t) = \sum_k a_k(t)\phi_k^{(0)}, \tag{1.16.5}$$

where the expansion coefficients $a_k(t)$ are functions of the time t. Substituting Eq.(1.16.5) into Eq.(1.16.4) and using the zero approximation equation Eq.(1.16.2), one obtains

$$i\hbar\sum_k \phi_k^{(0)}\frac{da_k(t)}{dt} = \sum_k a_k(t)\hat{H}^{(1)}\phi_k^{(0)}.$$

Multiplying by $\phi_m^{*(0)}$ from the left, integrating, and taking into account the orthogonality relation (1.15.6) we find

$$i\hbar\frac{da_m(t)}{dt} = \sum_k a_k\int \phi_m^{*(0)}\hat{H}^{(1)}\phi_k^{(0)}d\vec{r} = \sum_k a_k < \phi_m^{(0)}\,|\,\hat{H}^{(1)}(t)\,|\,\phi_k^{(0)} > . \tag{1.16.6}$$

The quantity $\int\phi_m^{*(0)}\hat{H}^{(1)}(t)\phi_k^{(0)}d\vec{r} = <\phi_m^{(0)}|\,\hat{H}^{(1)}(t)|\,\phi_k^{(0)}>$ is the time dependent matrix element of the perturbation. The substitution of the functions of the zero approximation, Eq.(1.16.3), yields the result

$$<\phi_m^{(0)} \mid \hat{H}^{(1)}(t) \mid \phi_k^{(0)}> = e^{\frac{i}{\hbar}\left(E_m^{(0)}-E_k^{(0)}\right)t} \int \psi_m^{(0)} \hat{H}^{(1)} \psi_k^{(0)} d\vec{r} =$$

$$e^{i\omega_{mk}t} < \psi_m^{(0)} \mid \hat{H}^{(1)} \mid \psi_k^{(0)} >, \tag{1.16.7}$$

where $\omega_{mk} = (E_m - E_k)/\hbar$.

We consider the perturbation of the n-th stationary state that corresponds in Eq.(1.16.5) to the following values of the coefficients at $t = 0$

$$a_n^{(0)} = 1, \quad a_k^{(0)} = 0 \quad \text{for} \quad k \neq n.$$

This means that the initial condition for Eq.(1.16.6) has the form

$$a_k^{(0)}(t = 0) = \delta_{kn}. \tag{1.16.8}$$

at $t = 0$. Then, it is convenient to represent a_k in Eq. (1.16.6) in the form

$$a_k = a_k^{(0)} + a_k^{(1)}. \tag{1.16.9}$$

Since the matrix elements of the perturbation $\hat{H}^{(1)}$ are considered to be small quantities of the first order, a_k can be taken in the form (1.16.8). Substituting Eq.(1.16.9) into Eq.(1.16.6) and keeping terms of the first order only, we find

$$i\hbar\frac{da_k^{(1)}(t)}{dt} = < \psi_n^{(0)} \mid \hat{H}^{(1)} \mid \psi_k^{(0)} > e^{i\omega_{kn}t}. \tag{1.16.10}$$

In order to indicate that we are seeking a correction to the initial state n of the system, we add a second index n to the coefficient $a_k^{(1)}$. Integration of Eq.(1.16.10) results in

$$a_{kn}^{(1)} = -\frac{i}{\hbar}\int_{-\infty}^{t} < \psi_k^{(0)} \mid \hat{H}^{(1)} \mid \psi_n^{(0)} > e^{i\omega_{kn}t} dt. \tag{1.16.11}$$

The constant of integration is taken to be zero in order that $a_{kn}^{(1)}$ be zero before the perturbation is switched on at $t \to -\infty$.

In the case of a perturbation with a harmonic dependence on time

$$\hat{H}^{(1)} = \hat{F} e^{-i\omega t} + \hat{F}^* e^{i\omega t}, \tag{1.16.12}$$

where $\hat{F}$ and $\hat{F}^*$ are time-independent amplitudes, the matrix element of the perturbation, Eq.(1.16.7), equals

$$< \phi_k^{(0)} \mid \hat{H}^{(1)} \mid \phi_n^{(0)} > = < \psi_k^{(0)} \mid \hat{F} \mid \psi_n^{(0)} > e^{i(\omega_{kn}-\omega)t} +$$

$$< \psi_k^{(0)} \mid \hat{F}^* \mid \psi_n^{(0)} > e^{i(\omega_{kn}+\omega)t}. \tag{1.16.13}$$

Combining Eq.(1.16.13) with Eq.(1.16.11) and assuming that the perturbation is switched on at $t = 0$ and switched off at $t = t_0$, we obtain for any time $t > t_0$

$$a_{kn}^{(1)}(t) = -\frac{< \psi_k^{(0)} \mid \hat{F} \mid \psi_n^{(0)} > e^{i(\omega_{kn}-\omega)t}}{\hbar(\omega_{kn}-\omega)} - \frac{< \psi_k^{(0)} \mid \hat{F}^* \mid \psi_n^{(0)} > e^{i(\omega_{kn}+\omega)t}}{\hbar(\omega_{kn}+\omega)}. \tag{1.16.14}$$

The total wave function equals

$$\phi(r,t) = \phi^{(0)}(r,t) + \sum_k a_{kn}^{(1)} \phi_k^{(0)}(r,t). \tag{1.16.15}$$

The result is applicable to a discrete set of unperturbed levels, when $\omega \neq \omega_{kn} = (E_k - E_n)/\hbar$.

Another result is obtained when a transition due to the perturbation (1.16.12) occurs from a discrete initial state i into the state f of the continuous spectrum, with $E_f > E_i$. Then the first term in Eq.(1.6.14) contributes at $\omega \approx \omega_{kn}$. The integration leads to

$$a_{fi}^{(1)} = -\frac{i}{\hbar} \int_0^t < \phi_f^{(0)} | \hat{H}^{(1)}(t) | \phi_i^{(0)} > dt =$$

$$- < \psi_f^{(0)} | \hat{F} | \psi_i^{(0)} > \frac{e^{i(\omega_{fi} - \omega)t} - 1}{\hbar(\omega_{fi} - \omega)}. \tag{1.16.16}$$

The lower limit in the integral is taken to be 0 in order to have $a_{fi} = 0$ at $t = 0$. The probability of transition from the initial state i into the final state f equals

$$| a_{fi}^{(1)} |^2 = |< \psi_f^{(0)} | \hat{F} | \psi_i^{(0)} >|^2 \left[\frac{4 \sin^2 \frac{\omega_{fi} - \omega}{2} t}{\hbar^2 (\omega_{fi} - \omega)^2} \right]. \tag{1.16.17}$$

When $t \to \infty$ the function in square brackets becomes more and more sharply peaked. We introduce the quantity $\delta(x)$ that has the following properties

$$\lim_{t \to \infty} \frac{\sin^2 xt}{\pi t x^2} = \delta(x), \tag{1.16.18}$$

and

$$\delta(x) = \begin{cases} 0 & \text{for } x \neq 0 \\ \infty & \text{for } x = 0 \end{cases}. \tag{1.16.19}$$

$\delta(x)$ is called the Dirac δ-function. It is normalized in the following way

$$\frac{1}{\pi} \int_{-\infty}^{+\infty} \frac{\sin^2 xt}{tx^2} dx = \int_{-\infty}^{+\infty} \delta(x) dx = 1, \tag{1.16.20}$$

and for any continuous function $f(x)$ satisfies the relation

$$\int_{-\infty}^{+\infty} f(x') \delta(x - x') dx' = f(x). \tag{1.16.21}$$

In terms of the δ-function, the probability Eq.(1.16.17) of finding the system in the final state after the perturbation is removed takes the form

$$| a_{fi} |^2 = \frac{2\pi}{\hbar} |< \psi_f^{(0)} | \hat{F} | \psi_i^{(0)} >|^2 \delta(E_f - E_i - \hbar\omega) t, \tag{1.16.22}$$

where the relation $\delta(ax) = \delta(x)/a$ has been used. We consider the transition in the narrow interval of the continuous spectrum of final states from E_f up to $E_f + dE_f$. The probability of finding the system in one of these states given by Eq.(1.16.22) is proportional to t. Therefore, the transition probability per unit time and per energy interval dE_f equals

$$dW_{i\to f} = \frac{2\pi}{\hbar} |< \psi_f^{(0)} | \hat{F} | \psi_i^{(0)} >|^2 \delta(E_f - E_i - \hbar\omega) dE_f . \tag{1.16.23}$$

The probability $dW_{i\to f}$ is nonzero only for $E_f(0) = E_i(0) + \hbar\omega$. We see that the δ-function expresses the law of energy conservation.

In case of the time-dependent perturbation the transition probability W from the state with the energy E into the state with the energy E' is proportional to

$$W \sim \frac{\sin^2 \frac{E'-E}{2\hbar} t}{(E'-E)^2} \tag{1.16.24}$$

It follows from Eq.(1.16.24) that the most probable value of the observed quantity $(E'-E)$ corresponds to

$$(E'-E)\,\Delta t \sim \hbar, \tag{1.16.25}$$

where Δt is the time interval between measurements of E' and E. The less is the interval Δt, the larger is the measured energy interval $\Delta E = E' - E$. It is important that the quantity $\hbar/\Delta t$ does not depend on the magnitude of the perturbation.

Eq.(1.16.25) means that in quantum mechanics the conservation law of the energy is verified for two measurements separated by Δt with accuracy $\hbar/\Delta t$. The relation (1.16.25) is known as ***uncertainty principle for energy-time:***

$$\Delta E \Delta t \sim \hbar. \tag{1.16.26}$$

We apply Eq.(1.16.26) to the system with the energy E_0 which decays in the time interval τ due to an external perturbation into two parts with energies E_1 and E_2. Time interval τ is reverse proportional to the probability of decay. Applying Eq.(1.16.26), one can find:

$$(E_0 - E_1 - E_2) \sim \frac{\hbar}{\tau}. \tag{1.16.27}$$

Equation (1.16.27) means that the energy of quasistationary decaying system is defined with the accuracy of the order of $\hbar/\tau$. The quantity $\Gamma = \hbar/\tau$ is called the width of the energy level E_0.

BIBLIOGRAPHY

1. R.P. Feynman, R.B. Leighton, and M. Sands, *Radiation, Waves, Quanta* In: *The Feynman Lectures on Physics* (Addison-Wesley, Menlo Park, 1963).

2. R. Eisberg and R. Resnik, *Quantum Physics* (John Wiley & Sons, New York, 1985).

CHAPTER 2

Introduction to Solid State Physics

2.1 CHEMICAL BONDS IN CRYSTALS

Knowledge of the structure of atomic levels permits classifying the chemical elements in nature and understanding the background of the chemical bond in solids. Hydrogen is the first element in the Periodic Table. We have presented its energy levels and wave functions in Eqs.(1.13.39) and (1.13.40). There are four quantum numbers that define an electronic state in hydrogen:

principle $n = 1, 2, 3, \ldots\ldots.$;

orbital $l = 0, 1, 2, \ldots\ldots. n-1$; or $l = s, p, d \ldots..$;

magnetic $m = 0, \pm 1, \pm 2 \ldots. \pm l$;

spin $s_z = +1/2, -1/2$.

A similar classification is possible for many-electron atoms taking that the nuclear charge is Ze', where Z is the atomic number of the atom in the Periodic Table of elements (the number of charged protons in the nucleus).

According to the Pauli exclusion principle, two electrons are not allowed to occupy a state with the same four quantum numbers. The electronic structure of complex atoms can be understood in terms of filling up higher and higher energy levels. The chemical behavior (valence) of complex atoms is determined mainly by the outermost ***valence*** electrons. The electrons of the internal shells and the nuclei make up the ***ionic cores*** of atoms. They are of no importance for the chemical bond.

The energy levels of the hydrogen atom Eq.(1.13.40) are highly degenerate, being independent of l, m, and s_z. The energy levels in many electron atoms depend not only on the principal quantum number n, but also on the orbital quantum number l. Therefore electrons are subdivided into "subshells" that correspond to different values of l, $l = 0, 1, 2, 3, 4....$ The spectroscopic notations for these subshells are $s, p, d, f, ...,$ respectively.

For a given value of l there are $(2l + 1)$ values of m, and for each of these there are two possible values of the spin quantum number s_z.

Periods of the Periodic Table of elements are constructed according to this system of subshells. Hydrogen is the first element in the Periodic Table. Having one electron, it is in the $1s$ state. The first index "1" means that the principal quantum number is $n = 1$. "s" means that the orbital quantum number is $l = 0$. The next element is helium, which has a second electron with opposite spin in the same subshell. Helium is said to have the $1s^2$ ***electronic configuration***. The superscript "2" denotes that there are two electrons in the subshell. These two electrons should have opposite spin quantum numbers. The s-shell with two electrons is complete. A helium atom with its closed electronic subshell has a low chemical activity.

The next period starts with Li, which has an electronic configuration with an extra electron in the next subshell (Li: $1s^2 2s$). The following element is Be. It possesses one more electron. The corresponding configuration is (Be: $1s^2 2s^2$). In the next element boron (B), an additional, fifth electron starts the next electronic subshell, p-shell. The electronic configuration of B is $1s^2 2s^2 2p^1$. In the p-shell, the orbital quantum number $l = 1$. Therefore, for elements following boron there are three corresponding values of the magnetic quantum number $m = 0, \pm 1$. Taking into account the two possible values of spin, one can find six possible electrons in the p-shell, and so on.

Some possible shell and subshell configurations are listed in Table 2.1.1. For the important semiconductor materials, silicon (Si, atomic number 14), germanium (Ge, atomic number 32), gallium (Ga, atomic number 31), and arsenic (As, atomic number 33), the electronic configurations are the following:

IV Si: $1s^2 2s^2 2p^6 3s^2 3p^2$, Ge: $1s^2 2s^2 2p^6 3s^2 3p^6 3d^{10} 4s^2 4p^2$,

III Ga: $1s^2 2s^2 2p^6 3s^2 3p^6 3d^{10} 4s^2 4p^1$,

V As: $1s^2 2s^2 2p^6 3s^2 3p^6 3d^{10} 4s^2 4p^3$.

Since the outer valence electron configurations are very similar for Si and Ga, we can expect that their chemical and physical properties are similar too. In fact, both elements are semiconductors.

The most stable electronic configurations exist in atoms with completed s- and p subshells of the valence electrons. Inert gases have closed subshells and they are chemically inactive:

Ne: $1s^2 2s^2 2p^6$,

Ar: $1s^2 2s^2 2p^6 3s^2 3p^6$,

Kr: $1s^2 2s^2 2p^6 3s^2 3p^6 3d^{10} 4s^2 4p^6$.

Shell (n)	l	m	Electronic configuration	Number of electrons
1	0	0	$1s$	2
2	0	0	$2s$	2
	1	$0, \pm 1$	$2p$	6
3	0	0	$3s$	2
	1	$0, \pm 1$	$3p$	6
	2	$0, \pm 1, \pm 2$	$3d$	10

Table 2.1.1 Possible shell and subshell configurations.

Chemical bonds in a crystal are formed in such a way that neighboring atoms share their valence electrons trying to complete the s and p shells of the valence electrons.

A. Ionic Crystals

Ionic crystals are made up of positive and negative ions. The ionic bond results mainly from the attractive electrostatic interaction of neighboring ions with opposite charges. There is also a repulsive interaction with second neighbors with the same charge. Attraction and repulsion together make a balance which results in equilibrium of an ionic crystal. The resulting electronic configuration in a crystal corresponds to closed electronic shells. Consider the example of the LiF crystal. We see from the Periodic Table that neutral Li and F atoms have the following electronic configurations:

Li: $1s^2 2s$,

F: $1s^2 2s^2 2p^5$.

To achieve a stable electronic configuration the Li atom gives an electron to an F atom. Both become ions with the closed shell stable configurations of inert gases, i.e., helium (Li^+) and neon (F^-)

Li^+: $1s^2$,

F^-: $1s^2 2s^2 2p^6$.

For all elements with almost closed shells, there exists a tendency to form an ionic bond. These compounds are dielectrics, and their crystal lattice is a closed packed one. The ionic bond is shown schematically in Fig.2.1.4a.

B. Covalent crystals

Covalent bonding is typical for atoms with a low level of the outer shell filling. Consider the example of a Si crystal. Silicon has the configuration $(Core + 3s^2 3p^2)$. In order to complete the outer $3p^2$ shell, a silicon atom forms four bonds with four other neighboring silicon atoms situated in corners of the tetrahedron. It shares two valence electrons with each of four silicon neighbors (8 valence electrons altogether).

The wave functions of these bonds can be represented as linear combinations of atomic wave functions of s- and p-type. These functions are often called ***atomic orbitals***. One s-orbital and three p-orbitals, having two electrons with opposite spin each, form the outer subshells of a Si atom. It is shown in Section 1.13 that an s-orbital of an atom has spherical symmetry, see Fig.2.1.1a. The angular dependence of each of the three p-orbitals, taken from Section 1.13, is shown in Figs. 2.1.1b.

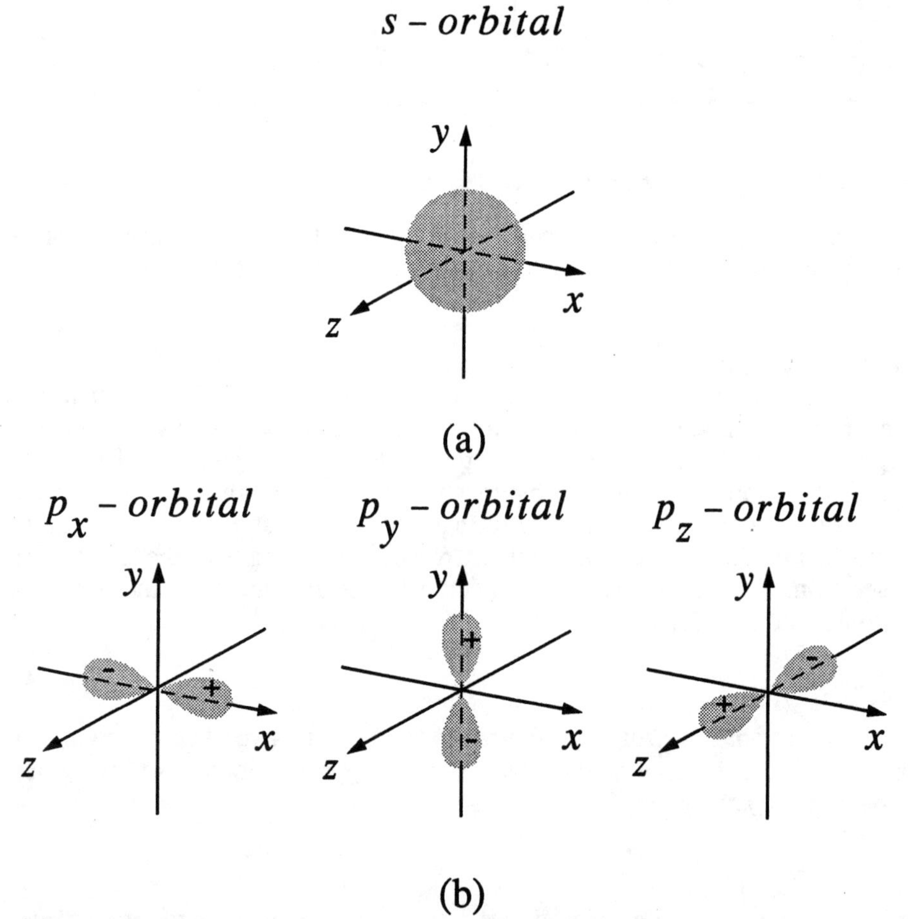

Fig. 2.1.1 (a) Depiction of an *s*-orbital. (b) Three *p*-orbitals.

The amplitude of each of the three p-orbitals is positive in the positive direction of the corresponding axis, and is negative in the negative direction of that axis. Four linear combinations of these atomic orbitals can be formed, which are called sp^3-***hybridized orbitals*** (or sp^3-hybrids):

$$\psi_1 = \frac{1}{2} | s > + | p_x > + | p_y > + | p_z >, \qquad (2.1.1a)$$

$$\psi_2 = \frac{1}{2} | s > + | p_x > - | p_y > - | p_z >, \qquad (2.1.1b)$$

$$\psi_3 = \frac{1}{2} | s > - | p_x > + | p_y > - | p_z >, \qquad (2.1.1c)$$

$$\psi_4 = \frac{1}{2} | s > - | p_x > - | p_y > + | p_z > . \qquad (2.1.1d)$$

When the three directed $| p >$ orbitals are hybridized with the $| s >$ orbital, which is positive in all directions, the negative parts of the p-orbitals are cancelled, so that only the positive parts remain in the directions of the diagonals of a cube, see Fig.2.1.2.

Hybridized directed orbitals do not correspond to the ground state of the crystal. Their energy is 5-10 eV higher. This energy is compensated in a crystal by the interaction between the atoms. sp^3- hybrids of two neighboring atoms form symmetric and antisymmetric combinations. The constructive interference of these two sp^3- hybrids results in the symmetric function shown in Fig.2.1.3a. This function gives a finite probability of finding an electron between two neighboring atoms. This situation is known from molecular physics, where the presence of the electron between two nuclei holds them together. The corresponding state is called a ***bonding state***. The valence band of a semiconductor crystal originates from the bonding states of the atoms which build up the crystal. The resulting bonding energy is lower

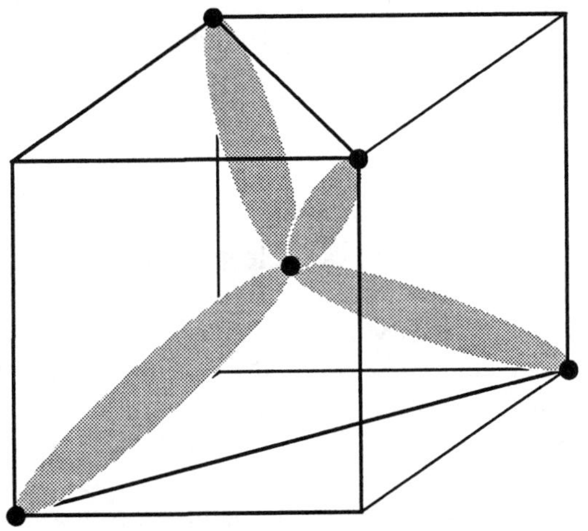

Fig. 2.1.2 sp^3 - hybridized orbitals.

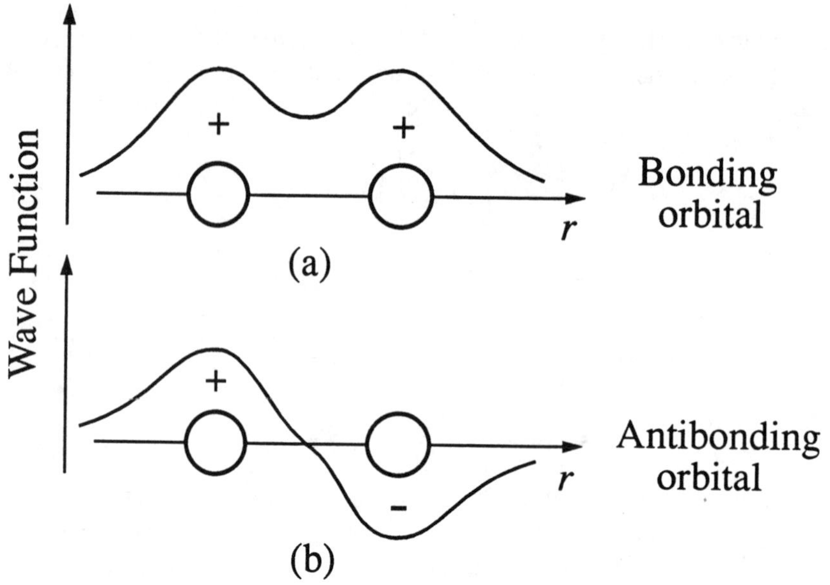

Fig. 2.1.3 (a) Bonding orbital. (b) Antibonding orbital.

than that of the ground atomic state. The gain in the energy caused by the formation of the crystal is called the ***cohesive energy***.

The destructive interference of two sp^3- orbitals on neighboring atoms results in the antisymmetric combination shown in Fig.2.1.3b. The probability of finding an electron between the atoms vanishes. This state is called an ***antibonding state***. The conduction band of semiconductor crystal originates from the antibonding states of the atoms which build up the crystal. The binding energy of these materials is $1-10\ eV/atom$.

A chemical bond of this type plays a significant role in Si, Ge, and in all III-V compounds, e.g. in GaAs. The electronic configurations of Ga and As atoms are

Ga: $\quad \text{Core} + 4s^2 4p^1,$

As: $\quad \text{Core} + 4s^2 4p^3.$

When making a crystal of GaAs, As gives an electron to Ga making them both ions. The Coulomb interaction of these ions contributes to the ionic bonding in III-V compounds. But each ion has only two $4s$ electrons and two $4p$ electrons, which is quite far from a closed subshell. Therefore, the rest of the bonding goes through the sp^3- hybridized orbitals. There is a covalent contribution to the bonding energy. Therefore, III-V compounds are materials with mixed bonding, partly ionic and partly covalent.

In a similar way, we can form many semiconductor compounds by combining elements from the column III with elements from the column V in the Periodic Table. Some examples are well known:

InAs, InP, AlAs, GaP, AlP, InSb, BN, AlN, GaSb, GaN.

Elements of the II-nd and the VI-th columns can also be combined in II - VI semiconductors with very similar properties:

CdS, ZnS, CdTe, CdSe.

Moreover, many III-V and II-VI compounds form semiconductor alloys or mixed crystals

$$Al_xGa_{1-x}As, \quad In_xGa_{1-x}As, \quad GaIn_xP_{1-x}.$$

where x is the molar fraction of Al or In. These materials are called ***ternary compounds***. By varying x from 0 to 1, one can change the properties of $A_xB_{1-x}C$ crystals from those of BC to those of AC. ***Quaternary compounds*** are also known

$$A_xB_{1-x}C_yD_{1-y}, \text{ e.g., } In_xGa_{1-x}As_yP_{1-y},$$

where x and y are molar fractional concentrations. This "materials engineering" allows designing semiconductor materials with desired properties. The covalent bond is shown in Fig.2.1.4b.

C. Metals

Metals (e.g. Na, K, Ca) consist of metal ions regularly situated in space. Each atom contributes an electron to a kind of electron "see", in which the ions are imbedded as shown in Fig.2.1.4c. The system as a whole is neutral. The electrons contribute significantly to the binding energy of metals. The binding energy is $1-10\ eV/atom$.

D. Molecular crystals and inert gas crystals

Atoms of these materials are bound by the interaction of the fluctuating dipole moments which exist in each molecule. This weak interaction is called the van der Waals interaction. The potential energy of this dipole-dipole interaction has the following dependence on the distance r between the atoms

$$U(r) \sim -\frac{1}{r^6}.$$

Minus means that it is attractive interaction. There is also a repulsive interaction which prevents atoms from falling into each other and the lattice from collapsing. It results from Pauli principle. When electronic clouds overlap the energies of electrons increase. This results in a very sharp repulsion of atoms. These crystals have a very low binding energy of the order of $0.1-1\ eV/atom$. An example of this bond is given in Fig.2.1.4d.

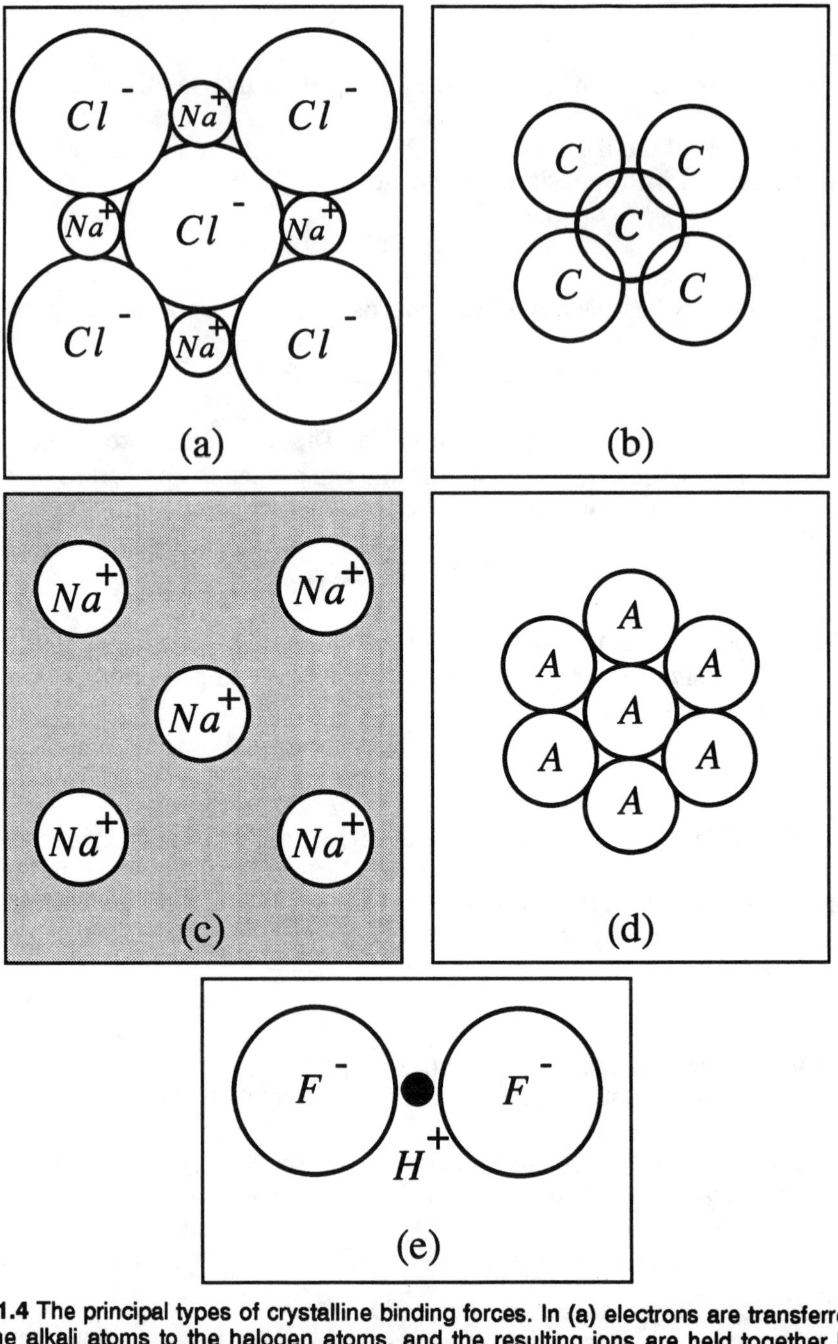

Fig. 2.1.4 The principal types of crystalline binding forces. In (a) electrons are transferred from the alkali atoms to the halogen atoms, and the resulting ions are held together by attractive electrostatic forces between the positive and negative ions. In (b) the neutral atoms of covalent material appear to be bound together by the overlapping parts of their electron distributions. In (c) the valence electrons in metal are taken away from each alkali atom to form a common electron sea in which the positive ions are imbedded. In (d) neutral atoms of inert gas crystal are bound together weakly by the van der Waals forces associated with fluctuations in their charge distributions. (e) hydrogen bond. (After Ch. Kittel. "Introduction to Solid State Physics". John Wiley & Sons, N.Y., 1986, p. 98).

E. Crystals with "hydrogen bonds"

Neutral hydrogen has only one electron. If it gives the electron to the neighboring atom, there remain the proton that form the hydrogen bond with proton of the neighboring atom with the bond energy of the order of $0.1\ eV$. The hydrogen bond connects only two atoms. An example of hydrogen fluoride atom stabilized by hydrogen bond is given in Fig.2.1.4e.

This example completes the review of principle types of crystalline binding shown in Fig.2.1.4.

2.2 CRYSTAL LATTICE

The most important feature of solids is that the atoms in them perform small-amplitude vibrations about certain positions regularly distributed in space. These positions are called ***sites*** of the crystal lattice. With a certain accuracy we can neglect these vibrations and consider the solid as a system (array) of regularly distributed atoms. In thermodynamic equilibrium solids are stable crystalline systems. Together with crystals, there are also amorphous solids, the atoms of which vibrate about randomly distributed positions. Amorphous solids are metastable systems. Their lifetimes are often very long, and we feel them to be stable. Nevertheless, if one waits long enough, amorphous solids will crystallize. An example of a long-lived amorphous solid is the usual glass in our windows.

We consider crystalline solids that have perfectly regular crystal lattices. The most important characteristic of a crystal is the spatial periodicity of the atomic positions, which is described by the ***translational symmetry***. Translational symmetry is the property of the crystal to go into itself under ***parallel transfers*** in certain directions and over certain distances. Any three-dimensional lattice is defined by three fundamental noncomplanar ***primitive translation vectors***

$$\vec{a}_1, \vec{a}_2, \vec{a}_3, \tag{2.2.1}$$

such that the position of any lattice site is determined by the ***translation vector***

$$\vec{R}_l = l_1\vec{a}_1 + l_2\vec{a}_2 + l_3\vec{a}_3, \tag{2.2.2}$$

where l_1, l_2, and l_3 are independent arbitrary positive or negative integers, including zero.

The parallelepiped whose edges are the basis vectors $\vec{a}_1$, $\vec{a}_2$, $\vec{a}_3$ is called the ***primitive cell,*** if its volume $v=\vec{a}_1\cdot(\vec{a}_2\times\vec{a}_3)$ is the smallest with respect to the volumes obtained with all other possible choices of basis vectors.

We note that the choice of basis vectors $\vec{a}_1$, $\vec{a}_2$, $\vec{a}_3$ and the corresponding choice of the primitive cell is arbitrary to some extent. An example of an oblique two-dimensional lattice is shown in Fig.2.2.1. It is seen that there are several ways to choose the primitive cell. The parallelograms 1, 2, 3 in Fig.2.2.1 have the meaning of two-dimensional primitive cell and they have equal areas. Any of them can be taken as the primitive cell.

In order to see translational symmetry, one can start from an arbitrary point of

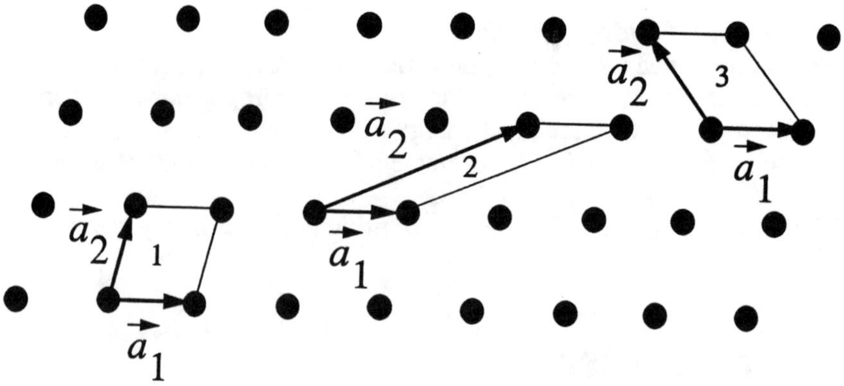

Fig. 2.2.1 Alternative choice of primitive cell in the crystal lattice.

the lattice as the origin, and find all the other equivalent positions in space by the application of translations from Eq.(2.2.1). We call points ***equivalent*** if they are indistinguishable from physical and geometrical points of view.

The total set of equivalent lattice points that can be obtained by the translations (2.2.1) is called a ***Bravais lattice***. Several possible equivalent Bravais lattices with different origins are shown in Fig.2.2.2 for the two-dimensional rectangular lattice. It should be noted that the points of a Bravais lattice are, in general, not the real sites of a crystal, i.e. the locations of atoms. If there is only one atom in the primitive cell, it is convenient to take the points of a Bravais lattice as the positions of the atoms. The Bravais lattice coincides with the real lattice in this case. This is the case for crystals such as copper or alkali metals, the structural unit of which contains a single atom.

In compound crystals, the Bravais lattice does not contain all the sites of the crystal. For example, in the two-dimensional crystal shown in Fig.2.2.3a, there is

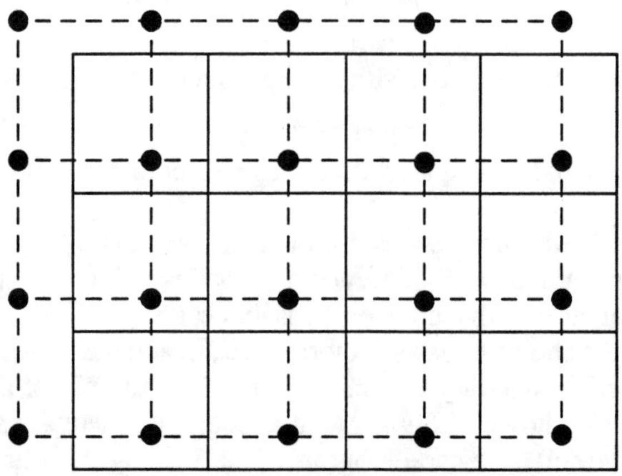

Fig. 2.2.2 Two possible choices of Bravais lattices with different origins for two-dimensional rectangular lattice.

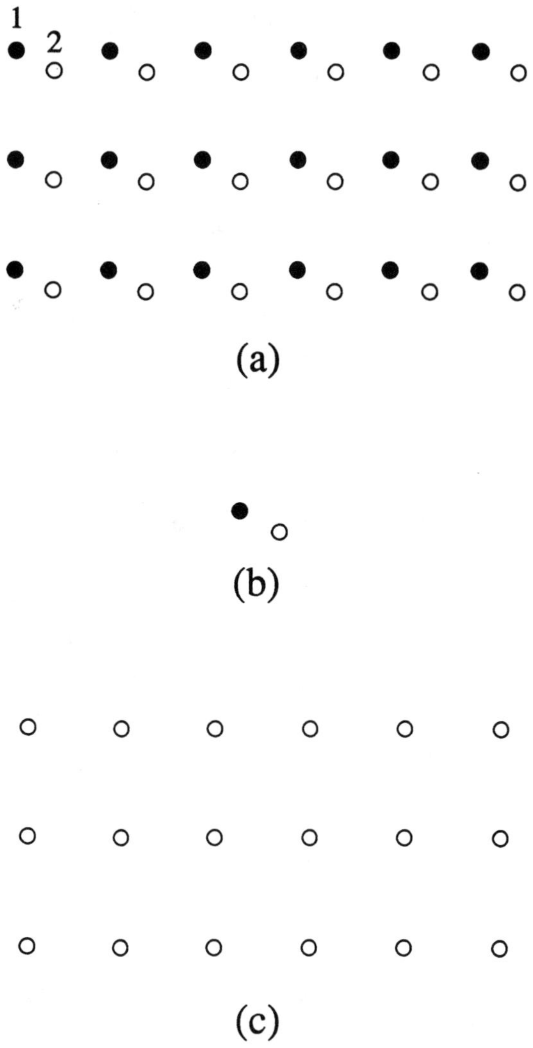

Fig. 2.2.3 (a) A two-dimensional lattice with two atoms in a primitive cell. (b) Basis consisting of two atoms. (c) Bravais lattice of the crystal.

no way to get to position 2 from position 1 by application of a translation vector $\vec{R}_l$ from Eq.(2.2.2). In order to reach position 2 starting from position 1 additional symmetry operations (rotations, reflections, non-primitive translations) are needed, in addition to the translations. Sites 1 and 2 are not connected by a translation. They comprise the ***basis*** of the crystal which contains two atoms shown in Fig.2.2.3b. The Bravais lattice for the crystal from Fig.2.2.3a is shown in Fig.2.2.3c.

Figure 2.2.4 shows another example of a two-dimensional crystal in which the lattice site E is not reached by a translation $\vec{R}_l$ from the lattice site A. The primitive cell of this lattice is the parallelepiped $ABCD$, which contains two atoms. These

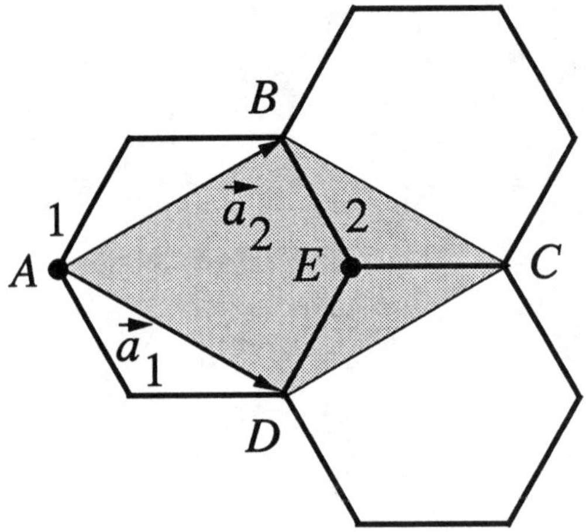

Fig. 2.2.4 Lattice with basis. The lattice site E cannot be reached by any translations from A.

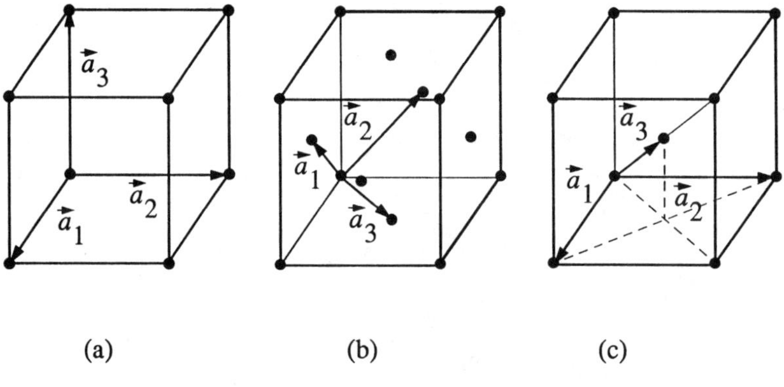

Fig. 2.2.5 Primitive translation vectors of (a) a simple cubic lattice, (b) face centered cubic lattice, (c) body centered cubic lattice.

two atoms constitute the basis of the crystal. The above mentioned extra symmetry operations which are needed to get to position 2 from position 1 reflect the symmetry of the basis or, in other words, the internal symmetry of the primitive cell.

We see that the crystal lattice has different symmetries. One of them is translational symmetry, and the other defines the structure of the basis. We shall discuss this second symmetry below in Chapter 3. The consequences of the translational symmetry represented by the vectors $\vec{R}_l$ will be considered here in Chapter 2.

In a lattice with the symmetry of a cube, three orthogonal edges of the cube are primitive translation vectors only in the simple cubic lattice shown in Fig.2.2.5a. In the face centered cubic lattice, there is no way to obtain the lattice sites at the centers of the cube faces with the help of the primitive translation vectors $\vec{a}_1, \vec{a}_2, \vec{a}_3$ presented in Fig.2.2.5a. Three other primitive vectors shown in Fig.2.2.5b are taken as primitive translations for this lattice. We see that primitive translation vectors are not always orthogonal. The volume of the parallelepiped (the volume of the primitive cell) defined by these vectors is only one fourth of the volume of the cube a^3, where a is the edge of the cube. Primitive translation vectors for the body centered cubic lattice are also different. They are shown in Fig.2.2.5c.

2.3 FREE ELECTRONS IN A CRYSTAL

The majority of the physical properties of crystals is defined by the outer valence electrons in the constituent atoms. The positively charged atomic cores create a periodic potential energy $U(\vec{r})$ for the electrons. The total time-independent Schrödinger equation of the electrons in the crystal has a very general form

$$\left\{\sum_i\left[-\frac{\hbar^2}{2m_0}\nabla_i^2+U(\vec{r}_i)\right]+\sum_{i\neq j}\frac{e'^2}{|\vec{r}_i-\vec{r}_j|}\right\}\Psi(\ldots\vec{r}_i\ldots)=\tilde{E}\,\Psi(\ldots\vec{r}_i\ldots). \qquad (2.3.1)$$

Here i, j run over all electrons. The first term in the square brackets represents the operator of the total kinetic energy of the electrons. The second term is the potential energy of the electrons created by ionic cores

$$U(\vec{r}_i)=\sum_l U(\vec{r}_i-\vec{R}_l), \qquad (2.3.2)$$

where $U(\vec{r}-\vec{R}_l)$ is the potential energy of the atom at the site $\vec{R}_l$ and the sum over l runs over all the crystal lattice sites. The third term in Eq.(2.3.1) represents the Coulomb potential energy of interelectron interaction. $\tilde{E}$ is the total energy of the electrons.

We start with the simplest case of noninteracting free electrons in a crystal. There is neither a potential energy U nor the Coulomb interaction in this simplified case. The problem, Eq.(2.3.1), reduces to the solution of the equation

$$\sum_i\left[-\frac{\hbar^2}{2m_0}\nabla_i^2\right]\Psi(\ldots\vec{r}_i\ldots)=\tilde{E}\,\Psi(\ldots\vec{r}_i\ldots). \qquad (2.3.3)$$

Since the Hamiltonian in Eq.(2.3.3) is a sum over independent electrons, equation (2.3.3) is solvable by separation of variables. The wave function Ψ is represented in the form of the product of wave functions $\psi(r_i)$ for each electron

$$\Psi(\ldots\vec{r}_i\ldots)=\psi(\vec{r}_1)\ldots\psi(\vec{r}_2)\ldots\psi(\vec{r}_i)\ldots, \qquad (2.3.4)$$

and the total energy of the system is the sum of energies of the individual electrons E_i

$$\tilde{E}=\sum_i E_i. \qquad (2.3.5)$$

The substitution of Eq.(2.3.4), (2.3.5) in Eq.(2.3.3) allows to reduce the problem to the solution of the Schrödinger equation for a single electron

$$\left[-\frac{\hbar^2}{2m_0}\nabla_i^2\right]\psi(\vec{r}_i)=E_i\psi(\vec{r}_i) \qquad (2.3.6)$$

where the label i can be omitted below.

Equation (2.3.6) has been already solved in Section 1.9. The eigenfunctions and eigenvalues are defined by Eq.(1.9.5). The conserved momentum of the free electron $\vec{p}$ is the quantum number that labels the energy and the ψ function of an electron.

We consider a crystal of finite size with volume $V = L^3$, where L is the edge of the cube, as shown in Fig.2.3.1. In principle, the electron wave function, Eq.(1.9.5), should vanish on the boundaries of the crystal. To simplify calculations, Born and von Karman suggested replacing zero boundary conditions for wave functions by ***periodic boundary conditions***. The wave function of the electron is assumed to be a periodic function with the periodicity of the sample, which has a length L in the $x-$, $y-$, and $z-$ directions, see Fig.2.3.1. The periodic boundary conditions thus have the form

$$\psi_{\vec{p}}(x,y,z) = \psi_{\vec{p}}(x+L,y,z), \tag{2.3.7a}$$

$$\psi_{\vec{p}}(x,y,z) = \psi_{\vec{p}}(x,y+L,z), \tag{2.3.7b}$$

$$\psi_{\vec{p}}(x,y,z) = \psi_{\vec{p}}(x,y,z+L). \tag{2.3.7c}$$

The values of the wave functions $\psi_{\vec{p}}$ are supposed to be equal on opposite faces of the cube. A boundary condition of this type avoids standing waves that could appear due to the reflections from the faces of the crystal. Applying Eq.(2.3.7) to the wave function from Eq.(1.9.5), we find

$$e^{\frac{i}{\hbar}p_xL} = 1, \qquad e^{\frac{i}{\hbar}p_yL} = 1, \qquad e^{\frac{i}{\hbar}p_zL} = 1. \tag{2.3.8}$$

The conditions (2.3.8) mean that the components of the momentum $\vec{p}$ take the discrete values

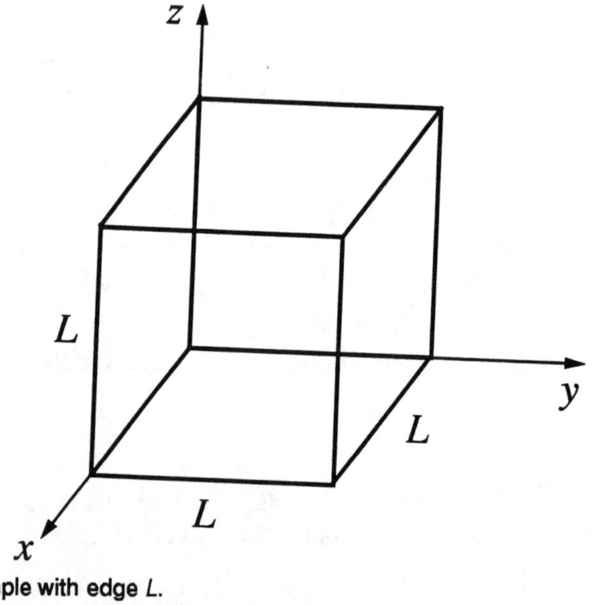

Fig. 2.3.1 Cubic sample with edge L.

$$\frac{p_x}{\hbar}=\frac{2\pi}{L}m_x, \qquad \frac{p_y}{\hbar}=\frac{2\pi}{L}m_y, \qquad \frac{p_z}{\hbar}=\frac{2\pi}{L}m_z, \tag{2.3.9}$$

where m_x, m_y, and m_z are integers, which are the ***quantum numbers*** of the problem. Equation (2.3.9) means that each component of momentum is a multiple of $(2\pi\hbar/L)$. Since the value of L is of macroscopic scale, the spectrum of $\vec{p}$-values is very dense.

The electron wave function $\psi_{\vec{p}}$, Eq.(1.9.5), should be normalized

$$\int_V |\psi_{\vec{p}}|^2\, d\vec{r}=1. \tag{2.3.10}$$

Here the integral is taken over the volume V of the crystal. The substitution of Eq.(1.9.5) in Eq.(2.3.10) yields the normalization constant $A=1/\sqrt{V}$, and one obtains, finally, the wave function of a free electron in a crystal with a finite volume V

$$\psi_{\vec{p}}=\frac{1}{\sqrt{V}}e^{\frac{i}{\hbar}\vec{p}\cdot\vec{r}}. \tag{2.3.11a}$$

The substitution of the momentum from Eq.(2.3.9) into the electron energy defined by Eq.(1.9.5) results in

$$E_{m_x m_y m_z}=\frac{\hbar^2}{2m_0}\left(\frac{2\pi}{L}\right)^2(m_x^2+m_y^2+m_z^2)=E_1(m_x^2+m_y^2+m_z^2). \tag{2.3.11b}$$

We see that the most of the eigenvalues Eq.(2.3.11b) are ***degenerate***. Different quantum numbers m_x, m_y, m_z corresponding to a different eigenfunctions, and therefore to different electron states, have the same value of the energy. For example, the six sets of quantum numbers m_x, m_y, m_z,

$$(m_x,m_y,m_z)=\begin{Bmatrix}(1,2,3) & (1,3,2) & (2,1,3)\\ (2,3,1) & (3,1,2) & (3,2,1)\end{Bmatrix},$$

generate the same energy eigenvalue, $14E_1$. This degeneracy is associated with the equivalence of three coordinate axes in space. Degeneracy of this type is called ***symmetry degeneracy***.

There are three ways to represent the energy spectrum of an electron. One of them is a direct plot of the energy $E(\vec{p})$ as a function of $\vec{p}$. Since the spectrum of $\vec{p}$ values is very dense, we can represent the dependence Eq.(1.9.5) as parabolic function $E=\vec{p}^2/2m_0$ for a certain direction of $\vec{p}$. See, for example, Fig.2.3.2a. Another possibility is connected with the representation of the surfaces of constant energy

$$E(\vec{p})=\frac{\vec{p}^2}{2m_0}=Constant.$$

In the case of free electrons, this constant energy surface is a sphere, as shown in Fig.2.3.2b. The second possibility is more informative than the first, since $E(\vec{p})$ is shown in Fig.2.3.2b for many directions in $\vec{p}$-space at the same time.

If the energy is highly degenerate, it is convenient to represent the energy spectrum using the concept of the density of states. The ***density of states*** is the

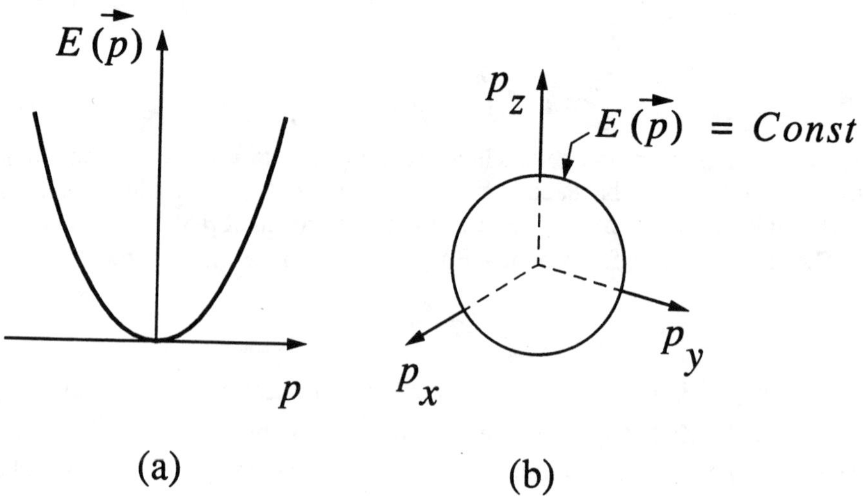

Fig. 2.3.2 Energy as function of quasimomentum. (a) Plot of energy E versus $\vec{p}$. (b) Surface of constant energy.

number of quantum states per unit energy interval per unit volume. The energy Eq.(1.9.5) is usually plotted in $\vec{p}$-space, the Cartesian coordinates of which are p_x, p_y, p_z (see Fig.2.3.3). Since $\vec{p}$ takes the discrete values (2.3.9), the ends of the $\vec{p}$ vectors form a regular lattice of $\vec{p}$ values, each corresponding to a certain state (wave function) of the electron. The volume per single state is determined from Eq.(2.3.9) by the element of volume

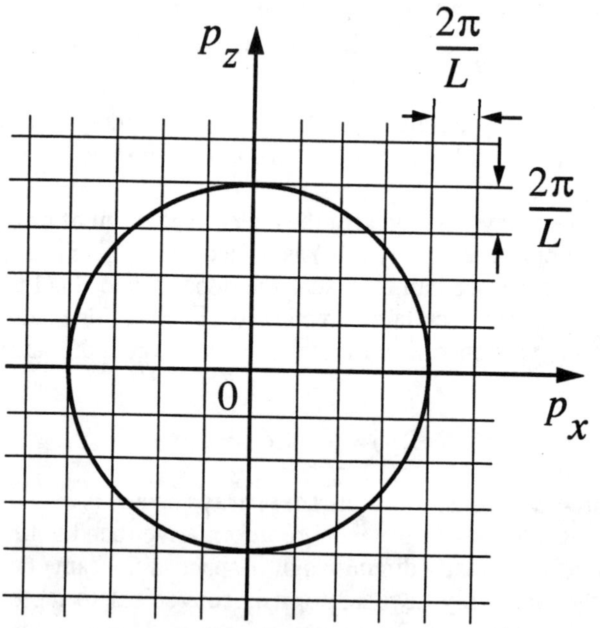

Fig. 2.3.3 Allowed values of $\vec{p}$ in two dimensions.

$$\Delta p_x \Delta p_y \Delta p_z = \left(\frac{2\pi\hbar}{L}\right)^3 \Delta m_x \Delta m_y \Delta m_z = \frac{(2\pi\hbar)^3}{V}, \qquad (2.3.12)$$

where Δm_x, Δm_y, Δm_z each equals to unity, and $V = L^3$ is the volume of the crystal. The number of states below a certain energy $E(\vec{p}) = Constant$ equals the number of states inside a sphere of radius p. The volume of the sphere is equal to$(4/3)\pi p^3$. The total number of states inside of the sphere therefore equals

$$N = 2\frac{4}{3}\pi p^3 \frac{V}{(2\pi\hbar)^3}.$$

The factor of "2" on the right hand side results from the two allowed spin values for each state. Substituting p^3 in terms of E, from Eq.(1.9.5), we get

$$N \;=\; 2\frac{4}{3}\pi\left(\frac{2m_0E}{\hbar^2}\right)^{3/2}\frac{V}{(2\pi)^3} \;=\; \frac{V}{3\pi^2}\left(\frac{2m_0E}{\hbar^2}\right)^{3/2}. \qquad (2.3.13)$$

We are interested in the increase of N when the energy E is increased by dE. Taking the derivative, one can find

$$\frac{dN}{dE}dE = \frac{V}{3\pi^2}\left(\frac{2m_0}{\hbar^2}\right)^{3/2}\frac{3}{2}E^{1/2}dE = \frac{V}{2\pi^2}\left(\frac{2m_0}{\hbar^2}\right)^{3/2}E^{1/2}dE = VD(E)dE.$$

The quantity $D(E)$ is the density of states of a three dimensional electron gas

$$D(E) = \frac{1}{2\pi^2}\left(\frac{2m_0}{\hbar^2}\right)^{3/2}E^{1/2} = \frac{3N}{2EV}. \qquad (2.3.14)$$

The density of states $D(E)$ in the case of free electrons is plotted in Fig.2.3.4. It increases proportionally to $E^{1/2}$. We note here that the factor V is quite often omitted in the equations like Eq.(2.3.14), because the density of states per unite volume is introduced instead of the density of states of a crystal with volume V.

The density of states represents the distribution of possible electron states in energy. The question of whether these states are occupied by electrons or not is answered by the statistical behavior of the collection of many electrons. The collective behavior is given by a probabilistic statistical approach based on the concept of temperature.

Two following properties of electrons are used in the statistical description: electrons are indistinguishable due to the spread of electron wave packets, the Pauli exclusion principle does not allow two electrons to be in the same quantum state.

The Fermi-Dirac distribution function $f(E)$ gives the probability that a state of energy E is occupied when the system is in thermal equilibrium. $f(E)$ is given by

$$f(E) = \frac{1}{e^{\frac{E(p)-\mu}{k_BT}} + 1}. \qquad (2.3.15)$$

The quantity $\mu(T)$ is called the ***chemical potential*** of an electron. Chemical potential $\mu(T)$ is to be chosen in such a way that the total number of electrons at any given temperature remains equal to

$$N = 2\sum_p f(E(\vec{p})). \qquad (2.3.16)$$

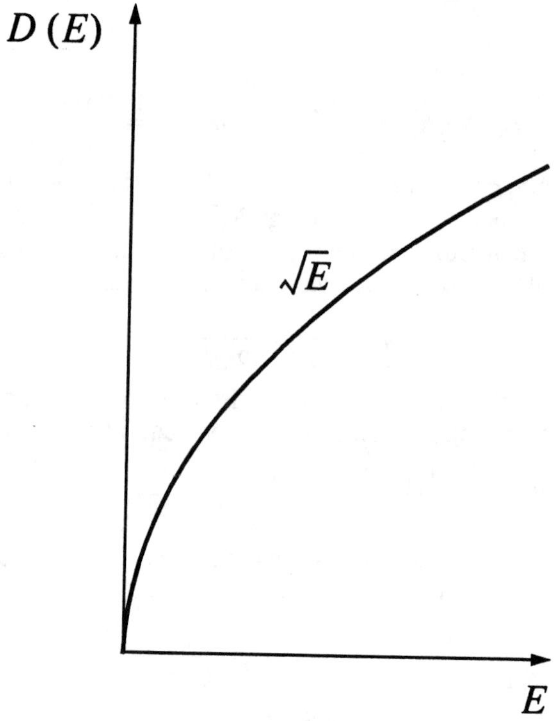

Fig. 2.3.4 Density of single-particle states as a function of energy for a free electron gas in three dimensions.

At temperature $T = 0$, $f(E)$ has the form of the step-like function shown in Fig.2.3.5 by the solid line. When $E > \mu$, $f(E) = 0$. When $E < \mu$, $f(E) = 1$. The step occurs at $E = \mu(0)$. This energy is called the ***Fermi energy***, $E_F = \mu(0)$. All the electron states with $E < E_F$ are occupied by electrons. All the states with $E > E_F$ are empty. The Fermi energy is the energy of the topmost occupied state at absolute zero. It can be given in terms of Fermi momentum $\vec{p}_F$

$$E_F = \frac{\vec{p}_F^{\,2}}{2m_0}, \tag{2.3.17}$$

The constant energy surface which corresponds to Eq.(2.3.16) is called the ***Fermi surface***.

The step in the Fermi-Dirac distribution Eq. (2.3.15) is well pronounced when the temperature remains below $\mu(0) \equiv E_F$, i.e.

$$T \ll \mu(0) \equiv E_F. \tag{2.3.18}$$

Under this condition the exponential factor $e^{\mu(0)/k_BT}$ is large:

$$e^{\frac{\mu(0)}{k_BT}} \gg 1. \tag{2.3.19}$$

Note that unequality (2.3.19) holds even when $\mu(0)/k_BT \sim 5-7$. The electron gas described by the well pronounced step is called ***degenerate*** electron gas.

In an opposite case of high temperatures

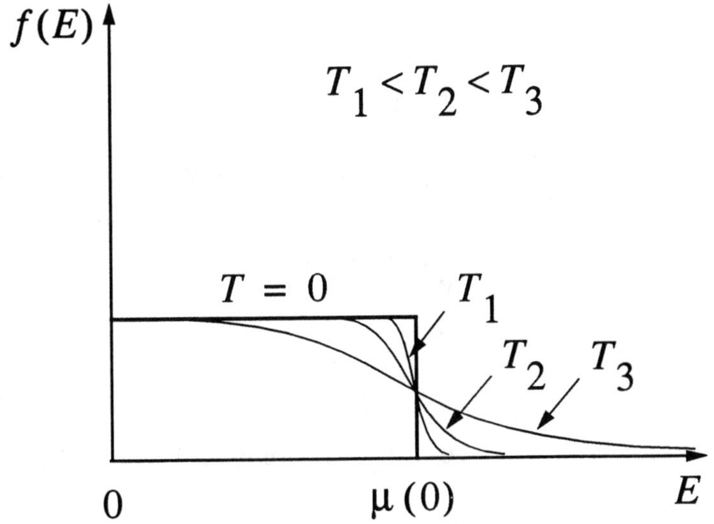

Fig. 2.3.5 Fermi-Dirac distribution as function of energy.

$$k_BT > \mu(0) \equiv E_F\ , \tag{2.3.20}$$

the exponential factor $e^{-\mu(T)/k_BT}$ is large:

$$e^{-\frac{\mu(T)}{k_BT}} \gg 1\ , \tag{2.3.21}$$

$\mu(T)$ being the negative quantity $\mu(T) < 0$.

Eq. (2.3.21) allows to neglect the unity in distribution (2.3.15) and to find the classical Boltzmann distribution function

$$f(E) = e^{\frac{\mu(T)}{k_BT}} e^{-\frac{E}{k_BT}} = Ae^{-\frac{E}{k_BT}}. \tag{2.3.22}$$

Here $A = e^{\mu(T)/T}$ is the normalization factor. The electron gas, which is described by Boltzmann distribution (2.3.22) is called the ***nondegenerated*** electron gas.

The condition (2.3.16) allows to find the concentration of electrons per unit volume, n:

$$n = \frac{N}{V} = 2\frac{1}{V}\sum_{\vec{p}} f(\vec{E}(\vec{p})). \tag{2.3.23}$$

For macroscope sample the spectrum of allowed values of $\vec{p}$ is very dense. It makes possible to replace the summation over allowed values of $\vec{p}$ in Eq. (2.3.23) by integration in $\vec{p}$ – space. The evaluation of the integral is much easier than taking the sum.

The volume of $\vec{p}$ – space per single allowed state equals $(2\pi\hbar)^3/V$. The density of allowed $\vec{p}$ – states is defined by the reciprocal quantity $V/(2\pi\hbar)^3$. Using this density one can write

$$\frac{1}{V}\sum_{\vec{p}}(\ldots) = \frac{1}{V}\frac{V}{(2\pi\hbar)^3}\int d^3p(\ldots). \tag{2.3.24}$$

Taking the $\lim\limits_{V\to\infty}$ of Eq. (2.3.23) makes it an exact relation.

Applying Eq. (2.3.24) to Eq. (2.3.23) results in

$$n = 2\frac{1}{(2\pi\hbar)^3}\int d^3p f(E(\vec{p})). \qquad (2.3.25)$$

Since $f(E)$ depends on the energy of electron, it is convenient to transfer to integration over the energy E. In spherical coordinates Eq. (2.3.25) takes the form

$$n = 2\frac{1}{(2\pi\hbar)^3}4\pi\int f(E(\vec{p}))p^2 dp .$$

Using the dependence $E = p^2/2m_0$, one can find

$$n = \int f(E)\frac{1}{2\pi^2}\frac{(2m_0)^{3/2}}{\hbar^2}\sqrt{E}\,dE . \qquad (2.3.26a)$$

The comparison with Eq. (2.3.14) shows that n here is given in terms of the density of states

$$n = \int f(E)D(E)dE . \qquad (2.3.26b)$$

The concentration of electrons in the degenerate electron gas at $T = 0$ can be found by integrating from 0 up to E_F and replacing $f(E) = 1$ within this interval

$$n = \int_0^{E_F} D(E)dE = \frac{1}{3\pi^2}\left(\frac{2m_0E_F}{\hbar^2}\right)^{3/2} . \qquad (2.3.27)$$

The Fermi energy E_F is

$$\mu(0) = E_F = (3\pi^2)^{2/3}\frac{\hbar^2}{2m_0}n^{2/3} . \qquad (2.3.28)$$

At low and finite temperatures satisfying condition (2.3.18) one can calculate the temperature dependence of μ in a degenerate electron gas

$$\mu(T) = E_F\left[1 - \frac{\pi^2}{12}\left(\frac{T}{E_F}\right)^2\right] . \qquad (2.3.29)$$

Because $T < E_F$, the second term is very small.

The electron concentration of nondegenerated electron gas is defined by Eq. (2.3.26a) with $f(E)$ from Eq. (2.3.22)

$$n = \frac{1}{2\pi^2}\left(\frac{2m_0}{\hbar^2}\right)^{3/2} e^{\frac{\mu(T)}{k_BT}} \int e^{-\frac{E}{k_BT}} E^{1/2} dE .$$

The integral is evaluated with substitution $x = E^{1/2}$ and using the integral

$$\int_{-\infty}^{+\infty} e^{-\alpha x^2} x^2 dx = \frac{1}{2}\frac{\sqrt{\pi}}{\alpha^{3/2}} .$$

The final formula for n is

$$n = 2\left(\frac{2\pi m_0 k_BT}{(2\pi\hbar)^2}\right)^{3/2} e^{\frac{\mu(T)}{k_BT}} = N_c e^{\frac{\mu(T)}{k_BT}} , N_C = 2\left(\frac{2\pi m_0 k_BT}{(2\pi\hbar)^2}\right)^{3/2} . \qquad (2.3.30)$$

The chemical potential of nondegenerated electron gas is

$$\mu(T) = k_B T \ln \frac{n}{N_c}. \tag{2.3.31}$$

In classical case, $\mu(T)$ depends strongly on the temperature T.

Transition from a degenerate to a nondegenerate electron gas is characterized by the temperature $T_D = E_F / k_B$ which is called the degeneracy temperature. In metals, where the concentration of electrons is very high, $n \approx 10^{22}\ cm^{-3}$, it follows from Eq. (2.3.28) that the Fermi energy E_F is about $5\,eV$. The corresponding degeneracy temperature equals $T_D = 60{,}000\,K$. This means that at room temperature $T = 300\,K$ the electron gas in metals is highly degenerate. In semiconductors, the electron concentrations are considerably lower: $n \approx 10^{17} - 10^{18}\ cm^{-3}$. The degeneracy temperature is, correspondingly, lower. (Note that this degeneracy of the electron gas has nothing in common with the degeneracy of energy levels in quantum mechanics).

The degenerate electron gas is an ideal, if the Coulomb potential energy of the electrons in the gas $e'^2/\bar{r}$, where $\bar{r}$ is the average separation of electrons, is less than the kinetic energy of electrons on the Fermi surface

$$\frac{e'^2}{\bar{r}} < \frac{\vec{p}_F^{\,2}}{2m_0}, \tag{2.3.31}$$

where the Fermi momentum $\vec{p}_F$ is defined by Eq.(2.3.16).

The average separation $\bar{r}$ is usually estimated from the volume per electron

$$\frac{1}{n} = \frac{V}{N} = \frac{4}{3}\pi(\bar{r})^3, \quad \text{or} \quad \bar{r} = \left(\frac{3}{4\pi}\right)^{\frac{1}{3}} n^{-\frac{1}{3}}, \tag{2.3.32}$$

It follows from Eq.(2.3.21) that $\bar{r}$ is proportional to $n^{-1/3}$. Substituting this value of $\bar{r}$ and the value of p_F from Eq.(2.3.17) into Eq.(2.3.20) and neglecting all numerical factors, we obtain the "gaseous" parameter for the electron gas in a crystal

$$n a_0^3 > 1, \tag{2.3.33}$$

where $a_0 = \hbar^2/m_0 e^2$ is the Bohr radius. We see from Eq.(2.3.22) that the ***degenerate electron gas is more ideal when it is denser.***

2.4 AN ELECTRON IN THE PERIODIC FIELD OF THE LATTICE

We now keep in the Schrödinger equation Eq.(2.3.1) both the first and the second terms and consider the behavior of an electron in the periodic field of the crystal lattice. The Schrödinger equation has the form

$$\sum_i \left[-\frac{\hbar^2}{2m_0}\nabla_i^2 + U(\vec{r}_i)\right]\Psi(\ldots\vec{r}_i\ldots) = \tilde{E}\,\Psi(\ldots\vec{r}_i\ldots).$$

Since the Hamiltonian is the sum of Hamiltonians of independent electrons, the procedure of separation of variables, which has been used in solving Eq.(2.3.3), works in this case and leads to the one-electron equation

$$\left[-\frac{\hbar^2}{2m_0}\nabla^2+U(\vec{r})\right]\psi(\vec{r})=E\psi(\vec{r}), \tag{2.4.1}$$

where E is the energy of the electron. The periodic potential of the lattice, $U(\vec{r})$, satisfies the periodicity condition (translational symmetry) of the crystal lattice

$$U(\vec{r})=U(\vec{r}+\vec{R}_l), \tag{2.4.2}$$

where the translation vector $\vec{R}_l$ is defined by Eq.(2.2.2). The translational symmetry of the lattice potential allows us to specify the electron wave function $\psi(\vec{r})$ to a certain extent. To see this we apply a translation operation to the Schrödinger equation Eq.(2.4.1) by making the coordinate substitution $\vec{r}\rightarrow(\vec{r}+\vec{R}_l)$, i.e. a translation through the vector $\vec{R}_l$,

$$\left[-\frac{\hbar^2}{2m_0}\nabla^2+U(\vec{r}+\vec{R}_l)\right]\psi(\vec{r}+\vec{R}_l)=E\psi(\vec{r}+\vec{R}_l). \tag{2.4.3}$$

Using the condition of translational symmetry Eq.(2.4.2), we see that Eq.(2.4.3) is transformed into the equation

$$\left[-\frac{\hbar^2}{2m_0}\nabla^2+U(\vec{r})\right]\psi(\vec{r}+\vec{R}_l)=E\psi(\vec{r}+\vec{R}_l), \tag{2.4.4}$$

which is equivalent to Eq.(2.4.1)

Since the Hamiltonian operators in Eqs.(2.4.4) and Eq.(2.4.1) are the same, the wave functions that satisfy these equations can differ only by a constant factor $C(\vec{R}_l)$,

$$\psi(\vec{r}+\vec{R}_l)=C(\vec{R}_l)\psi(\vec{r}). \tag{2.4.5}$$

A wave function should satisfy the normalization condition $\int|\psi|^2 d\vec{r}=1$. Substituting Eq.(2.4.5) into the normalization condition, we obtain

$$|C(\vec{R}_l)|^2=1,$$

or

$$C(\vec{R}_l)=e^{i\alpha(\vec{R}_l)}, \tag{2.4.6}$$

where $\alpha(\vec{R}_l)$ is a real function of $\vec{R}_l$ only.

In order to specify the function $C(\vec{R}_l)$ further, we now use a method that is typical for symmetry considerations. We make two successive translations of the coordinate $\vec{r}$ in the wave function $\psi(\vec{r})$. The first translation is by the vector $\vec{R}_l$, the second is by $\vec{R}_{l'}$. We have from Eq.(2.4.5) that the successively translated wave function is given by

$$\psi(\vec{r}+\vec{R}_l+\vec{R}_{l'})=C(\vec{R}_{l'})C(\vec{R}_l)\psi(\vec{r}). \tag{2.4.7}$$

The combined shift $(\vec{R}_l + \vec{R}_{l'})$ should result in the same wave function which, according to Eq.(2.4.5) is given by

$$\psi(\vec{r} + \vec{R}_l + \vec{R}_{l'}) = C(\vec{R}_l + \vec{R}_{l'})\psi(\vec{r}). \tag{2.4.8}$$

Equating wave functions (2.4.7) and (2.4.8), we find

$$C(\vec{R}_{l'})C(\vec{R}_l) = C(\vec{R}_l + \vec{R}_{l'})$$

or, substituting Eq.(2.4.6),

$$e^{i\alpha(\vec{R}_l + \vec{R}_{l'})} = e^{i\alpha(\vec{R}_l)} e^{i\alpha(\vec{R}_{l'})}.$$

It is seen that when the functions $e^{i\alpha(\vec{R}_l)}$ are multiplied, their arguments $\vec{R}_l$ add. The only known function which satisfies this condition is the exponential with a linear dependence of α on $\vec{R}_l$, i.e.

$$\alpha(\vec{R}_l) = \frac{\vec{p}}{\hbar} \cdot \vec{R}_l, \tag{2.4.9}$$

where $\vec{p}/\hbar$ is the proportionality coefficient that has the physical meaning of the quantum number of the electron in the periodic field of the lattice. Substituting Eq.(2.4.9) into Eq.(2.4.6), results in the final expression for $C(\vec{R}_l)$

$$C(\vec{R}_l) = e^{\frac{i}{\hbar}\vec{p} \cdot \vec{R}_l}. \tag{2.4.10}$$

The coefficient $\vec{p}$ here has the dimensions of momentum. But it is not a momentum. Because the momentum operator $(-i\hbar\nabla)$ does not commute with the Hamiltonian (2.4.1), momentum is not a conserved quantity for an electron in the periodic field of the crystal lattice. The new quantum number $\vec{p}$ is called the ***quasimomentum***. We shall see below that there are aspects in which the quasimomentum is similar to the momentum, but there are also facets, when it differs from the momentum.

The substitution of Eq.(2.4.10) into Eq.(2.4.5) results in the equation

$$\psi(\vec{r} + \vec{R}_l) = e^{\frac{i}{\hbar}\vec{p} \cdot \vec{R}_l}\psi(\vec{r}), \tag{2.4.11}$$

which shows that the translation $\vec{r} \to (\vec{r} + \vec{R}_l)$ of the wave function $\psi(\vec{r})$ results in its multiplication by the factor $e^{\frac{i}{\hbar}\vec{p} \cdot \vec{R}_l}$. Multiplying Eq.(2.4.11) by $e^{-\frac{i}{\hbar}\vec{p} \cdot (\vec{r} + \vec{R}_l)}$, we find that the resulting function $u(\vec{r})$ is a periodic function of $\vec{r}$ with the period of the lattice:

$$e^{-\frac{i}{\hbar}\vec{p} \cdot (\vec{r} + \vec{R}_l)}\psi(\vec{r} + \vec{R}_l) = e^{-\frac{i}{\hbar}\vec{p} \cdot \vec{r}}\psi(\vec{r}) = u(\vec{r}). \tag{2.4.12}$$

Finally, we see from Eq.(2.4.12) that the wave function of electron in the periodic field of the lattice has the form

$$\psi_{\vec{p}}(\vec{r}) = e^{\frac{i}{\hbar}\vec{p} \cdot \vec{r}} u_{\vec{p}}(\vec{r}), \tag{2.4.13}$$

where $u_{\vec{p}}(\vec{r})$ is given a label $\vec{p}$ that corresponds to the quasimomentum $\vec{p}$ in the exponential factor. The function $\psi_{\vec{p}}(\vec{r})$ satisfies the Schrödinger equation

$$\left[-\frac{\hbar^2}{2m_0}\nabla^2 + U(\vec{r})\right]\psi_{\vec{p}}(\vec{r}) = E(\vec{p})\psi_{\vec{p}}(r). \tag{2.4.14}$$

Both the eigenfunctions $\psi_{\vec{p}}(\vec{r})$ and the eigenvalues $E(\vec{p})$ are labeled by the quantum number $\vec{p}$ which results from the translational symmetry of the lattice. Equation (2.4.13) states that the wave function of an electron in a periodic lattice, $\psi_{\vec{p}}(\vec{r})$, is a plane wave modulated by the factor $u_{\vec{p}}(\vec{r})$. This result was first obtained by Bloch. The functions $\psi_{\vec{p}}(\vec{r})$ and $u_{\vec{p}}(\vec{r})$ are called the ***Bloch function*** and the ***modulating Bloch function***, respectively.

The possible values of the quasimomentum, $\vec{p}$, in a crystal of finite volume V are restricted by the ***periodic boundary conditions*** that are imposed on the Bloch wave function just as in the case of a free electron in the crystal considered in Section 2.3.

We consider the crystal slab shown in Fig.2.4.1. The slab has linear dimensions $L_1 = G_1a_1$, $L_2 = G_2a_2$, $L_3 = G_3a_3$, where G_1, G_2, G_3 are the numbers of elementary cells along each edge of the slab. The periodic boundary conditions for a Bloch function are the same as those assumed for a free electron in the crystal, see Eq.(2.3.7),

$$\psi_{\vec{p}}(x,y,z) = \psi_{\vec{p}}(x + G_1a_1, y, z), \tag{2.4.15a}$$

$$\psi_{\vec{p}}(x,y,z) = \psi_{\vec{p}}(x, y + G_2a_2), \tag{2.4.15b}$$

$$\psi_{\vec{p}}(x,y,z) = \psi_{\vec{p}}(x, y, z + G_3a_3). \tag{2.4.15c}$$

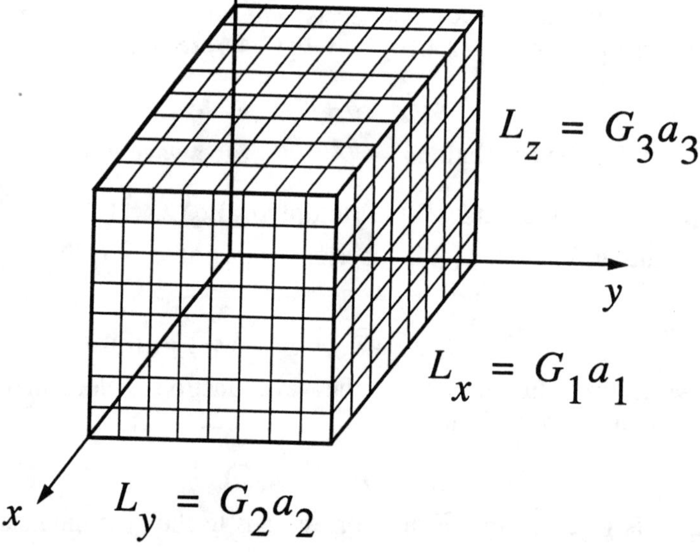

Fig. 2.4.1 A rectangular crystalline slab.

These conditions mean that wave functions on opposite faces of the slab are identical. Substituting the wave function (2.4.13) into the first boundary condition (2.4.15a), we find

$$e^{\frac{i}{\hbar}[p_x(x+G_1a_1)+p_yy+p_zz]}u_{\vec{p}}(\vec{r})=e^{\frac{i}{\hbar}(p_xx+p_yy+p_zz)}u_{\vec{p}}(\vec{r}).$$

Cancelling equal factors, we obtain

$$e^{\frac{i}{\hbar}p_xG_1a_1}=1. \qquad (2.4.16a)$$

Similar conditions follow from the other two equations (2.4.15b,c).

$$e^{\frac{i}{\hbar}p_yG_2a_2}=1, \quad e^{\frac{i}{\hbar}p_zG_3a_3}=1. \qquad (2.4.16b).$$

It follows from Eqs.(2.4.16) that the components of quasimomentum take the values

$$\frac{p_x}{\hbar}=\frac{2\pi}{G_1a_1}m_x, \quad \frac{p_y}{\hbar}=\frac{2\pi}{G_2a_2}m_y, \quad \frac{p_z}{\hbar}=\frac{2\pi}{G_3a_3}m_z, \qquad (2.4.17)$$

where (m_x, m_y, m_z) are positive and negative integers with absolute values up to the corresponding G_i, $i=1,2,3$. Since Eqs.(2.4.17) have the large numbers G_1, G_2, G_3 in their denominators, the spectrum of possible values of the quasimomentum $\vec{p}$ is very dense. We see that the size quantization of quasimomentum defined by Eq.(2.4.17) is similar to the size quantization of the free electron momentum given in Eq.(2.3.9).

The quasimomentum is not the only quantum number of the electron in the crystal. There are some other quantum numbers which are inherited from the orbitals of the atoms that constitute the crystal. We call these quantum numbers j and label the energy $E_j(\vec{p})$ and the wave function $\psi_{j\vec{p}}(\vec{r})$ by two quantum numbers $\vec{p}$ and j:

$$\psi_{j\vec{p}}=Ae^{\frac{i}{\hbar}\vec{p}\cdot\vec{r}}u_{j\vec{p}}(\vec{r}). \qquad (2.4.18)$$

Here A is an arbitrary amplitude of the wave function that is defined by the normalization and orthogonality conditions. In case of Bloch wave functions they have the form

$$\int_V \psi^*_{j'\vec{p}'}(\vec{r})\psi_{j\vec{p}}(\vec{r})d\vec{r}=\delta_{jj'}\delta_{\vec{p}\vec{p}'}, \qquad (2.4.19)$$

where the integral is taken over the volume of the crystal; $\delta_{jj'}$ and $\delta_{\vec{p}\vec{p}'}$ are Kronecker deltas.

2.5 PROPERTIES OF QUASIMOMENTUM

a. We consider now some aspects of quasimomentum which make it different from momentum. First of all, ***quasimomentum is not a single-valued quantity.*** To show this, we consider the behavior of the Bloch wave function (2.4.11) under the translation $\vec{r}\rightarrow\vec{r}+\vec{R}_l$. Equation (2.4.11) is a strong condition imposed on the wave function by the fact that the lattice is translationally invariant. The modulation function $u_{j\vec{p}}(\vec{r})$ is not changed by this translation. The Bloch function $\psi_{jp}(r)$ is multiplied by the factor $e^{\frac{i}{\hbar}\vec{p}\cdot\vec{R}_l}$ as a result of the translation.

We define a new vector $\vec{K}$ which satisfies the equation

$$(\vec{K}\cdot\vec{R}_l) = 2\pi m \tag{2.5.1}$$

for an arbitrary $\vec{R}_l$ from the set (2.2.2), where m is an integer. Equation (2.5.1) means that the scalar product $(\vec{K}\cdot\vec{R}_l)$ is a multiple of 2π.

The Bloch wave function $\psi_{\vec{p}+\hbar\vec{K}}(\vec{r})$, translated according to $\vec{r}\rightarrow\vec{r}+\vec{R}_l$, is multiplied by the same exponential $e^{\frac{i}{\hbar}\vec{p}\cdot\vec{R}_l}$ as $\psi_{\vec{p}}(\vec{r})$,

$$\psi_{\vec{p}+\hbar\vec{K}}(\vec{r}+\vec{R}_l) = e^{\frac{i}{\hbar}(\vec{p}+\hbar\vec{K})(\vec{r}+\vec{R}_l)}u_{\vec{p}+\hbar\vec{k}}(\vec{r}) = e^{\frac{i}{\hbar}\vec{p}\cdot\vec{R}_l}\psi_{\vec{p}+\hbar\vec{K}}(\vec{r}), \tag{2.5.2}$$

where Eq.(2.5.1) has been taken into account. The functions $\psi_{\vec{p}}(\vec{r})$ and $\psi_{\vec{p}+\hbar\vec{K}}(\vec{r})$ behave similarly under translations. This means that they are equivalent and equal:

$$\psi_{\vec{p}}(\vec{r}) = \psi_{\vec{p}+\hbar\vec{K}}(\vec{r}). \tag{2.5.3}$$

One can say that the quasimomenta $\vec{p}$ and $(\vec{p}+\hbar\vec{K})$ are equivalent in the sense that they correspond to the same state of the electron. The Bloch wave function $\psi_{\vec{p}}(\vec{r})$ satisfies the Schrödinger equation (2.4.14). The Bloch wave function $\psi_{\vec{p}+\hbar\vec{K}}(\vec{r})$ obeys the equation

$$\left[-\frac{\hbar^2}{2m_0}\nabla^2 + U(\vec{r})\right]\psi_{\vec{p}+\hbar\vec{K}}(\vec{r}) = E(\vec{p}+\hbar\vec{K})\psi_{\vec{p}+\hbar\vec{K}}(\vec{r}). \tag{2.5.4}$$

Comparing Eq.(2.4.14) and Eq.(2.5.4), and taking into account Eq.(2.5.3), we obtain one of the basic equations of solid state physics

$$E(\vec{p}) = E(\vec{p}+\hbar\vec{K}). \tag{2.5.5}$$

We see from Eq.(2.5.5) that due to the translational symmetry of the lattice, ***the electron energy $E(\vec{p})$ is a periodic function of the quasimomentum $\vec{p}$ with period $\hbar\vec{K}$.***

To clarify the physical meaning of the vector $\vec{K}$, we represent it in a form similar to that for $\vec{R}_l$ in Eq.(2.2.2),

$$\vec{K} = m_1\vec{K}_1 + m_2\vec{K}_2 + m_3\vec{K}_3, \tag{2.5.6}$$

where m_1, m_2, m_3 are integers. Each of the three vectors $\vec{K}_i$, $i = 1, 2, 3$ is defined to be perpendicular to the plane formed by two of the primitive translation vectors $\vec{a}_i$. The length of $\vec{K}_i$ is chosen such that its scalar product with the third vector $\vec{a}_i$ is simply 2π. Therefore,

$$\vec{a}_i\cdot\vec{K}_j = 2\pi\delta_{ij}, \tag{2.5.7}$$

where δ_{ij} is the Kronecker delta. When $i \neq j$, the product $\vec{a}_i\vec{K}_j = 0$, and if $i = j$, we have $\vec{a}_i\vec{K}_j = 2\pi$. Equation (2.5.7) gives an implicit dependence of $\vec{K}_1, \vec{K}_2, \vec{K}_3$ on $\vec{a}_1$, $\vec{a}_2, \vec{a}_3$. Direct substitution of Eqs.(2.2.2), (2.5.6), and (2.5.7) shows that Eq.(2.5.7) is equivalent to Eq.(2.5.1)

$$\vec{K} \cdot \vec{R}_l = (m_1\vec{K}_1 + m_2\vec{K}_2 + m_3\vec{K}_3) \cdot (l_1\vec{a}_1 + l_2\vec{a}_2 + l_3\vec{a}_3) = (m_1 l_1 + m_2 l_2 + m_3 l_3) 2\pi = 2\pi m.$$

Equations (2.5.7) can be solved for $\vec{K}_1$, $\vec{K}_2$, and $\vec{K}_3$, with the result that

$$\vec{K}_1 = \frac{2\pi(\vec{a}_2 \times \vec{a}_3)}{\vec{a}_1 \cdot (\vec{a}_2 \times \vec{a}_3)}, \quad \vec{K}_2 = \frac{2\pi(\vec{a}_3 \times \vec{a}_1)}{\vec{a}_1 \cdot (\vec{a}_2 \times \vec{a}_3)}, \quad \vec{K}_3 = \frac{2\pi(\vec{a}_1 \times \vec{a}_2)}{\vec{a}_1 \cdot (\vec{a}_2 \times \vec{a}_3)}. \tag{2.5.8}$$

We see that vector $\vec{K}$ defines a periodic lattice of points with primitive translation vector $\vec{K}_1$, $\vec{K}_2$, $\vec{K}_3$. This lattice is called the ***reciprocal lattice***. The vector $\vec{K}$ is a ***reciprocal lattice vector***. In the case of the simple cubic lattice, when $\vec{a}_1$, $\vec{a}_2$, $\vec{a}_3$ are perpendicular to each other, the primitive translation vectors (2.5.8) have the very simple forms

$$K_1 = \frac{2\pi}{a_1}(1,0,0), \quad K_2 = \frac{2\pi}{a_2}(0,1,0), \quad K_3 = \frac{2\pi}{a_3}(0,0,1). \tag{2.5.9}$$

The two-dimensional rectangular reciprocal lattice is shown in Fig.2.5.1 and its primitive cell is hatched. Taking into account the allowed values of the quasimomentum from Eqs.(2.4.17), and comparing them with Eq.(2.5.9), we see that the entire volume of the primitive cell of the reciprocal lattice is uniformly filled with

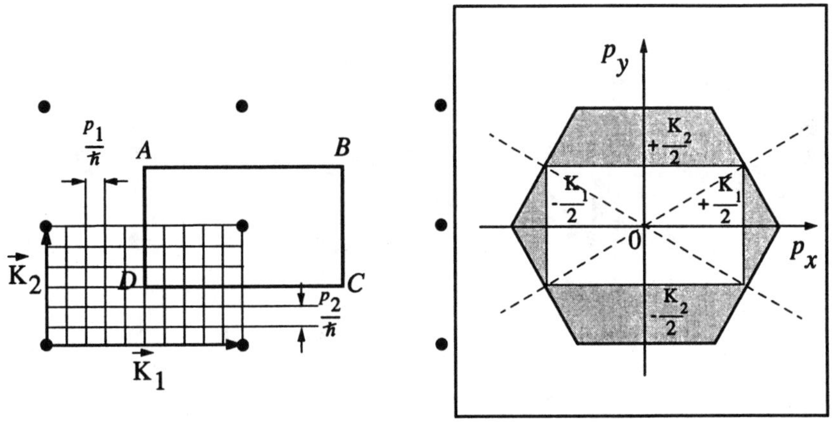

(a) (b)

Fig. 2.5.1 (a) A two-dimensional rectangular reciprocal lattice is shown and the symmetric first Brillouin zone ABCD is indicated by the heavy solid line. (b) The first and second Brillouin zones of a rectangular crystal lattice.

the allowed values of $\vec{p}$. The endpoints of all the possible $\vec{p}$ vectors form a very fine mesh in the primitive cell of the reciprocal lattice, see Fig.2.5.1a.

Since the energy of an electron is a function of the quasimomentum $\vec{p}$, we should consider it in the $\vec{p}$ space of the reciprocal lattice. Because the energy is a periodic function of $\vec{p}$, see Eq.(2.5.5), it is sufficient to consider it only inside the primitive cell of the reciprocal lattice.

The choice of primitive cell in reciprocal lattice is not unique, just as it is not in the direct space. As a rule, it is constructed in a symmetric way. One of the sites of the reciprocal lattice is taken as an origin. Vectors are drawn from this origin to nearby sites of the reciprocal lattice. The smallest volume entirely enclosed by planes that are perpendicular bisectors of these vectors is a primitive cell of the reciprocal lattice. This symmetric primitive cell of the reciprocal lattice is called the ***1st Brillouin zone.***

Following this rule, the construction of the 1st Brillouin zone for the two-dimensional rectangular lattice is shown in Fig.2.5.1a. The primitive cell is chosen by drawing vectors to connect a given lattice point to all nearby points. At their midpoints and normal to these vectors, new lines are drawn. The smallest area enclosed in this way is the 1st Brillouin zone of the two-dimensional rectangular lattice. The electron energy $E(\vec{p})$ is usually considered for p-values in this symmetric Brillouin zone.

The geometry of the 1st Brillouin zone is defined by Eq.(2.5.1) with $m = 1$, which is the equation of the crystal plane shown in Fig.2.5.2. All vectors $\vec{R}_l$ with their endpoints in the same crystal plane have the same projection on the normal to the plane. This projection is equal to the minimum of the inverse value of $\vec{K}$ corresponding to $m = 1$. We see that the crystal plane is defined by the vector $\vec{K}$ that is normal to it. This is the basic idea for constructing the reciprocal lattice.

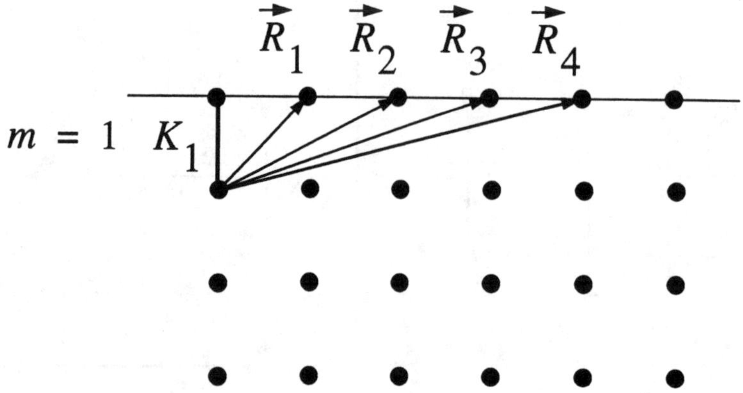

Fig. 2.5.2 A crystalline plane. The reciprocal lattice vector $\vec{K}_1$ is normal to it.

If $m=2$ in Eq.(2.5.1), we have to draw lines from the origin to the next neighboring sites of the reciprocal lattice. The perpendicular bisectors of these lines produce a new area that consist of four hatched areas shown in Fig.2.5.1b. This area is called the ***2nd Brillouin zone***. Taking $m = 3,4,5....,$ we can continue the procedure and construct the 3rd, 4th,... Brillouin zones.

b. The difference between quasimomentum and momentum is well seen by considering the addition of quasimomenta. To demonstrate this, we consider a system of two noninteracting electrons. The wave function of the system is the product of the wave function of the individual electrons:

$$\psi_{\vec{p}}(\vec{r}_1,\vec{r}_2) = \psi_{\vec{p}_1}(\vec{r}_1)\psi_{\vec{p}_2}(\vec{r}_2) = e^{\frac{i}{\hbar}\vec{p}_1\cdot\vec{r}_1} e^{\frac{i}{\hbar}\vec{p}_2\cdot\vec{r}_2} u_{\vec{p}_1}(\vec{r}_1)u_{\vec{p}_2}(\vec{r}_2), \qquad (2.5.10)$$

where $\vec{p}$ is the quasimomentum of the total system. Performing the symmetry operation of translation, we find that the separate translations $\vec{r}_1 \to \vec{r}_1 + \vec{R}_l$ and $\vec{r}_2 \to \vec{r}_2 + \vec{R}_l$ of the two electrons results in the multiplication of the right hand side of Eq.(1.5.10) by the factor $e^{\frac{i}{\hbar}\vec{p}_1\cdot\vec{R}_l} e^{\frac{i}{\hbar}\vec{p}_2\cdot\vec{R}_l}$. The same translations of the arguments of the wave function $\psi_{\vec{p}}$ for the total system of two electrons results in its multiplication by $e^{\frac{i}{\hbar}\vec{p}\cdot\vec{R}_l}$. Since the simultaneous translation of two particles is equivalent to the two successive translations we have

$$e^{\frac{i}{\hbar}\vec{p}\cdot\vec{R}_l} = e^{\frac{i}{\hbar}\vec{p}_1\cdot\vec{R}_l} e^{\frac{i}{\hbar}\vec{p}_2\cdot\vec{R}_l}. \qquad (2.5.11)$$

Equation (2.5.11) holds, if the total quasimomentum of the system equals

$$\vec{p} = \vec{p}_1 + \vec{p}_2 + \hbar\vec{K}, \qquad (2.5.12)$$

where $\vec{K}$ is a reciprocal lattice vector. Equation (2.5.11) shows that the vector $\hbar\vec{K}$ may occur in all the possible quasimomentum conservation laws in a crystal. For example, it can play an important role in electronic collisions. If two colliding electrons have initially the total quasimomentum $(\vec{p}_1 + \vec{p}_2)$ which extends outside the 1st Brillouin zone, the addition of $\hbar\vec{K} = (-2\pi\hbar/a, 0, 0)$ transforms $(\vec{p}_1 + \vec{p}_2)$ into the final total quasimomentum $\vec{p}$ inside the first Brillouin zone, see Fig.2.5.3. The final quasimomenta of electrons $\vec{p}_1$ and $\vec{p}_2$ have the directions shown in Fig.2.5.3. They are different from what one obtains from the momentum conservation law of classical mechanics. In the process of collision with the participation of $\hbar\vec{K}$, the component p_x acquires a negative value, which is why the process is called an ***Umclapp process*** (process with flip over). The presence of $\hbar\vec{K}$ in Eq.(2.5.11) indicates that in the process of electronic collision a part of the quasimomentum could be transferred to or taken from the crystal as a whole. The collision processes with $\vec{K} = 0$ are called ***normal processes.***

c. The distinction between quasi-momentum and momentum is also well seen when the motion of an electron in an external field is considered. The group velocity of the wave packet built of Bloch wave functions with quasimomenta near a certain value of $\vec{p}$ in accordance with Eq.(1.3.12) equals

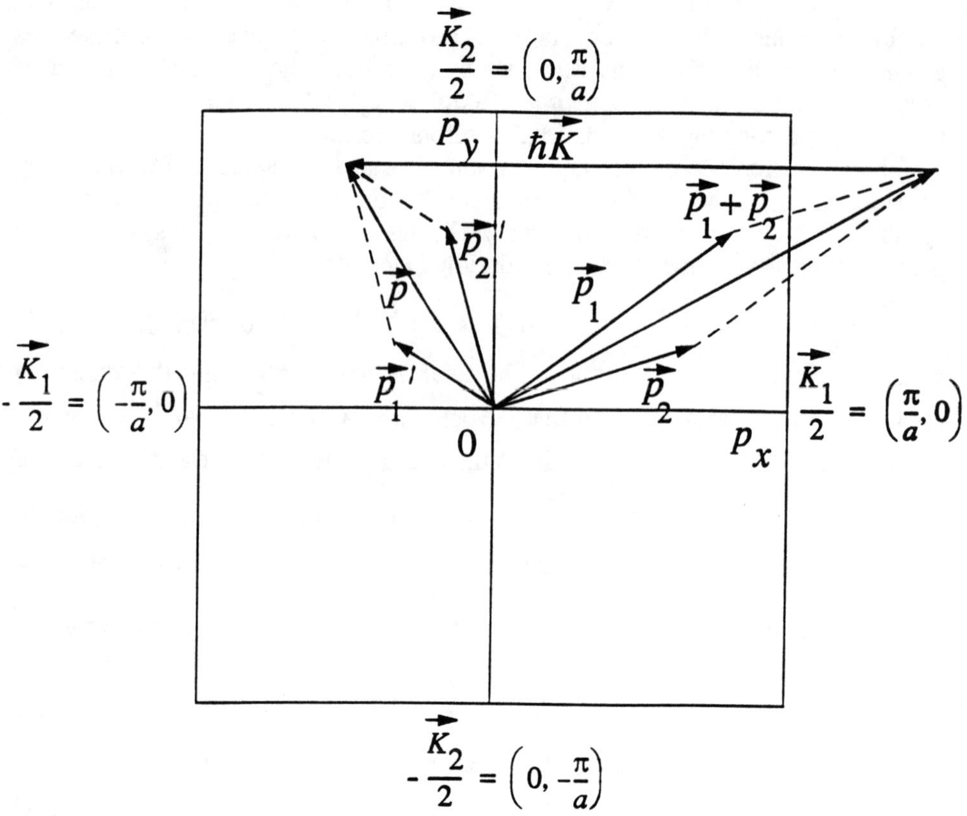

Fig. 2.5.3 An example of Umclapp-Process.

$$v = \frac{d\hbar\omega(\vec{k})}{d\hbar\vec{k}} = \frac{dE(\vec{p})}{d\vec{p}} = \nabla_{\vec{p}} E(\vec{p}), \tag{2.5.13}$$

where the quantum mechanical equations (1.3.3) and (1.3.4) have been used. The effect of the periodic potential of the lattice on the electron is taken into account through the dependence of $E(\vec{p})$ on the quasimomentum $\vec{p}$.

In the presence of an external force $\vec{F}$ the rate of change of the quasimomentum and the corresponding rate of change of the electronic energy appear. The latter equals the work done on the electron by $\vec{F}$ per unit time

$$\frac{dE(p)}{dt} = \vec{F} \cdot \vec{v}. \tag{2.5.14}$$

Using the chain rule for derivatives one obtains

$$\frac{dE(\vec{p})}{dt} = \sum_{\alpha} \frac{dE(\vec{p})}{dp_\alpha}\frac{dp_\alpha}{dt} = \nabla_{\vec{p}} E(\vec{p}) \frac{d\vec{p}}{dt} = \vec{v} \cdot \frac{d\vec{p}}{dt}, \quad \alpha = x, y, z. \tag{2.5.15}$$

Combining Eq.(2.5.14) with Eq.(2.5.15), we find

$$\vec{v}\cdot\frac{d\vec{p}}{dt}=\vec{v}\cdot\vec{F},$$

or

$$\frac{d\vec{p}}{dt}=\vec{F}. \tag{2.5.16}$$

This equation looks like the classical equation of motion. But there is a significant difference. For a free electron, the time derivative of the momentum would be defined by the total force on the electron. For an electron in a crystal, we have in Eq.(2.5.14) the external force only. Effects of the periodic lattice potential are already taken into account in the quasimomentum $\vec{p}$. For example, the equation of motion of an electron in a crystal in a uniform magnetic field $\vec{B}$ is

$$\frac{d\vec{p}}{dt}=(-e)(\vec{v}\times\vec{B}),$$

where $\vec{v}$ is the electron group velocity. Substituting $\vec{v}$ from Eq.(2.5.13), one finds

$$\frac{d\vec{p}}{dt}=(-e)(\nabla_{\vec{p}}E\times\vec{B}). \tag{2.5.17}$$

It follows from the vector cross-product that in a magnetic field, an electron moves in $\vec{p}$ space in a direction that is normal to $\nabla_{\vec{p}}E(p)$. This means that the electron moves on a surface of constant energy. On the other hand, this motion also occurs in a plane normal to the direction of $\vec{B}$. The electron orbit is defined by the intersection of this plane with the constant energy surface as shown in Fig.2.5.4.

2.6 TIME REVERSAL SYMMETRY

The equations of motion in classical mechanics contain the second time derivative. Therefore, the reversal of time

$$t\rightarrow -t \tag{2.6.1}$$

does not change classical equations of motion. It is said that the equations of motion are ***invariant against time reversal.***

The time-dependent Schrödinger equation in quantum mechanics contains the first time derivative

$$i\hbar\frac{\partial\phi(\vec{r},t)}{\partial t}=\hat{H}\,\phi(\vec{r},t), \tag{2.6.2}$$

where according to Eq.(1.5.5)

$$\phi(\vec{r},t)=e^{-\frac{i}{\hbar}E(\vec{p})t}\,\psi_{\vec{p}}(\vec{r}). \tag{2.6.3}$$

Here $\psi_{\vec{p}}(\vec{r})$ is a Bloch wave function from Eq.(2.4.13). The time-dependent Schrödinger equation (2.6.2) remains ***invariant*** when the procedure of time reversal Eq.(2.6.1) is accompanied by the complex conjugation of the equation (2.6.2)

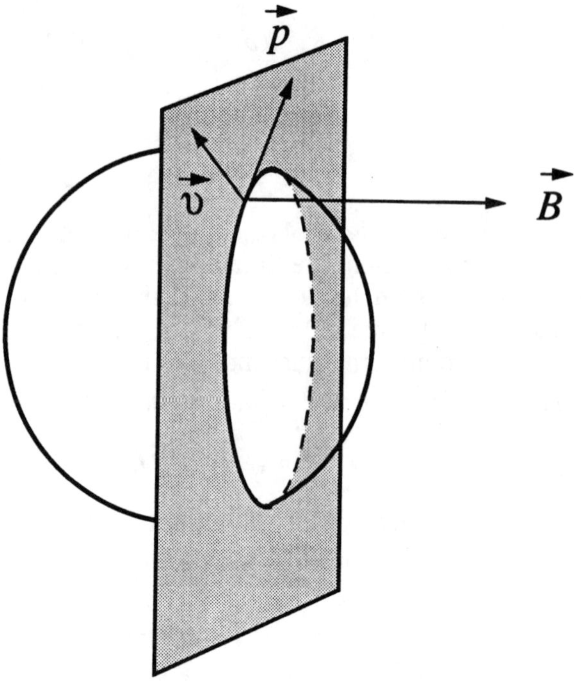

Fig. 2.5.4 Trajectory ("orbit") of an electron in a magnetic field.

$$-i\hbar\frac{\partial\phi^*(\vec{r},t)}{\partial(-t)} = \hat{H}\,\phi^*(\vec{r},t), \tag{2.6.4}$$

where

$$\phi^*(\vec{r},t) = e^{+\frac{i}{\hbar}E(\vec{p})\cdot(-t)}\,\psi^*_{\vec{p}}(\vec{r}). \tag{2.6.5}$$

Comparison of Eq.(2.6.3) and Eq.(2.6.5) shows that the evolution of $\phi(\vec{r},t)$ in time is the same as the evolution of $\phi^*(\vec{r},t)$ in the direction $(-t)$.

Substituting the wave function (2.6.3) into Eq.(2.6.2) and the wave function (2.6.5) into Eq.(2.6.4) results in similar time-independent Schrödinger equations

$$\hat{H}\psi_{\vec{p}} = E(\vec{p})\psi_{\vec{p}}, \tag{2.6.6}$$

and

$$\hat{H}\psi^*_{\vec{p}} = E(\vec{p})\psi^*_{\vec{p}}, \tag{2.6.7}$$

respectively. We see from these equations that $\psi_{\vec{p}}$ and $\psi^*_{\vec{p}}$ belong to the same eigenvalue $E(\vec{p})$. Since we have accepted in Eq.(2.4.13) the notation that the modulation Bloch function $u_{\vec{p}}(\vec{r})$ carries the same quantum number $\vec{p}$ as in the exponential factor of the Bloch wave function, the complex conjugate function

$$\psi^*_{\vec{p}} = e^{-\frac{i}{\hbar}\vec{p}\cdot\vec{r}}\,u^*_{\vec{p}}(\vec{r})$$

can be rewritten in the identical form

$$\psi^*_{\vec{p}} = e^{\frac{i}{\hbar}(-\vec{p})\cdot\vec{r}} u_{-\vec{p}}(r) = \psi_{-\vec{p}}. \tag{2.6.8}$$

Equation (2.6.8) means that $\phi^*_{\vec{p}}$ satisfies the equation

$$\hat{H}\psi^*_{\vec{p}} = E(-\vec{p})\psi^*_{\vec{p}} \tag{2.6.9}$$

and belong to the eigenvalue $E(-\vec{p})$. To make Eqs.(2.6.7) and (2.6.9) consistent, one must put

$$E(\vec{p}) = E(-\vec{p}). \tag{2.6.10}$$

According to Eq.(2.6.10), the energy of an electron in a crystal is always an ***even function of the quasimomentum***. The invariance of time-dependent Schrödinger equation (2.6.2) with respect to time-reversal results in the symmetry of the energy stated by Eq.(2.6.10).

2.7 NEARLY FREE ELECTRONS

The general ability of electron de Broglie waves to interfere leads to the conclusion that the energy spectrum of electron in the periodic potential of the lattice should have ***allowed*** and ***forbidden energy bands***. The latter are called ***energy gaps.*** We demonstrate the origin of the energy gaps by considering the periodic potential $U(\vec{r})$ as small perturbation on the free electron motion. Since we are interested in the stationary states of the electron, we use the time-independent Schrödinger equation (1.5.6).

$$\left[\hat{H}^{(0)} + U(\vec{r})\right]\psi_{\vec{p}} = E(\vec{p})\psi_{\vec{p}}, \tag{2.7.1}$$

where $\hat{H}^{(0)} = -\frac{\hbar^2}{2m_0}\nabla^2$ is the free electron Hamiltonian and $U(\vec{r})$ is a small perturbation.

In the zeroth approximation, when $U = 0$, Eq.(2.7.1) reduces to the free electron Schrödinger equation

$$\hat{H}^{(0)}\psi^{(0)}_{\vec{p}} = E^{(0)}(\vec{p})\psi^{(0)}_{\vec{p}}. \tag{2.7.2}$$

The solution of Eq.(2.7.2) has been obtained in Section 1.9, Eqs.(1.9.5), and has been discussed in detail in Sec.2.3.

In order to consider the effects of $U(\vec{r})$ on the free electron motion, it is convenient to represent the periodic potential $U(\vec{r})$ in the form of a Fourier expansion

$$U(\vec{r}) = \sum_{\vec{K}} U_{\vec{K}} e^{i\vec{K}\cdot\vec{r}}, \tag{2.7.3}$$

where $\vec{K}$ is a reciprocal lattice vector, and $U_{\vec{K}}$ is the expansion coefficient. We can easily check that $U(\vec{r})$ in the form of Eq.(2.7.3) is a periodic function of $\vec{r}$. It follows from Eq.(2.5.1) that

$$U(\vec{r}+\vec{R}_l)=\sum_{\vec{K}} U_{\vec{K}} e^{i\vec{K}\cdot(\vec{r}+\vec{R}_l)}=U(\vec{r}).$$

In order to find corrections to $\psi_{\vec{p}}^{(0)}$ and $E^{(0)}(\vec{p})$ due to the periodic potential $U(\vec{r})$, we use Eqs.(1.15.11) and (1.15.14) of the perturbation approximation. Using Eq.(2.7.3) and the wave functions (1.9.5), one can calculate the matrix elements

$$<\psi_{\vec{p}'}^{(0)} \mid U(\vec{r}) \mid \psi_{\vec{p}}^{(0)}>=\sum_{\vec{K}} U_{\vec{K}} \frac{1}{V}\int e^{\frac{i}{\hbar}(-\vec{p}'+\hbar\vec{K}+\vec{p})\cdot\vec{r}} d\vec{r}=\sum_{\vec{K}} U_{\vec{K}} \delta_{\vec{p}',\vec{p}+\hbar\vec{K}}, \qquad (2.7.4)$$

where $\delta_{\vec{p}',\vec{p}+\hbar\vec{K}}$ is the Kronecker delta. The diagonal matrix element with $\vec{p}=\vec{p}'$ has the constant value

$$<\psi_{\vec{p}}^{(0)} \mid U(\vec{r}) \mid \psi_{\vec{p}}^{(0)}>=\sum_{\vec{K}} U_{\vec{K}} \frac{1}{V}\int e^{i\vec{K}\cdot\vec{r}} d\vec{r}=\sum_{\vec{K}} U_{\vec{K}} \delta_{\vec{K},0}=U_0 \quad . \qquad (2.7.5)$$

The constant U_0 shifts the origin of the energy and does not give any contribution to the dependence of the energy on the quasimomentum $\vec{p}$.

We are seeking a correction to $E^{(0)}(\vec{p})$ which depends on the quasimomentum $\vec{p}$. Therefore, we have to find the second order correction to the energy from Eq.(1.15.15) of the perturbation approximation where, in accordance with Eq.(2.7.4), nondiagonal matrix elements with $\vec{p}'=\vec{p}+\hbar\vec{K}$ contribute to the electron energy. The substitution of these matrix elements into Eq.(1.5.15) gives

$$E^{(2)}(\vec{p})=\sum_{\vec{p}'}\sum_{\vec{K},\vec{K}'} U_{\vec{K}'}^{*} U_{K} \frac{\delta_{\vec{p}',\vec{p}+\hbar\vec{K}}\delta_{\vec{p}',\vec{p}+\hbar\vec{K}'}}{E^{(0)}(\vec{p})-E^{(0)}(\vec{p}')}.$$

Using the Kronecker deltas in order to carry out the sums over $\vec{p}'$ and $\vec{K}'$, we obtain

$$E^{(2)}(\vec{p})=\sum_{\vec{K}} \mid U_{\vec{K}} \mid^2 \frac{1}{E^{(0)}(\vec{p})-E^{(0)}(\vec{p}+\hbar\vec{K})}. \qquad (2.7.6)$$

One can see from the final Eq.(2.7.6) that there are points of $\vec{p}$space where the denominator vanishes

$$E^{(0)}(\vec{p})=E^{(0)}(\vec{p}+\hbar\vec{K}), \qquad (2.7.7)$$

and $E^{(2)}(\vec{p})$ becomes infinite. The perturbation approximation, being an expansion in the parameter

$$\frac{\mid U_{\vec{K}} \mid^2}{E^{(0)}(\vec{p})-E^{(0)}(\vec{p}+\hbar\vec{K})} \ll 1 \qquad (2.7.8)$$

is not valid at these points.

Equation (2.7.7) means that the vectors $\vec{p}$ and $(\vec{p}+\hbar\vec{K})$ are connected by geometry in the first Brillouin zone shown in Fig.2.7.1. The dashed line AB is transverse to the vector $\vec{K}$ at its midpoint. The faces of the first Brillouin zone are similar bisectors. This means that the condition expressed by Eq.(2.7.7) holds on the Brillouin zone boundaries. It follows from Eq.(2.7.7) that the absolute values of the vectors $\vec{p}$ and $\vec{p}+\hbar\vec{K}$ should be equal, which leads to:

$$\vec{p}^2=(\vec{p}+\hbar\vec{K})^2.$$

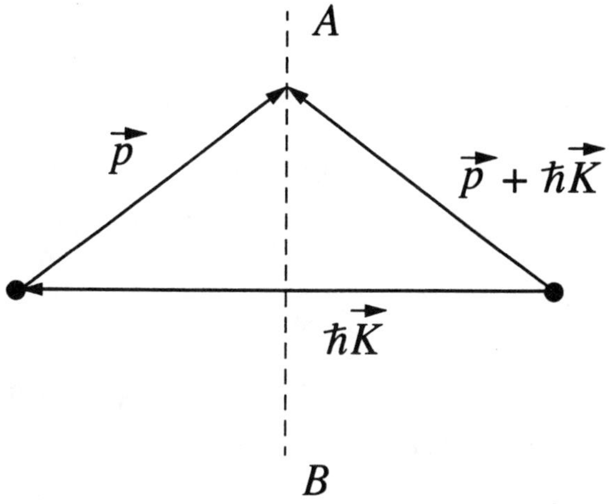

Fig. 2.7.1 Geometry of condition (2.7.7)

Because $(\vec{p}+\hbar\vec{K})^2 = p^2 + 2(\vec{p}\cdot\hbar\vec{K}) + \hbar^2K^2$, one finds that

$$|p|\cos\theta = \frac{|\hbar K|}{2},$$

where θ is the angle between $\vec{p}$ and $\vec{K}$. For the smallest value of $\vec{K}$, which corresponds to $m = 1$ in Eq.(2.5.1), the last equation takes the form

$$p\cos\theta = \hbar\pi/a. \tag{2.7.9}$$

This is a Bragg-type condition for the reflection of the electron from an atomic plane. We see that the periodic potential $U(\vec{r})$ results in the Bragg reflection of electrons from the plane transverse to the vector $\hbar\vec{K}$. If the endpoint of $\vec{p}$ lies close to a Brillouin zone boundary, the electron experiences strong Bragg reflection from the Brillouin zone boundary. This reflection is specific for an electron in the crystal lattice potential.

In order to find the energy of an electron near the Brillouin zone boundary, we must use a modified perturbation approximation which takes into account the double degeneracy of the electronic levels expressed by Eq.(2.7.7).

We introduce the following notations for degenerate eigenvalues: $E_1^{(0)} = E^{(0)}(\vec{p})$ and $E_2^{(0)} = E^{(0)}(\vec{p}+\hbar\vec{K})$. The corresponding eigenfunctions are

$$\psi_1^{(0)} = \frac{1}{\sqrt{V}}e^{\frac{i}{\hbar}\vec{p}\cdot\vec{r}} \quad \text{and} \quad \psi_2^{(0)} = \frac{1}{\sqrt{V}}e^{\frac{i}{\hbar}(\vec{p}+\hbar\vec{K})\cdot\vec{r}} \tag{2.7.10}$$

In order to take into account the degeneracy of levels, we take $\psi_{\vec{p}}^{(0)}$ in the form of a linear combination of the functions (2.7.10),

$$\psi_{\vec{p}}^{(0)} = C_1\psi_1^{(0)} + C_2\psi_2^{(0)}. \tag{2.7.11}$$

The substitution of Eq.(2.7.11) into Eq.(2.7.1) results in

$$\left[\hat{H}^{(0)} + U(\vec{r})\right](C_1\psi_1^{(0)} + C_2\psi_2^{(0)}) = E(C_1\psi_1^{(0)} + C_2\psi_2^{(0)}). \tag{2.7.12}$$

Multiplying Eq.(2.7.12) from the left by $\psi_1^{*(0)}$, and integrating over the volume of the crystal, we obtain

$$\int \psi_1^{*(0)}\left[\hat{H}^{(0)} + U(\vec{r})\right](C_1\psi_1^{(0)} + C_2\psi_2^{(0)})d\vec{r} = E\int \psi_1^{*(0)}(C_1\psi_1^{(0)} + C_2\psi_2^{(0)})d\vec{r}.$$

Substituting $\hat{H}^{(0)}\psi_1^{(0)} = E_1^{(0)}\psi_1^{(0)}$ and taking into account the normalization and orthogonality of the wave functions $\psi_1^{(0)}$ and $\psi_2^{(0)}$ one obtains the equation

$$E_1^{(0)}C_1 + <\psi_1^{(0)}|U|\psi_1^{(0)}> C_1 + <\psi_1^{(0)}|U|\psi_2^{(0)}> C_2 = E(\vec{p})C_1. \qquad (2.7.13)$$

The multiplication of Eq.(2.7.2) by $\psi_2^{*(0)}$, integration over the volume and using $\hat{H}^{(0)}\psi_2^{(0)} = E_2^{(0)}\psi_2^{(0)}$ leads to a second equation

$$E_2^{(0)}C_2 + <\psi_2^{(0)}|U|\psi_2^{(0)}> C_2 + <\psi_2^{(0)}|U|\psi_1^{(0)}> C_1 = E(\vec{p})C_2. \qquad (2.7.14)$$

Equations (2.7.13) and (2.7.14) are a linear system for the determination of the coefficients C_1 and C_2. These equations are consistent if the determinant of the coefficients vanishes

$$\begin{bmatrix} E_1^{(0)} + <\psi_1^{(0)}|U|\psi_1^{(0)}> - E(\vec{p}), & <\psi_1^{(0)}|U|\psi_2^{(0)}> \\ <\psi_2^{(0)}|U|\psi_1^{(0)}>, & E_2^{(0)} + <\psi_2^{(0)}|U|\psi_2^{(0)}> - E(\vec{p}) \end{bmatrix} = 0 \qquad (2.7.15)$$

Equation (2.7.15) is the characteristic equation for the determination of the allowed values of the energy $E(\vec{p})$.

Using the wave functions Eqs.(2.7.10) and the expansion (2.7.3) for the periodic potential $U(\vec{r})$, we can calculate the matrix elements appearing in Eq.(2.7.15) in an explicit form

$$<\psi_1^{(0)}|U(\vec{r})|\psi_1^{(0)}> = <\psi_2^{(0)}|U(\vec{r})|\psi_2^{(0)}> = U_0,$$

$$<\psi_1^{(0)}|U(\vec{r})|\psi_2^{(0)}> = \sum_{\vec{K}'} U_{\vec{K}'}\delta_{\vec{K}',\vec{K}} = U_{\vec{K}},$$

$$<\psi_2^{(0)}|U(\vec{r})|\psi_1^{(0)}> = U_{-\vec{K}} = U_{\vec{K}}^*,$$

where $\delta_{\vec{K},\vec{K}'}$ is the Kronecker delta. The transformation of the determinant into an algebraic equation leads to the quadratic equation

$$E^2 - E(E_2^{(0)} + E_1^{(0)} + 2U_0) + (E_1^{(0)} + U_0)(E_2^{(0)} + U_0) - |U_{\vec{K}}|^2 = 0, \qquad (2.7.16)$$

whose solution takes the form

$$E_\pm = U_0 + \frac{1}{2}[E^{(0)}(\vec{p}) + E^{(0)}(\vec{p} + \hbar\vec{K})] \pm \left\{\frac{1}{4}[E^{(0)}(\vec{p}) - E^{(0)}(\vec{p} + \hbar\vec{K})]^2 + |U_{\vec{K}}|^2\right\}^{1/2}. \qquad (2.7.17)$$

The quantity U_0 that defines the origin of the energy is omitted everywhere below.

We investigate the behavior of the solution (2.7.17) near the point $\vec{p} \to 0$ of the Brillouin zone. When $p \to 0$ the difference of the unperturbed energies is so large $[E^{(0)}(\vec{p}) - E^{(0)}(\vec{p} + \hbar\vec{K})]^2 \gg |U_{\vec{K}}|^2$ that we can neglect $|U_{\vec{K}}|^2$ under the square root and find two solutions

$$E_+ = E^{(0)}(\vec{p} + \hbar\vec{K})\,|_{\vec{p}\to 0} = E^{(0)}(\hbar\vec{K}), \tag{2.7.18}$$

$$E_- = E^{(0)}(\vec{p})\,|_{p\to 0} = 0. \tag{2.7.19}$$

The coefficient C_1 is found by substitution of E_- from Eq.(2.7.19) into Eq.(2.7.13):

$$E^{(0)}(\vec{p})C_1 + <\psi_1^{(0)} \mid U \mid \psi_2^{(0)}> C_2 = E^{(0)}(\vec{p})C_1.$$

It follows from this equation that $C_2 = 0$. The normalization condition $|C_1|^2 + |C_2|^2 = 1$ enables us to find C_1. Since $C_2 = 0$, we have $C_1 = 1$. Substituting C_1 and C_2 into the linear combination Eq.(2.7.11) results in

$$\psi_-^{(0)} = \psi_1^{(0)} = e^{\frac{i}{\hbar}\vec{p}\cdot\vec{r}}.$$

The substitution of E_+ from Eq(2.7.18) into Eq.(2.7.14) leads to

$$E^{(0)}(\vec{p} + \hbar\vec{K})C_2 + <\psi_2^{(0)} \mid U \mid \psi_1^{(0)}> C_1 = E^{(0)}(\vec{p} + \hbar\vec{K})C_2.$$

It follows from this equation that $C_1 = 0$ and it then follows from the normalization condition that $C_2 = 1$. Substitution of these results into the linear combination Eq.(2.7.11) gives the second wave function

$$\psi_+^{(0)} = \psi_2^{(0)} = e^{\frac{i}{\hbar}(\vec{p}+\hbar\vec{K})\cdot\vec{r}}.$$

Both values of the energy are shown in Fig.2.7.2a at $\vec{p} = 0$ in the diagram where the electron energy is plotted as a function of the quasimomentum.

We now calculate the energy of the electron near the boundary of the Brillouin zone $\vec{p} = \hbar\vec{K}/2$ where Bragg reflection occurs. In this case the difference of unperturbed energies becomes smaller than $|U_{\vec{K}}|$. Only $|U_{\vec{K}}|^2$ remains in the square root of Eq.(2.7.17). One obtains finally

$$E_+ = E^{(0)}\left(\frac{\hbar\vec{K}}{2}\right) + |U_{\vec{K}}|, \quad E_- = E^{(0)}\left(\frac{\hbar\vec{K}}{2}\right) - |U|_{\vec{K}}. \tag{2.7.20}$$

These values are very close and split by the small energy $2|U_{\vec{K}}|$, see Fig.2.7.2a. E_+ is higher than the unperturbed value $E^{(0)}(\hbar\vec{K}/2)$ and E_- is lower than this value. A splitting of the energy at the boundary of Brillouin zone appears, i.e. there appears a ***forbidden energy gap***.

It follows from the general considerations presented in Sections 2.5 and 2.6 that the electron energy is a periodic and even function of the quasimomentum, see Eqs.(2.5.5) and (2.6.10). By making the energy depicted in Fig.2.7.2a even and periodic, one obtains the total complex band structure of the electron energy with its allowed and forbidden bands, which is shown in Fig.2.7.2b. The forbidden gap is indicated by the dashed lines.

The periodic structure of the electron energy gives rise to several ways of plotting the band structure.

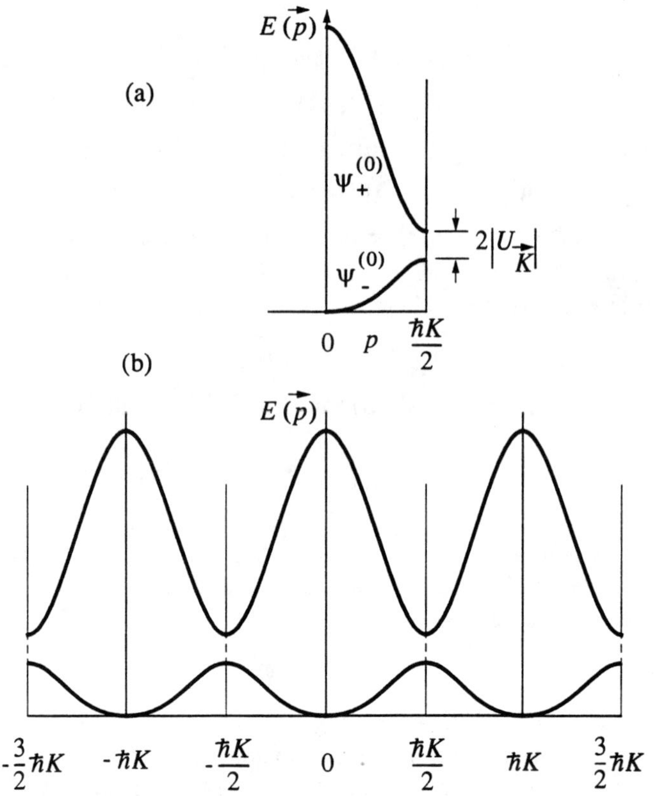

Fig. 2.7.2 Total electron band structure for nearly free electrons.

1. The ***reduced zone scheme***. Because the energy is a periodic function of $\vec{p}$ with period $\hbar\vec{K}$, there is the possibility of plotting representatives of all allowed bands in the reduced interval of quasimomentum in the 1st Brillouin zone.

$$-\frac{\pi}{a_i} \leq \frac{p_i}{\hbar} \leq +\frac{\pi}{a_i}, \quad i = 1,2,3. \tag{2.7.21}$$

Since band gaps are well seen in this scheme, it is used in the physics of semiconductors. The reduced zone scheme is shown in Fig.2.7.3a.

2. The ***extended zone scheme***. The representatives of each allowed energy band are taken in different Brillouin zones. There is the possibility to choose the representative of the lowest energy band in the 1st Brillouin zone, the representative of the second allowed energy band in the 2nd Brillouin zone, that is within the intervals

$$-\frac{2\pi}{a_i} \leq \frac{p_i}{\hbar} \leq -\frac{\pi}{a_i} \quad \text{and} \quad +\frac{\pi}{a_i} \leq \frac{p_i}{\hbar} \leq +\frac{2\pi}{a_i}, \tag{2.7.22}$$

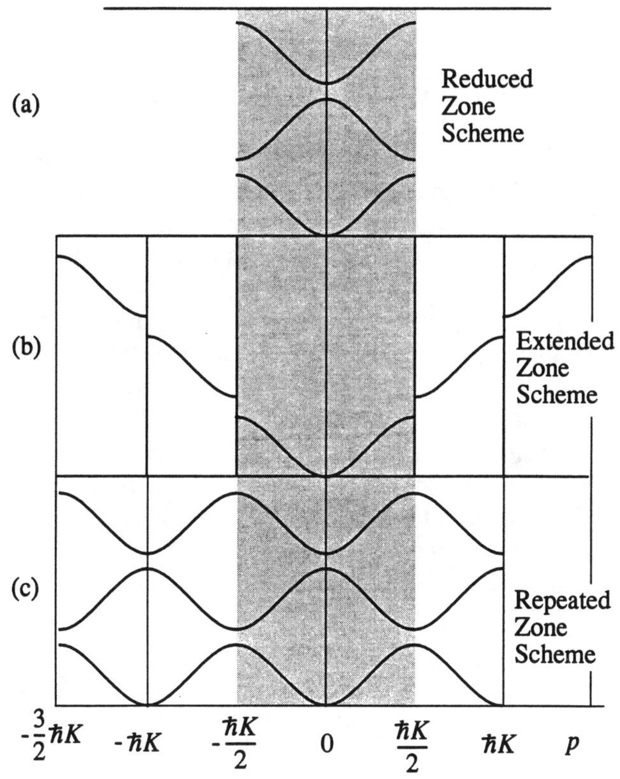

Fig. 2.7.3 Three energy bands plotted in the reduced (a), extended (b), and periodic (c) zone schemes.

and so on. This scheme is shown in Fig.2.7.3b. Inside each Brillouin zone, the resulting band structure has a certain similarity to the parabolic energy dependence of a free electron. But it differs considerably from the free electron parabolic spectrum near the Brillouin zone boundaries.

3. The ***repeated zone scheme***. Because the Brillouin zone is the unit cell of the reciprocal lattice, it may be repeated by translations through reciprocal lattice vectors. The energy in each allowed band is a periodic continuous function of the quasimomentum $\vec{p}$, which is taken in the interval

$$-\infty \leq p_i \leq +\infty.$$

The repeated zone scheme is shown in Fig.2.7.3c. It is used often in metals where the intraband behavior of electrons is of the most interest.

2.8 ELECTRON IN A SUPERLATTICE

We now discuss the energy spectrum of electrons in semiconductor layered structures consisting of materials with different physical properties laid regularly one upon another. The recent crystal growth technologies, like Molecular Beam Epitaxy (MBE), Liquid Phase Epitaxy (LPE) or Metallo-Organic Chemical Vapor Deposition (MOCVD) opened the way to growing such semiconductor layered structures. The thickness of the layers can be varied from 10 to 1000 *nm*. There are special cases of these layered structures, in which the layers are repeated in a periodic manner. These structures are called ***superlattices***. Structures have been realized with $A_xB_{1-x}C$ semiconductor alloys in which the alternate layers have different values of the fractional composition x, as is shown in Fig.2.8.1 for a GaAs/AlGaAs structure. Some examples of devices exploiting these structures are semiconductor lasers, field effect transistors, and nonlinear optical elements.

In a superlattice with alternating layers, there is an extra periodic potential $U(z)$, where z is the direction normal to the layers, which has the macroscopic period d that is much larger than the crystal lattice period a

$$d \gg a. \tag{2.8.1}$$

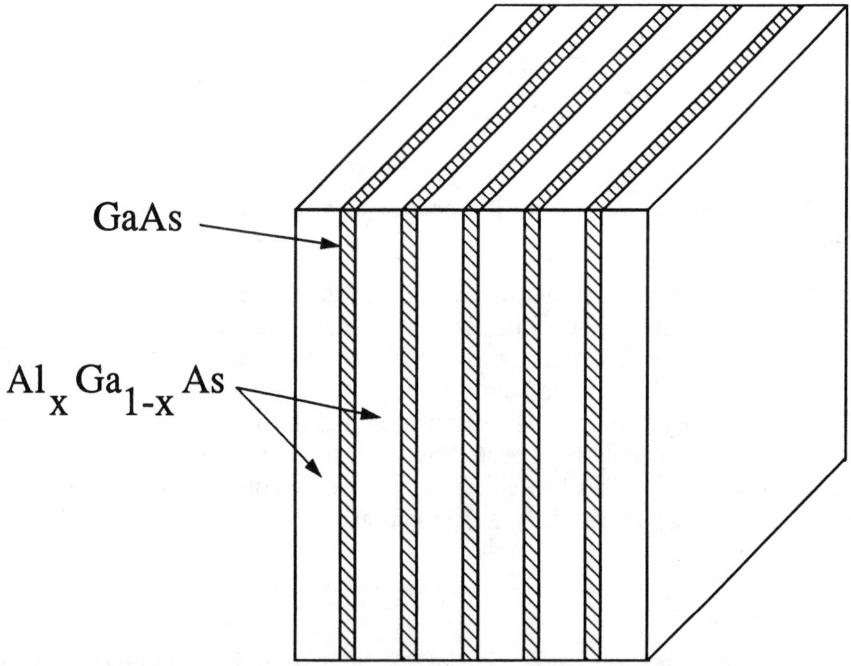

Fig. 2.8.1 Superlattice fabricated from alternating layers of GaAs and $Al_xGa_{1-x}As$.

This extra potential $U(z)$ changes the conditions of quasimomentum quantization (2.7.21). The component of quasimomentum in the z-direction p_z is specified by a new size-quantization rule which is defined by the period of the superlattice d,

$$\frac{p_z}{\hbar} = \frac{2\pi}{Nd} m_z, \tag{2.8.2}$$

where N is the total number of repeating layers in the superlattice.

The superlattice periodicity results in new Brillouin zone boundaries. The new 1st Brillouin zone corresponds to

$$-\frac{\pi}{d} \le \frac{p_x}{\hbar} \le \frac{\pi}{d}. \tag{2.8.3}$$

Comparing the interval (2.8.3) with the Brillouin zone structure from Section 2.7, one can see that the 1st Brillouin zone is reduced from the interval (2.7.21) to considerably narrower quasimomentum interval (2.8.3).

By treating the superlattice potential $U(z)$ as a small perturbation, one can follow the procedure of the calculations in Section 2.7 and obtain Bragg-type reflections of electron from the new Brillouin zone boundaries and new energy gaps. The electron energy $E(\vec{p})$ from the first Brillouin zone of an infinite crystal breaks up into new ***mini-bands*** separated by new ***mini-gaps***. The new band structure in the p_z-direction is illustrated schematically in Fig.2.8.2.

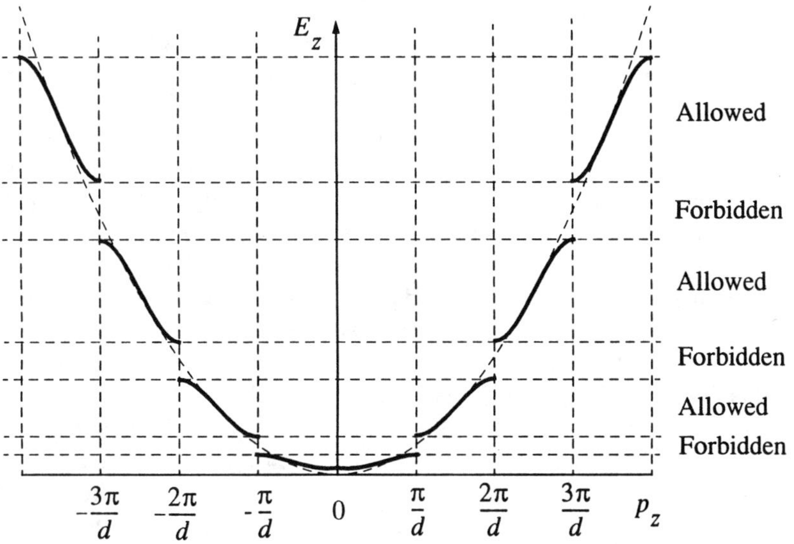

Fig. 2.8.2 Mini-gaps in the energy spectrum of an electron in semiconductor superlattice.

2.9 METALS, INSULATORS, AND SEMICONDUCTORS

The electronic wave functions (orbitals) in each energy band are labeled by the allowed values of the quasimomentum $\vec{p}$ given by Eq.(2.4.17). Using the reduced band scheme, we should take values of $\vec{p}$ within the 1st Brillouin zone, see condition Eq.(2.7.21). Substituting the allowed values of p_i from Eq.(2.4.17) into Eq.(2.7.21) and the values of $\hbar K_i$ from Eq.(2.5.9), one finds that

$$-\frac{G_i}{2} \le m_i \le \frac{G_i}{2}, \; i = 1,2,3 . \tag{2.9.1}$$

Equation (2.9.1) means that each component p_i has G_i different values. The total number of quasimomentum values within the energy band is $G_1G_2G_3$ or G^3, if $G_1 = G_2 = G_3 = G$.

Each orbital, according to the Pauli principle can be occupied by two electrons only. There are $2G^3$ independent orbitals (states) in each energy band. In the case of monovalent metals (Na, K, Li, Au ...) there is only one atom in each primitive cell of the crystal. This atom has one valence electron. Altogether, there are as many electrons as elementary cells: G^3. If G^3 electrons are distributed over $2G^3$ orbitals in the energy band, we find one half-filled energy band. These materials are ***metals*** with a typical band structure shown in Fig.2.9.1a. If the temperature $T \to 0$, the electrons occupy the lowest possible levels up to the Fermi energy, which is situated in the middle of the conduction band. Metals are good conductors.

If there is an odd number of electrons in the primitive cell, we have always a metal. Such are Al, Ga, In, Te, which have three electrons per prmitive cell. It is important to note that some elements of the fifth group crystalize in materials with two atoms per primitive ceell. This results in ten electrons per a primitive cell. $10G^3$ electrons of the crystal fill in almost five energy bands. The sixth band overlaps slightly with the fifth band. Electrons transfer from the fifth to the sixth band creates empty levels in the fifth band as shown in the Fig.2.9.2. These materials are called ***semimetals*** and they conduct current. The example is As, Sb, Bi.

Now, consider the III-V compounds, e.g. GaAs. There are two atoms (Ga and As) in each elementary cell. Ga has three valence electrons and As has five valence electrons. Together they have 8 electrons per elementary cell. The total number of electrons is $8G^3$. Dividing the number of electrons $8G^3$ by the number of states in each band $2G^3$, we see that electrons occupy the top four filled energy bands. At zero temperature these materials should be ***insulators***. An external electric field does not cause an electric current to flow at zero temperature. At finite temperatures some electrons are excited from the upper filled valence band into the lowest empty conduction band. These materials behave as ***semiconductors***. The band structure of a semiconductor is shown in Fig.2.9.1b.

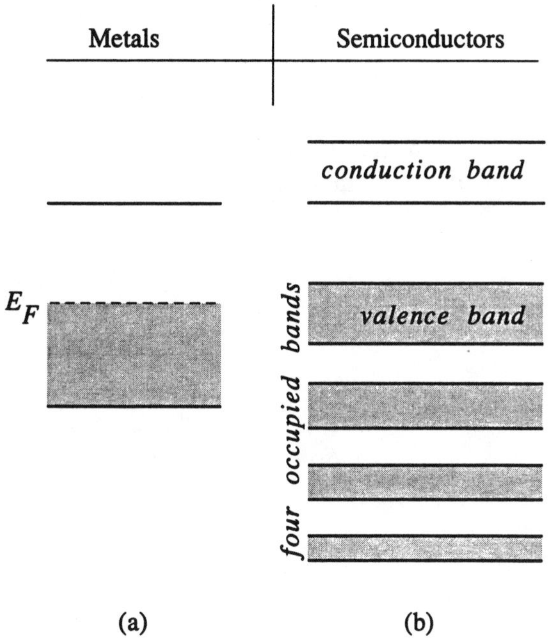

Fig. 2.9.1 Band structure of metals (a), and semiconductors (b).

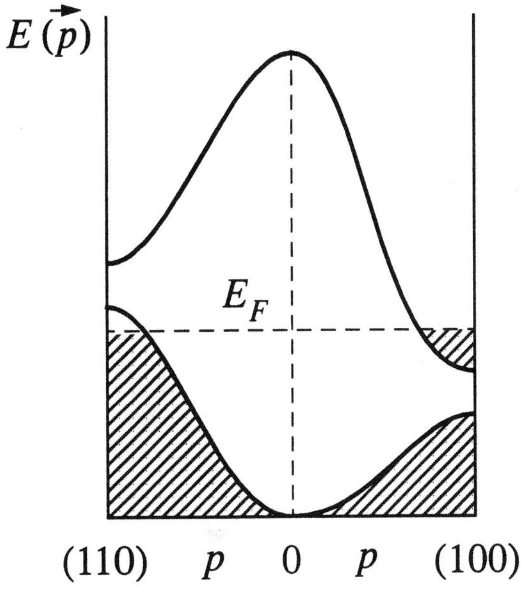

Fig. 2.9.2 Overlapping bands for semimetals. Energy as function of quasimomentum $\vec{p}$ for two directions: (110) and (100).

BIBLIOGRAPHY

W.A. Harrison *Electronic Structure and Properties of Solids: the Physics of Chemical Bond* (Freeman, San Francisco, 1980).

CHAPTER 3

Applications of Symmetry to Band Structure of Solids

3.1 ELEMENTS OF SYMMETRY

The application of symmetry simplifies significantly the determination of the electronic band structures. Symmetry allows to find the number of electron energy levels in the Brillouin zone, the degree of degeneracy of these levels, different types of selection rules.

We have already discussed in Section 2.2 the translational symmetry of a crystal lattice. But the translational symmetry does not give the total symmetry a crystal. In each crystal, there are some equivalent directions. Since it is a parallel transfer, a translation does not change directions. Some operations, which are rotations and mirror-reflections, should be added to change directions in a crystal. If a finite crystal is carried into itself due to rotation or mirror-reflection, we speak about the ***point symmetry*** of directions in the crystal. Such displacements can involve rotations and reflections whose axes and planes intersect in one common point. The existence of one point of intersection guarantees that no linear displacement of the crystal accompanies these reflections and rotations. Successive rotations about two non-intersecting axes lead to a translation of the body, whose effects have been considered earlier.

The macroscopic properties of crystals depend only on directions in the crystal. The point symmetry shows the equivalency of directions in the crystal. Therefore, the point symmetry is often called the ***macroscopic symmetry of directions***.

Let us consider basic elements of symmetry.

1. Rotation axis.

If the crystal goes into itself on rotation through an angle $2\pi/n$ about some axis that passes through the crystal, then this axis is said to be an n-fold axis C_n. The number n can take on any integer values, $n = 1, 2, 3....$ The value $n = 1$ corresponds to rotation through the angle 2π which is the same as 0. Therefore, C_1 is the identity operation usually called E.

Repeating C_n two, three, ..., n –times, one obtains rotations through the angles

$$C_n^2 = 2\frac{2\pi}{n}, \quad C_n^3 = 3\frac{2\pi}{n}, \ldots. \quad C_n^n = n\frac{2\pi}{n} = 2\pi. \tag{3.1.1a}$$

Performing the rotation n times, we return to the initial position of the body. This means that C_n^n is the identity operation E also.

Consider, for example, C_3. The prism shown in Fig.3.1.1 is a body having C_3 symmetry. C_3 corresponds to three types of rotations

$$C_3 = \tfrac{2\pi}{3} = 120°, \quad C_3^2 = 2\tfrac{2\pi}{3} = 240°, \quad C_3^3 = 3\tfrac{2\pi}{3} \equiv E. \tag{3.1.1b}$$

We see that the three-fold rotation C_3^3 carries the prism into itself. C_3^3 is the identity operation E.

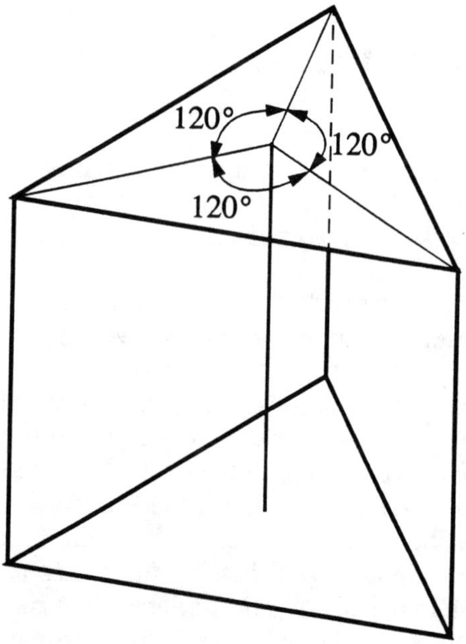

Figure 3.1.1 A prism which has C_3- symmetry.

2. Mirror reflections in a plane.

If the body is carried into itself by reflection in some plane, this plane is said to be a ***mirror reflection plane***. It is denoted by σ. A double reflection is the identity operation $\sigma^2 = E$, since it sends the body into itself.

There are two types of reflection planes. A reflection plane that contains a rotation axis is named σ_v; a reflection plane perpendicular to such an axis is called σ_h. Planes of both types are shown in Fig.3.1.2.

3. Rotary-reflection axis.

The successive application of a rotation C_n and mirror reflection σ_h yields a new type of transformation called an ***n-fold rotary-reflection axis*** S_n. A body has an n-fold rotary-reflection axis, if it goes into itself by a rotation through an angle $2\pi/n$ about this axis followed by a reflection in a plane perpendicular to it

$$S_n = \sigma_h C_n = C_n \sigma_h. \tag{3.1.2}$$

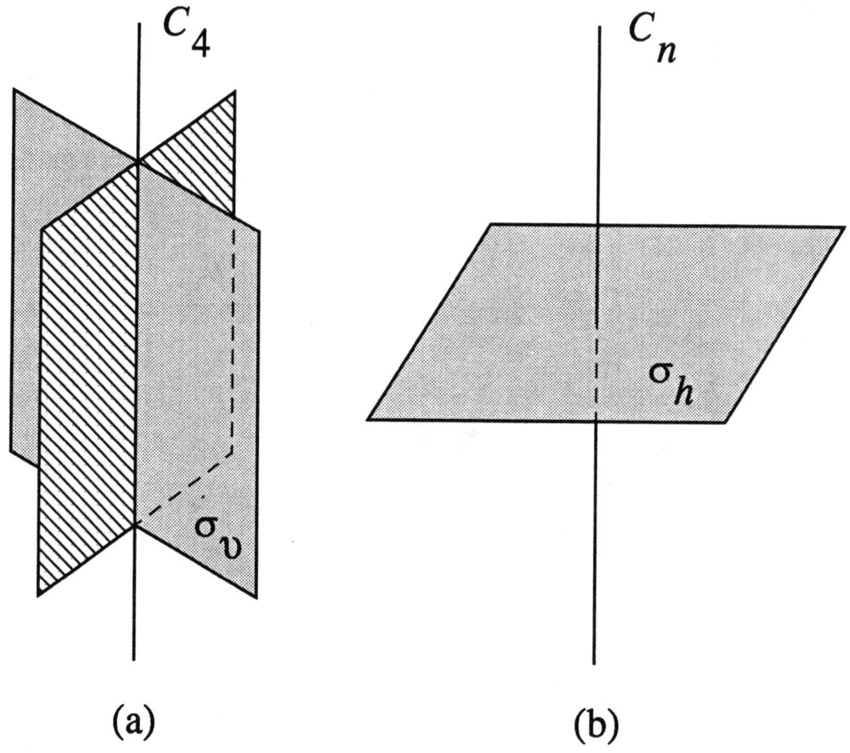

Figure 3.1.2 Two types of reflection planes: (a) the plane σ_v passes through the vertical axis; (b) the plane σ_h is perpendicular to the axis C_n

A structural building block of the semiconductor GaAs, consisting of atoms in the center and corners of a tetrahedron, is shown in Fig.3.1.3. It has the symmetry S_4. A rotation through the angle $2\pi/4$ about the vertical axis and a subsequent reflection σ_h take the block into itself.

An operation S_n is a new symmetry operation in the case of even n only. When repeated n times S_n^n reduces to the identity transformation $S_n^n = \sigma_h^n C_n^n = E$ for even n. In case of odd n, the operation $S_n^n = \sigma_n^n C_n^n = \sigma_h$ reduces to the simple mirror reflection discussed above.

An important special case of S_n is the rotary-reflection axis of the second order

$$S_2 = C_2 \sigma_h . \tag{3.1.3}$$

It is called the ***inversion operation*** $I \equiv S_2$. The inversion operation I consists of a rotation through π followed by reflection in a plane normal to the rotation axis. It is shown in Fig.3.1.4 that the inversion operation transforms an arbitrary point of the body A into the point A' that is on the straight line through A and the stationary center O, so that $OA' = OA$. When the inversion operation is applied to the radius-vector $\vec{r} = (x, y, z)$, all the components of this vector change their signs

$$I(x, y, z) = (-x, -y, -z),$$

or in vector form

$$I\vec{r} = -\vec{r}. \tag{3.1.4}$$

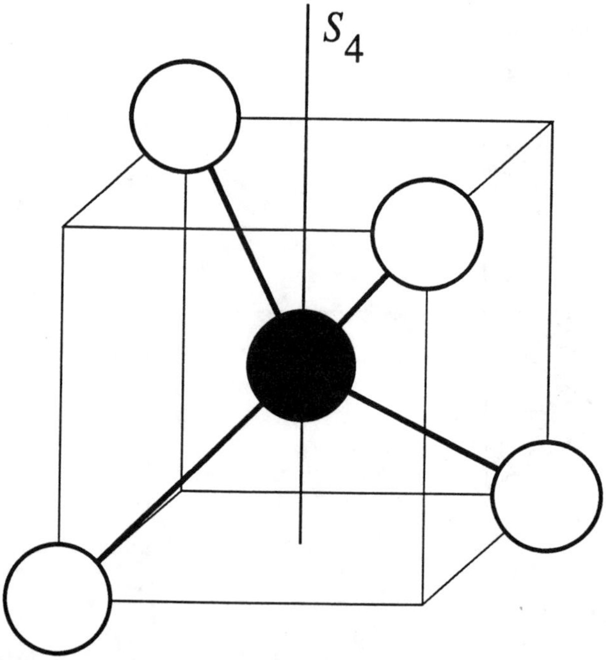

Figure 3.1.3 Structural block with S_4 - symmetry of the crystal structure of a semiconductor such as GaAs.

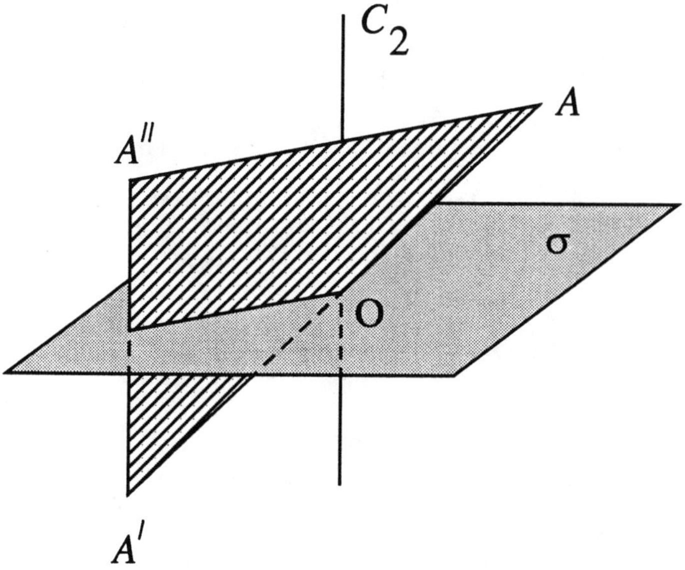

Figure 3.1.4 Point *O* is a center of inversion.

Therefore, a right hand coordinate system is transformed by the inversion I into a left hand coordinate system.

In crystals, these rotations and mirror reflections should coexist with translational symmetry. In three dimensional lattices only two-, three-, four-, and six-fold rotations are compatible with translational symmetry. A single free molecule can have any degree of rotational symmetry, in contrast to an infinite crystal. Fig.3.1.5 shows what happens in a two-dimensional periodic lattice of blocks with five-fold symmetry: pentagons do not fit together and some empty areas remain.

Types of Bravais lattice symmetry with respect to rotations and mirror reflections which are compatible with translational symmetry are classified according to ***crystal systems***. In three-dimensional space there are only seven crystal systems that are accomplished by fourteen Bravais lattices. In two-dimensional systems (the surface of a crystal is an example of a two-dimensional system) there are only 5 types of Bravais lattices.

The macroscopic point symmetry of crystals is classified according to ***crystal classes***. In the case of compound crystals with a basis, the primitive cell contains more than one atom, and the point symmetry may be, in fact, lower than the symmetry of the Bravais lattice. For example, the Bravais lattice of GaAs belong to the cubic crystal system called O_h. On the other hand, there are two different atoms in the primitive cell of GaAs. Due to this, the symmetry of a GaAs crystal belong to the crystal class of the tetrahedron called T_d. The symmetry T_d is a lower symmetry than O_h. There are altogether 32 crystallographic point groups.

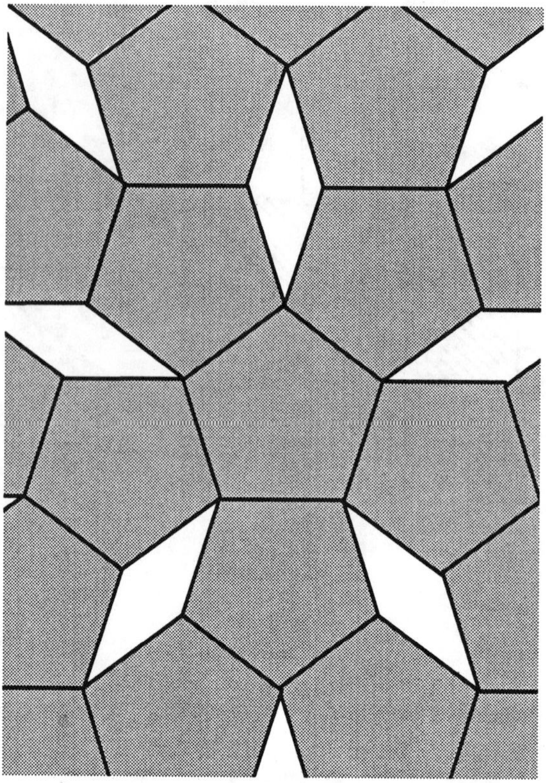

Figure 3.1.5 A five-fold axis of symmetry cannot exist in a crystal lattice because it is not possible to fill the entire plane with a connected array of pentagons (After C. Kittel "Introduction to Solid State Physics", J.Wiley & Sons, N.Y., 1986).

The knowledge of the Bravais lattice and crystal class still is not enough for a description of the symmetry of a crystal. Crystals also have ***microsymmetry***, which sends all the points of the crystal into themselves. This microsymmetry is called the ***space symmetry of the crystal***. In space symmetry, in addition to simple rotations and reflections, there appear compound symmetry operations which combine rotations with nonprimitive translation in such a way that the rotation itself is not a symmetry operation. For example, the crystal lattices of Ge and Si and some other typical semiconductors have a nontrivial translation through 1/4 of a cube diagonal. A combination of a rotation with a subsequent displacement along the same axis results in a ***screw axis***. A mirror reflection in a plane with subsequent translation in the same plane results in the operation of a ***glide plane***.

The total set of crystal symmetry elements is called the ***space group of the crystal***. Space symmetries of different crystals comprise 230 different space groups. We turn now to a consideration of consequences of the point symmetry of crystals.

3.2 SYMMETRY OF THE SQUARE

We shall find the total set of symmetry operations of such a simple body as a square. The symmetry of the square is the symmetry of some crystal lattice surfaces. The square that is shown in Fig.3.2.1, has obviously many symmetry operations which carry it into itself. It is possible to find the minimum set of symmetry operations for the square by applying them to the vector $\vec{r}$ which is directed from the center of the square to an ***arbitrary*** point of the square, see Fig 3.2.1.

There is the identity operation E which transforms the spatial components of the vector $\vec{r} = (x, y, z)$ into themselves. The vector $\vec{r}$ has the position 1 in Fig.3.2.1. There is the four-fold rotation axis C_4 that goes through the center of the square and is directed normal to the surface of the square. The operation C_4 means that there are three rotations C_4, C_4^2, C_4^3 which carry the square into itself. Operation C_4 transforms the components (y, z) of the vector $\vec{r}$ into $(z, -y)$, and $\vec{r}$ takes the position 2 in Fig.3.2.1. The operation $C_4^2 = C_2$ transforms (y, z) into $(-y, -z)$, and $\vec{r}$ takes the position 3 in Fig.3.2.1. C_4^3 transforms (y, z) into $(-z, y)$, and $\vec{r}$ takes the position 4.

The two-fold rotation axis called U_2 goes through the mid points of opposite edges of the square. There are two transverse axes of this type. The rotation about the first axis U_2 transforms (y, z) into $(y, -z)$, and position 1 goes into position 5. The second axis U_2 transforms (y, z) into $(-y, z)$, and position 1 goes into position 6.

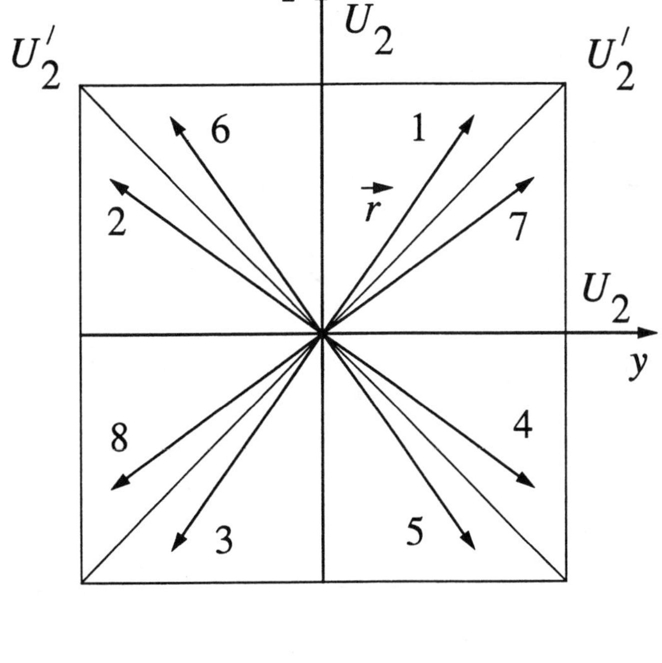

Figure 3.2.1 A square in the direct space.

There are also two two-fold axes along the diagonals of the square, called U'_2. One of these rotations carries (y,z) into (z,y), (point 1 goes into 7), and the other sends (y,z) into $(-z,-y)$, (point 1 goes into 8). One can see from Fig.3.2.1 that there are 8 equivalent positions of $\vec{r}$ which are obtained by 8 transformations, and there are no other equivalent positions for $\vec{r}$. This means that we have found the total set of symmetry operation of the square.

In principle, there are some other symmetry operations which carry the square into itself, for example, the mirror reflections in planes which are transverse to the square and go through diagonals of the square. But these are excess operations, since they do not send $\vec{r}$ into a new equivalent position on the square. The symmetry operations listed in Table 3.2.1 comprise the minimum set of possible symmetry operations of the square.

Similar procedures carried out for the tetrahedron lead to 24 equivalent positions of an arbitrary $\vec{r}$–vector inside the tetrahedron. This means that the group symmetry of a tetrahedron consists of 24 symmetry operations. In the case of the cube, we find 48 positions of an arbitrary vector $\vec{r}$ in the cube, which means that there are 48 symmetry operations in the point symmetry group of a cube.

	Symmetry operations	Transformation of coordinate axes	Positions
1	E	$yz \rightarrow yz$	1
2	C_4, C_4^3	$yz \rightarrow \begin{bmatrix} z & -y \\ -z & y \end{bmatrix}$	2 4
3	$C_4^2 \equiv C_2$	$yz \rightarrow -y, -z$	3
4	$2U_2$	$yz \rightarrow \begin{bmatrix} -y & z \\ y & -z \end{bmatrix}$	6 5
5	$2U'_2$	$yz \rightarrow \begin{bmatrix} -z & -y \\ z & y \end{bmatrix}$	8 7

Table 3.2.1 The symmetry operations of a square.

3.3 SYMMETRY TRANSFORMATIONS OF AN ELECTRON HAMILTONIAN IN A CRYSTAL

Consider, for example, the rotation of the body about axis z through the angle α in the right hand screw direction. Under the rotation through α the radius vector $\vec{r}' = (x', y', z')$ goes into the new vector $\vec{r} = (x, y, z)$ shown in Fig. 3.3.1.

Components (x, y) of $\vec{r}$ are equal to

$$x = r\cos(\alpha+\beta) = r(\cos\alpha\cos\beta - \sin\alpha\sin\beta),$$
$$y = r\sin(\alpha+\beta) = r(\sin\alpha\cos\beta + \cos\alpha\sin\beta).$$

Since

$$\cos\beta = \frac{x'}{|\vec{r}'|} \quad \text{and} \quad \sin\beta = \frac{y'}{|\vec{r}'|},$$

and $|\vec{r}'| = |\vec{r}|$, one can get

$$x = x'\cos\alpha - y'\sin\alpha,$$
$$y = x'\sin\alpha + y'\cos\alpha. \tag{3.3.1}$$

The real and orthogonal matrix of the coefficients in Eq.(3.3.1),

$$S(\alpha) = \begin{pmatrix} \cos\alpha, & -\sin\alpha, & 0 \\ \sin\alpha, & \cos\alpha, & 0 \\ 0, & 0, & 1 \end{pmatrix}, \tag{3.3.2}$$

gives a complete description of this rotation. The rotation in the opposite direction through the angle $(-\alpha)$ is given by the inverse matrix S^{-1}, which is defined by the relation $SS^{-1} = 1$, where 1 is the unit matrix. In a similar way, a reflection in the xy-plane is given by the matrix

$$S(\sigma) = \begin{pmatrix} 1, & 0, & 0 \\ 0, & 1, & 0 \\ 0, & 0, & -1 \end{pmatrix}. \tag{3.3.3}$$

One can write any symmetry operation in the form of a matrix S which transforms the space coordinate $\vec{r}'$ into the new coordinate $\vec{r}$

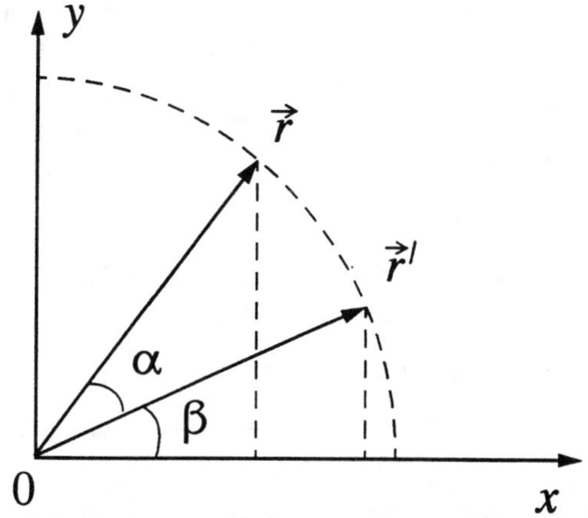

$$\vec{r} = S\vec{r}'. \tag{3.3.4}$$

In components, Eq.(3.3.4) has the form

$$\begin{pmatrix} x \\ y \\ z \end{pmatrix} = \begin{pmatrix} S_{11}S_{12}S_{13} \\ S_{21}S_{22}S_{23} \\ S_{31}S_{32}S_{33} \end{pmatrix} \begin{pmatrix} x' \\ y' \\ z' \end{pmatrix},$$

where S_{ij}, $i,j = 1,2,3$, are coefficients in the linear equations which carry $\vec{r}'$ into $\vec{r}$. The inverse transformation S^{-1} (a rotation in the opposite sense) is also a symmetry operation of the body. Therefore,

$$\vec{r}' = S^{-1}\vec{r}. \tag{3.3.5}$$

The convenience of the matrix representation of symmetry operations follows from the fact that successive applications of several symmetry operations is represented by the product of the corresponding matrices.

If $f(\vec{r})$ describes a physical property of the crystal, then the application of the symmetry operation S means that the symmetry operation is applied to the argument $\vec{r}$ of $f(\vec{r})$

$$Sf(\vec{r}) = f(S^{-1}\vec{r}). \tag{3.3.6}$$

We turn now to a consideration of the consequences of symmetry for an electron in the periodic field of the crystal lattice. The corresponding Schrödinger equation is given in Eq.(2.4.14)

$$\hat{H}(\vec{r})\psi_{j\vec{p}}(\vec{r}) = E_j(\vec{p})\psi_{j\vec{p}}(\vec{r}), \tag{3.3.7}$$

where $\hat{H} = -\frac{\hbar^2}{2m_0}\nabla^2 + U(\vec{r})$. The application of symmetry operation S^{-1} from Eq.(3.3.5) to equation (3.3.7) results in

$$\hat{H}(S^{-1}\vec{r})\psi_{j\vec{p}}(S^{-1}\vec{r}) = E_j(\vec{p})\psi_{j\vec{p}}(S^{-1}\vec{r}). \tag{3.3.8}$$

Consider first the Hamiltonian of Eq.(3.3.8). Since the kinetic energy operator in the Hamiltonian, $\left[-\frac{\hbar^2}{2m_0}\nabla^2\right]$, is proportional to the scalar product $(\nabla \cdot \nabla)$, it does not change its form under any rotation, reflections or translations, because any scalar product depends only on the angle between vectors and does not depend on the choice of the reference system. The potential energy, $U(\vec{r})$, which is the periodic potential of the crystal lattice, possesses all the symmetry of the lattice. It should remain unaltered when the symmetry operation S^{-1} from Eq.(3.3.5) is applied to it: $U(\vec{r}) = S^{-1}U(\vec{r}) = U(S^{-1}\vec{r}) = U(\vec{r}')$. Therefore, the total Hamiltonian $\hat{H}(\vec{r})$ is not changed under the symmetry operation S^{-1}. It is said that Hamiltonian is ***invariant*** under the symmetry operation of the crystal lattice S^{-1}

$$\hat{H}(\vec{r}) = S^{-1}\hat{H}(\vec{r}) = \hat{H}(S^{-1}\vec{r}) = \hat{H}(\vec{r}'). \tag{3.3.9}$$

Combining Eq.(3.3.9) with Eq.(3.3.8), one obtains the Schrödinger equation in the form

$$\hat{H}(\vec{r})\psi_{j\vec{p}}(S^{-1}\vec{r}) = E_j(\vec{p})\psi_{j\vec{p}}(S^{-1}\vec{r}) \tag{3.3.10}$$

with the same Hamiltonian as in Eq.(3.3.7). The same operation S^{-1} applied to the Bloch wave function given by Eq.(2.4.18), results in

$$\psi_{j\vec{p}}(S^{-1}\vec{r}) = A e^{\frac{i}{\hbar}\vec{p}\cdot S^{-1}\vec{r}} u_{j\vec{p}}(S^{-1}\vec{r}). \tag{3.3.11}$$

The exponent in the Bloch wave function Eq.(3.3.11) has the form of a scalar product, which is invariant against rotations and reflections. Applying the operation S to both vectors in the scalar product, we obtain it in the form

$$(\vec{p}\cdot S^{-1}\vec{r}) = (S\vec{p}\cdot SS^{-1}\vec{r}) = (S\vec{p}\cdot\vec{r}), \tag{3.3.12}$$

where the symmetry operation S is applied to vector $\vec{p}$. Here we have used the identity $SS^{-1} = 1$, where 1 is the unit matrix. The substitution of the relation (3.3.12) in the wave function Eq.(3.3.11) results in

$$\psi_{j\vec{p}}(S^{-1}\vec{r}) = A e^{\frac{i}{\hbar}S\vec{p}\cdot\vec{r}} u_{j\vec{p}}(S^{-1}\vec{r}). \tag{3.3.13}$$

According to the definition of the Bloch wave function in Section 2.4, the modulating Bloch wave function $u_{j\vec{p}}(\vec{r})$ should be labeled by the same index $\vec{p}$ which appears in the exponent of the factor $e^{\frac{i}{\hbar}\vec{p}\cdot\vec{r}}$. Therefore, we can change the notations in Eq.(3.3.13) in the following way

$$\psi_{j\vec{p}}(S^{-1}\vec{r}) = e^{\frac{i}{\hbar}S\vec{p}\cdot\vec{r}} u'_{jS\vec{p}}(\vec{r}) = \psi_{jS\vec{p}}(\vec{r}), \tag{3.3.14}$$

where the new modulating Bloch wave function $u'_{jS^{-1}\vec{p}}(\vec{r})$ has the same index $S\vec{p}$ that is present in the exponent. On the one hand, the wave function $\psi_{j\vec{p}}(S^{-1}\vec{r})$ satisfies Eq.(3.3.10) with eigenvalue $E_j(\vec{p})$. On the other hand, the Schrödinger equation for $\psi_{jS\vec{p}}(\vec{r})$ can be written in the alternative form

$$\hat{H}(\vec{r})\psi_{jS\vec{p}}(\vec{r}) = E_j(S\vec{p})\psi_{jS\vec{p}}(\vec{r}) \tag{3.3.15}$$

with eigenvalue $E_j(S\vec{p})$. Equations (3.3.10) and (3.3.15) are consistent only when

$$E_j(\vec{p}) = E_j(S\vec{p}). \tag{3.3.16}$$

Equation (3.3.16) means that the electron energy as a function of the quasimomentum $\vec{p}$ has the total point symmetry of the crystal lattice. It also means that the symmetry of the reciprocal lattice is the same as the symmetry of the lattice in direct space. Equation (3.3.16) is the third general relation for the electron energy which complements Eq.(2.5.5) and Eq.(2.6.10). Altogether, we have

$$E_j(\vec{p}) = E_j(\vec{p} + \hbar\vec{K}), \tag{3.3.17a}$$

$$E_j(\vec{p}) = E_j(-\vec{p}), \tag{3.3.17b}$$

$$E_j(\vec{p}) = E_j(S\vec{p}). \tag{3.3.17c}$$

These three relations constitute the basis for understanding the structure of electron energy levels in solids.

3.4 CLASSIFICATION OF ELECTRON ORBITALS IN A BRILLOUIN ZONE

Equation (3.3.17c) allows classifying electron energy levels and the corresponding wave functions at different points $\vec{p}$ in the Brillouin zone. As an example, we consider the 1st Brillouin zone of the two-dimensional square lattice, whose symmetry has been discussed in Section 3.2.

It follows from Eq.(3.3.17c) that both the reciprocal lattice and the 1st Brillouin zone have the symmetry of the square shown in Fig.3.4.1. An arbitrary vector $\vec{p}$ in the 1st Brillouin zone is transformed by the symmetry operations of the square from Table 3.2.1 into eight equivalent positions in $\vec{p}$ –space. The eight equivalent values of $\vec{p}$ define eight different orbitals belonging to the same energy $E(S\vec{p})$. This 8-fold degeneracy of the energy should be taken into account when solving the Schrödinger equation at an arbitrary point of the Brillouin zone. The characteristic determinant of the type of Eq.(2.7.15), which defines the energy of the electron, is (8×8). In the case of a crystal with the symmetry of the cube we would find a (48×48) characteristic determinant.

The problem is partly simplified if special points of the Brillouin zone are considered, which are located on the symmetry axes of the square. An example is given in Fig.3.4.2. There are only four equivalent positions of the vector $\vec{p}$, which results in a four-fold degeneracy of the energy. The characteristic determinant for

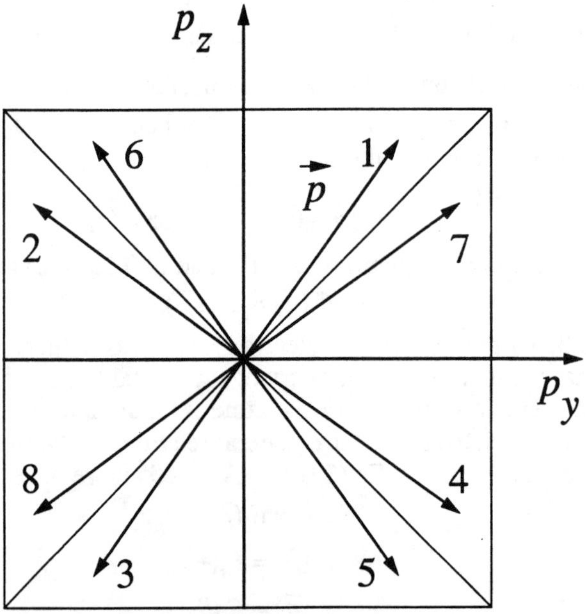

Figure 3.4.1 The first Brillouin zone of the two-dimensional crystal lattice with symmetry of the square. Eight equivalent positions of $\vec{p}$ are indicated.

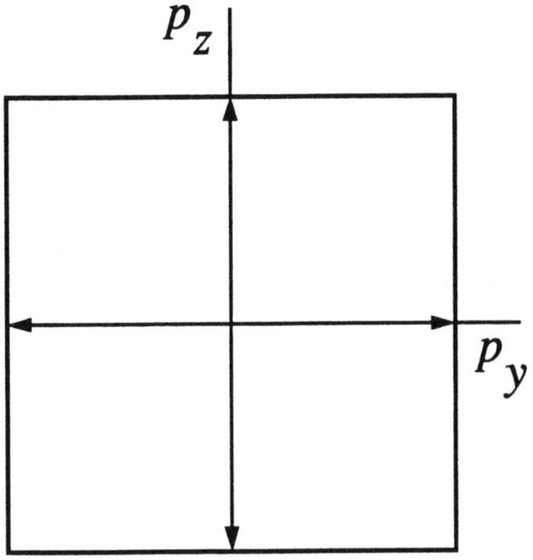

Figure 3.4.2 Four equivalent positions of $\vec{p}$ on the symmetry axes of square first Brillouin zone.

the calculation of the electron energy is (4×4).

The simplest case is that of the central point of the Brillouin zone $\vec{p} = 0$ depicted in Fig.3.4.1. It does not have any equivalent positions, and there is no degeneracy of the energy mentioned above. The more symmetric is the point of the Brillouin zone, the lower is the order of the characteristic equation, and the simpler are the calculations of the electron energy. This advantage of the symmetry points is always used in numerical calculations of the electron energy in a crystal. The symmetry points in the Brillouin zone have standard notations: Γ is the point in the center of the square, X is the mid point of the square edge, and so on. These symmetry points of the square are shown in Fig.3.4.3. The standard notations for the symmetry points for a cubic face centered lattice are shown in Fig.3.4.4.

As an example, we consider the classification of atomic orbitals at the point Γ of the Brillouin zone for a crystal with the point symmetry of the tetrahedron T_d. The classification at the Γ-point is the simplest, because this $\vec{p} = 0$ has no equivalent positions. The point symmetry operations of the tetrahedron are shown in Fig.3.4.5a. There are 6 reflection planes σ, four C_3 axes, three S_4 axes which connect midpoints of opposite edges of the tetrahedron, and the identity transformation E. There are altogether 24 operations. The same symmetry operations in a cube are shown in Fig.3.4.5b. These symmetry elements are also listed in the first column of Table 3.4.1. All 24 symmetry operations are represented in this table by 24 matrices S. The argument of S shows how these symmetry operations transform the components of the radius-vector $\vec{r} = (x, y, z)$. For example, $S_2(x, \bar{y}, \bar{z})$ means that S_2 changes the signs of the y- and z- components. $(-x)$ is indicated by $\bar{x}$, and $(-y)$ is shown as $\bar{y}$. The application of S_2 to $\vec{r}$ results in

$$S_2\vec{r} = S_2(x, y, z) = (x, -y, -z) \equiv (x, \bar{y}, \bar{z}). \tag{3.4.1}$$

The application of S_2 to an arbitrary function $f(\vec{r})$ results in

$$S_2 f(\vec{r}) = f(S_2(x, y, z)) = f(x, -y, -z) = f(x, \bar{y}, \bar{z}). \tag{3.4.2}$$

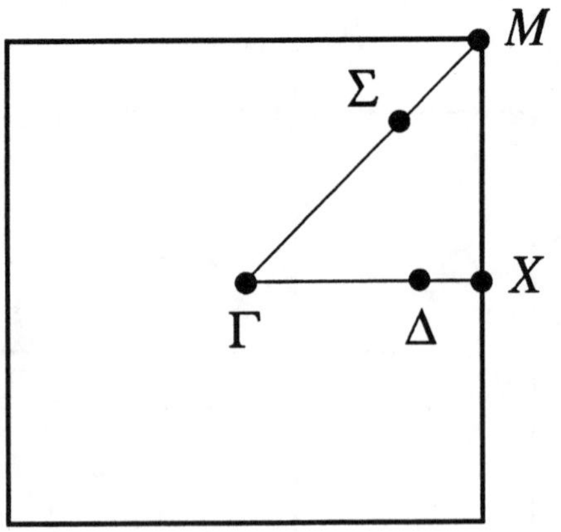

Figure 3.4.3 Symmetry points in the 1st Brillouin zone of a two-dimensional square lattice.

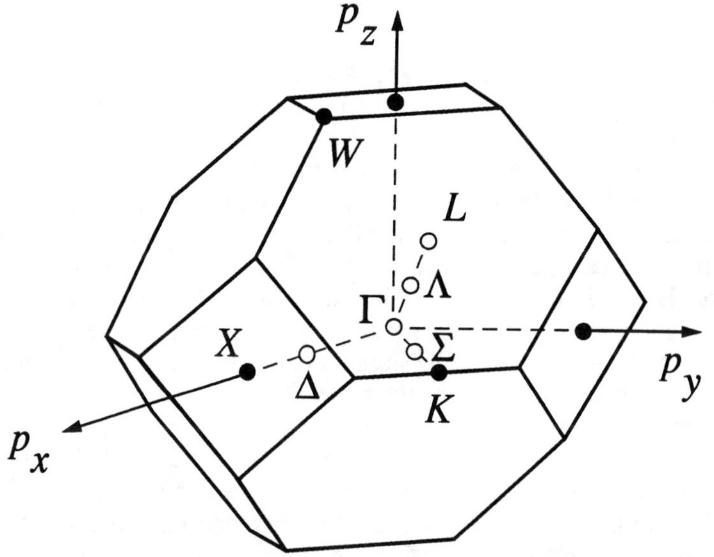

Figure 3.4.4 Symmetry points in the 1st Brillouin zone of the face centered cubic lattice.

The time-independent Schrödinger equation at the point Γ of the Brillouin zone ($\vec{p}=0$) has the form

$$\left[-\frac{\hbar^2}{2m_0}\nabla^2+U(\vec{r})\right]u_{j0}(\vec{r})=E_j(0)u_{j0}(\vec{r}). \tag{3.4.3}$$

The Bloch wave function defined by Eq.(2.4.13) reduces at this point to the modulation Bloch function $u_{j0}(\vec{r})$. Applying all symmetry operations from Table 3.4.1 to Eq.(3.4.3), we find that according to Eq.(3.3.9) the Hamiltonian $\hat{H}(\vec{r})$ remains unaltered under all operations S. The wave functions behave in a different

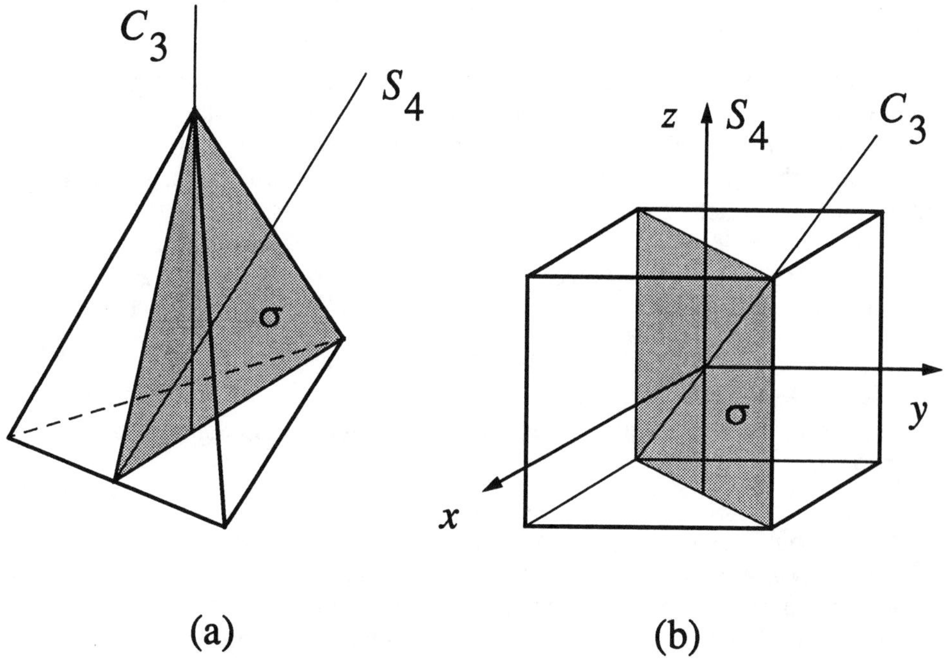

Figure 3.4.5 (a) Symmetry operations of a tetrahedron. (b) Some symmetry operations of a tetrahedron shown in a cube.

E	$S_1\,(xyz)$
$3S_4^2 \equiv 3C_4^2$	$S_2\,(x\bar{y}\bar{z})$, $S_3\,(\bar{x}y\bar{z})$, $S_4\,(\bar{x}\bar{y}z)$
$4C_3$, $4C_3^2$	$S_5\,(yzx)$, $S_6\,(\bar{y}z\bar{x})$, $S_7\,(\bar{y}\bar{z}x)$, $S_8\,(y\bar{z}\bar{x})$ $S_9\,(zxy)$, $S_{10}\,(\bar{z}\bar{x}y)$, $S_{11}\,(z\bar{x}\bar{y})$, $S_{12}\,(\bar{z}x\bar{y})$
$3S_4$, $3S_4^3$	$S_{13}\,(\bar{x}z\bar{y})$, $S_{14}\,(\bar{x}\bar{z}y)$, $S_{15}\,(\bar{z}\bar{y}x)$ $S_{16}\,(z\bar{y}\bar{x})$, $S_{17}\,(y\bar{x}\bar{z})$, $S_{18}\,(\bar{y}x\bar{z})$
6σ	$S_{19}\,(xzy)$, $S_{20}\,(x\bar{z}\bar{y})$, $S_{21}\,(zyx)$ $S_{22}\,(\bar{z}y\bar{x})$, $S_{23}\,(yxz)$, $S_{24}\,(\bar{y}\bar{x}z)$

Table 3.4.1 Symmetry operations of a tetrahedron.

way. In order to understand this behavior, a certain similarity between the possible angular dependence of the electron wave function and the angular dependence of the orthogonal polynomials of lowest degree can be used. This similarity follows from the fact that wave functions, just as the lowest degree orthogonal polynomials, are orthogonal to each other.

1. Let $E_j(0)$ be an energy level of electron in the crystal whose wave function remains unchanged under each transformation S or S^{-1} from Table 3.4.1

$$u_{j0}(\vec{r}) = S^{-1}u_{j0}(\vec{r}) = u_{j0}(S^{-1}\vec{r}) = u_{j0}(\vec{r}'). \tag{3.4.4}$$

Equation (3.4.4) means that $u_{j0}(\vec{r})$ is an invariant function which is not changed under rotations and reflections. A function of this type should depend on $r^2 = x^2 + y^2 + z^2$ only,

$$u_{j0} = f_1(r). \tag{3.4.5}$$

Because the only one wave function (3.4.5) corresponds to the energy level $E_j(0)$, this level is nondegenerate. The orbital given by Eq.(3.4.5) is called an orbital of Γ_1 type. We already know an example of a wave function that satisfies Eq.(3.4.4). It is the wave function of the s-state of the hydrogen atom ψ_{100}, given by Eq.(1.13.38).

2. Another level $E_{j_1}(0)$ can correspond to the wave function that transforms under symmetry transformations from Table 3.4.1 as the first degree polynomial x. This is a function of the type

$$u_{j_10}^{(1)}(\vec{r}) = f_2(r)x, \tag{3.4.6a}$$

where $f_2(r)$ depends only on $|\vec{r}\,|$. Consulting Table 3.4.1, we see that some symmetry operations transform x into y, and some other operations carry x into z. This means that along with the wave function Eq.(3.4.6a), we should consider two other functions

$$u_{j_10}^{(2)} = f_2(r)y \quad \text{and} \quad u_{j_10}^{(3)} = f_2(r)z. \tag{3.4.6b}$$

The wave functions (3.4.6) are orthogonal to the wave function Eq.(3.4.5). All three functions (3.4.6a), (3.4.6b) go one into another under operations S from the Table 3.4.1. These three orbitals belong to the same energy level $E_{j_1}(0)$, which is triply degenerate. Wave functions of this type are called wave functions of Γ_{15} type. Wave functions of the p –states in the hydrogen atom given by Eqs.(1.13.39) behave this way.

3. The level $E_{j_2}(0)$ may have a wave function that has an angular dependence represented by the next lowest degree polynomial x^2. It follows from the Table 3.4.1 that the transformations S create also the function y^2 and z^2. We should remember that wave functions are orthogonal to each other. This means that we can take only those second degree polynomials which are orthogonal to the radial combination $r^2 = x^2 + y^2 + z^2$, which has been already used in the wave function Eq.(3.4.5). The

Γ_1	1
Γ_2	$x^4(y^2-z^2)+y^4(z^2-x^2)+z^4(x^2-y^2)$
Γ_{12}	$z^2-\frac{1}{2}(x^2+y^2),\ x^2-y^2$
Γ_{15}	x, y, z
Γ_{25}	$x(y^2-z^2),\ y(z^2-x^2),\ z(x^2-y^2)$

Table 3.4.2 Five different orthogonal polynomials representing the symmetry of five possible orbitals for a crystal with T_d symmetry.

procedure of constructing orthogonal polynomials is known from mathematics. It leads to two independent second degree polynomials: $(x^2 - y^2)$ and $(z^2 - 1/2(x^2+y^2))$. The corresponding wave functions are

$$u^{(1)}_{j_20} = f_3(r)\,(x^2 - y^2), \tag{3.4.7a}$$

$$u^{(2)}_{j_20} = f_3(r)\,[z^2 - 1/2(x^2 + y^2)]. \tag{3.4.7b}$$

The energy level $E_{j_2}(0)$ is doubly degenerate. The wave functions (3.4.7) are called functions of Γ_{12} type.

Since in Table 3.4.1 five different types of symmetry operations are given, we can find five different types of orthogonal polynomials, which represent the five possible angular dependences of electronic wave functions. These functions are given in Table 3.4.2.

Applying the inversion I to the 24 operations S from the Table 3.4.1, leads to an additional 24 symmetry operations, which form, together with the operations of Table 3.4.1, 48 elements of the point symmetry of a cube. These 48 operations belong to 10 different types. The construction of orthogonal polynomials leads to five more types of wave functions which are given in Table 3.4.3. There are altogether 10 types of electronic wave functions in a cubic crystal, which are given in Table 3.4.2 and Table 3.4.3.

It is seen from Table 3.4.1 that the symmetry operations of a tetrahedron

Γ_1'	$xyz\left[x^4\left(y^2-z^2\right)+y^4\left(z^2-x^2\right)+z^4\left(x^2-y^2\right)\right]$
Γ_2'	xyz
Γ_{12}'	$xyz\left[z^2-\frac{1}{2}\left(x^2+y^2\right)\right]$, $xyz\left(x^2-y^2\right)$
Γ_{15}'	$yz\left(y^2-z^2\right)$, $zx\left(z^2-x^2\right)$, $xy\left(x^2-y^2\right)$
Γ_{25}'	yz, zx, xy

Table 3.4.3 Five additional orthogonal polynomials, representing the symmetry of five extra orbitals for a crystal with O_h symmetry.

(T_d-symmetry) always change the signs of the components of $\vec{r}$ in pairs. In crystals with the symmetry of a cube (O_h-symmetry), each component of $\vec{r}$ is allowed to change sign by itself due to the inversion I. This makes all the selection rules for quantum transitions of electrons in crystals with T_d and O_h symmetry different.

We turn now to a discussion of the band structures of typical semiconductors where symmetry arguments are of great help.

BIBLIOGRAPHY

1. H. Jones *The Theory of Brillouin Zones and Electronic States in Crystals* (North-Holland, Amsterdam, 1975).
2. L.D. Landau, E.M. Lifshits *Statistical Physics, 2nd ed.* (Addison-Wesley, Reading, Massachussetts, 1966).

CHAPTER 4

Effective Mass Approximation

4.1 EFFECTIVE MASS APPROXIMATION: NONDEGENERATE ENERGY BANDS

The complete determination of the electron energy throughout the entire Brillouin zone requires numerical calculations that employ computers. The main problem in these calculations is the proper choice of an adequate crystal lattice potential $U(\vec{r})$.

But the knowledge of the potential energy is not needed for most semiconductors. Because the number of current carriers is not large, most of them concentrate near the extremes of energy in the Brillouin zone. The possible dependence of $E_j(\vec{p})$ on $\vec{p}$ near the extrema of energy can be obtained without numerical calculations on the basis of symmetry arguments alone.

The expansion of the energy in the components of the quasimomentum near the possible extremum located at the Γ-point ($\vec{p}=0$) of the Brillouin zone has the form

$$E_j(\vec{p}) = E_j(0) + \sum_{\alpha}\left(\frac{\partial E_j(\vec{p})}{\partial p_\alpha}\right)_{p_\alpha=0} p_\alpha + \frac{1}{2}\sum_{\alpha,\beta}\left(\frac{\partial^2 E_j(\vec{p})}{\partial p_\alpha \partial p_\beta}\right)_{\vec{p}=0} p_\alpha p_\beta , \tag{4.1.1}$$

where $\alpha, \beta = x, y, z$. The necessary condition for an energy extremum is

$$\left(\frac{\partial E_j(\vec{p})}{\partial \vec{p}}\right)_{\vec{p}=0} = 0.$$

Therefore, the linear term in the expansion (4.1.1) vanishes. The third term in the expansion represents a quadratic dependence of the energy on the quasimomentum. The coefficient $\left(\frac{\partial^2 E_j(\vec{p})}{\partial p_\alpha \partial p_\beta}\right)_{\vec{p}=0}$ is the curvature of the parabolic dependence of $E_j(\vec{p})$ on $\vec{p}$ near the point of the extremum, see Fig.4.1.1. This coefficient has the dimensions of a reciprocal mass and is called the ***reciprocal effective mass*** of the electron

$$\left(\frac{1}{m}\right)_{\alpha\beta} = \left(\frac{\partial^2 E_j(\vec{p})}{\partial p_\alpha \partial p_\beta}\right)_{\vec{p}=0}. \tag{4.1.2}$$

The parabolic dependence of the energy on quasimomentum Eq.(4.1.1) differs from the free electron energy $E(\vec{p}) = \vec{p}^2/2m_0$ by the replacement of m_0^{-1} by the anisotropic reciprocal effective mass $m_{\alpha,\beta}^{-1}$.

Starting from the time-independent Schrödinger equation (2.4.1), we shall find the reciprocal mass $(1/m)_{\alpha\beta}$ in terms of electron band structure parameters. Substituting the Bloch wave function Eq.(2.4.13) into the Schrödinger equation and using the equation of vector calculus

$$\nabla^2 f\psi = \psi\nabla^2 f + f\nabla^2\psi + 2\nabla f \cdot \nabla\psi,$$

one finds

$$\left[-\frac{\hbar^2}{2m_0}\nabla^2 + U(\vec{r}) + \frac{\vec{p}^2}{2m_0} - i\frac{\hbar}{m_0}(\vec{p}\cdot\nabla)\right]u_{j\vec{p}}(\vec{r}) = E_j(\vec{p})u_{jp}(\vec{r}). \tag{4.1.3}$$

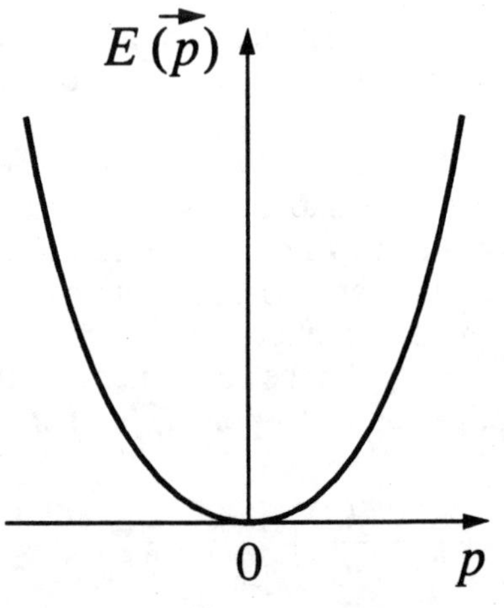

Figure 4.1.1 The parabolic dependence of $E(\vec{p})$ versus p.

If we are considering the extremum of energy at the Γ point of the Brillouin zone, $\vec{p}=0$, the Bloch wave functions reduce at this point to the Bloch modulating function $u_{j0}(\vec{r})$ which satisfies the equation

$$\left[-\frac{\hbar^2}{2m_0}\nabla^2+U(\vec{r})\right]u_{j0}(\vec{r})=E_j(0)u_{j0}(\vec{r}). \tag{4.1.4}$$

The functions $u_{j0}(\vec{r})$ being the solutions of Eq.(4.1.4), obey conditions of orthogonality and normalization

$$\int u_{j'0}^{*}u_{j0}(\vec{r})d\vec{r}=\delta_{jj'}, \tag{4.1.5}$$

where $\delta_{jj'}$ is the Kronecker delta.

The third and fourth terms in Eq.(4.1.3) depend on the quasimomentum $\vec{p}$. Close to the point Γ, they can be treated as a small perturbation

$$\hat{H}^{(1)}=\frac{\vec{p}^{\,2}}{2m_0}-i\frac{\hbar}{m_0}(\vec{p}\cdot\nabla) \tag{4.1.6}$$

to Eq.(4.1.4), which is considered as the equation of the zeroth approximation. The Bloch modulation functions $u_{j0}(\vec{r})$ and the energy $E_j(0)$ are zero approximation wave functions and energies, respectively.

The Hamiltonian $\hat{H}^{(1)}$ is not an external perturbation. But because the microsymmetry of an arbitrary point $\vec{p}$ in the 1st Brillouin zone is lower than the microsymmetry of the point Γ, $\hat{H}^{(1)}$ defined by Eq.(4.1.6) reduces the symmetry of the total Hamiltonian of Eq.(4.1.3). In this sense, $\hat{H}^{(1)}$ is very similar to an external perturbation.

We apply the perturbation approximation from Section 1.15 and find the first order perturbation to the energy $E_j(0)$ from $\hat{H}^{(1)}$

$$E_j^{(1)}(\vec{p})=\langle u_{j0}|\hat{H}^{(1)}|u_{j0}\rangle=\int u_{j0}^{*}\frac{\vec{p}^{\,2}}{2m_0}u_{j0}d\vec{r}+\int u_{j0}^{*}\left[-i\frac{\hbar}{m_0}(\vec{p}\cdot\nabla)\right]u_{j0}d\vec{r}.$$

Here the second term, linear in $\vec{p}$, vanishes, since the integrand is an odd function of $\vec{r}$. Using the orthogonality condition (4.1.5), we find the first order perturbation to the energy $E_j(0)$

$$E_j^{(1)}(\vec{p})=\frac{\vec{p}^{\,2}}{2m_0}. \tag{4.1.7}$$

There is another contribution to the energy $E_j(0)$ that is of second order in $\vec{p}$ and that should be taken into account. It comes from the second order perturbation approximation, Eq.(1.15.15)

$$E_j^{(2)}(\vec{p}) = \frac{1}{m_0^2}\sum_{j'}{}' \frac{|< u_{j0} | \vec{p}\cdot(-i\hbar\nabla) | u_{j'0} >|^2}{E_j(0) - E_{j'}(0)} =$$

$$\frac{1}{m_0^2}\sum_{j'}{}' \frac{< u_{j0} | \vec{p}\cdot(-i\hbar\nabla) | u_{j'0} >< u_{j'0} | \vec{p}\cdot(-i\hbar\nabla) | u_{j0} >}{E_j(0) - E_{j'}(0)}. \qquad (4.1.8)$$

The second line of Eq.(4.1.8) results from the operator $(-i\hbar\nabla)$ being hermitian. Representing the scalar product in the form $-i\hbar(\vec{p}\cdot\nabla) = \sum_{\alpha=x,y,z} p_\alpha(-i\hbar\nabla_\alpha)$ one finds the second order perturbation to $E_j(0)$

$$E_j^{(2)}(\vec{p}) \;=\; \frac{1}{m_0^2}\sum_{\alpha,\beta} p_\alpha p_\beta \sum_{j'}{}' \frac{< u_{j0} | -i\hbar\nabla_\alpha | u_{j'0} >< u_{j'0} | -i\hbar\nabla_\beta | u_{j0} >}{E_j(0) - E_{j'}(0)}. \qquad (4.1.9)$$

The total energy equals the sum of the energies from Eqs.(4.1.7) and (4.1.9),

$$E_j(\vec{p}) = E_j(0) + E_j^{(1)}(\vec{p}) + E_j^{(2)}(\vec{p}) = \frac{1}{2}\sum_{\alpha,\beta}\left(\frac{1}{m}\right)_{\alpha\beta} p_\alpha p_\beta\,, \qquad (4.1.10)$$

where

$$\left(\frac{1}{m}\right)_{\alpha\beta} = \frac{1}{m_0}\delta_{\alpha\beta} + \frac{2}{m_0^2}\sum_{j'}{}' \frac{< u_{j0} | -i\hbar\nabla_\alpha | u_{j'0} >< u_{j'0} | -i\hbar\nabla_\beta | u_{j0} >}{E_j(0) - E_{j'}(0)}. \qquad (4.1.11)$$

Here $\alpha, \beta = x, y, z$. Relation (4.1.10) represents the parabolic dependence of $E_j(\vec{p})$ on $\vec{p}$.

Equation (4.1.10) has the dimensions of the energy $E_j(\vec{p})$ and is a scalar function. On the right hand side of Eq.(4.1.10), we have the product of quasimomentum components $p_\alpha p_\beta$. The proportionality coefficient between the energy and the vector components product $p_\alpha p_\beta$ should be the second rank tensor $m_{\alpha\beta}^{-1}$ with dimensions of an inverse mass. It is the ***effective mass*** of the electron. Physical properties of tensors are discussed briefly in Appendix 1. Since $E_{j'}(0)$ may be larger or smaller than $E_j(0)$, the effective mass could be smaller or larger than the free electron mass. One can see from Eqs.(4.1.9) and (4.1.11) that the effective mass results from band-to-band transitions of the electron. The largest contribution comes from the band nearest to $E_j(0)$.

The second rank tensor $m_{\alpha\beta}^{-1}$ has the general form of the matrix

$$m_{\alpha\beta}^{-1} = \begin{pmatrix} m_{xx}^{-1} & m_{xy}^{-1} & m_{xz}^{-1} \\ m_{yx}^{-1} & m_{yy}^{-1} & m_{yz}^{-1} \\ m_{zx}^{-1} & m_{zy}^{-1} & m_{zz}^{-1} \end{pmatrix}. \qquad (4.1.12)$$

The elements of the matrix in Eq.(4.1.12) depend on the choice of the reference system in $\vec{p}$-space. In the reference system of the principal axes the matrix $m_{\alpha\beta}^{-1}$ has the diagonal form

$$m_{\alpha\beta}^{-1}=\begin{pmatrix} m_1^{-1} & 0 & 0 \\ 0 & m_2^{-1} & 0 \\ 0 & 0 & m_3^{-1} \end{pmatrix}, \tag{4.1.13}$$

where m_α^{-1}, $\alpha = 1,2,3$, are the principal values of the matrix. In the principal axes of the matrix, the energy Eq.(4.1.10) takes the form

$$E_j(\vec{p})=E_j(0)+\frac{1}{2}\left[\frac{p_x^2}{m_1}+\frac{p_y^2}{m_2}+\frac{p_z^2}{m_3}\right]. \tag{4.1.14}$$

One can see from Eq.(4.1.14) that the surface of constant energy $E_j(\vec{p}) = Constant$ in the vicinity of an extremal point is an ellipsoid.

According to Eq.(3.3.17c), the surface of constant energy should possess the total point symmetry of the crystal. For example, in crystals with the symmetry of a tetrahedron (an example is the semiconductor crystal GaAs), the surface of constant energy should go into itself as a result of C_4-rotations around the coordinate axes, see the second line in Table 3.4.1. The principal axes of the matrix $m_{\alpha\beta}^{-1}$ coincide in this case with the C_4-axes of the tetrahedron. Because three C_4 axes represent equivalent perpendicular directions in the tetrahedron, the principal values of the matrix $m_{\alpha\beta}^{-1}$ are equal, that is

$$m_1=m_2=m_3=m^*, \tag{4.1.15}$$

and $m_{\alpha\beta}^{-1} = (m^*)^{-1}\delta_{\alpha\beta}$. The reciprocal mass from Eq.(4.1.11) takes the form

$$\frac{1}{m^*}=\frac{1}{m_0}+\frac{2}{m_0^2}\sum_{j'}{}'\frac{|<u_{j0}|-i\hbar\nabla|u_{j'0}>|^2}{E_j(0)-E_{j'}(0)}. \tag{4.1.16}$$

The surface of constant energy reduces to a sphere with its center at the Γ-point of the Brillouin zone

$$E_j(\vec{p})=E_j(0)+\frac{1}{2m^*}(p_x^2+p_y^2+p_z^2)=E_j(0)+\frac{\vec{p}^2}{2m^*}=Constant. \tag{4.1.17}$$

The dependence of the energy on the quasimomentum is parabolic and isotropic, see Fig.4.1.2a. The spherical dependence of E on $\vec{p}$ is found in GaAs. The effective mass of GaAs is $m^* = 0.066\, m_0$.

In some materials with cubic symmetry, like Ge or Si, an extremum of energy happens to be not at the Γ-point, but at some other points $\vec{p}_0$ of the Brillouin zone. The electron energy has the form of the expansion

$$E_j(\vec{p}+\vec{p}_0)=E_j(\vec{p}_0)+\frac{1}{2}\sum_{\alpha\beta}m_{\alpha\beta}^{-1}(p-p_0)_\alpha(p-p_0)_\beta. \tag{4.1.18}$$

The point symmetry at $\vec{p}_0$ is lower than at the central point Γ. The surfaces of constant energy from Eq.(4.1.18) are not spheres: they are ellipsoids that have the symmetry of the point $\vec{p}_0$.

Consider the example of germanium (Ge), which has the point symmetry O_h. The energy extremum occurs in Ge at a point $\vec{p}_0$ located on the ***body diagonal*** of the cube. It is called the L- point of the Brillouin zone. The diagonal of the cube is

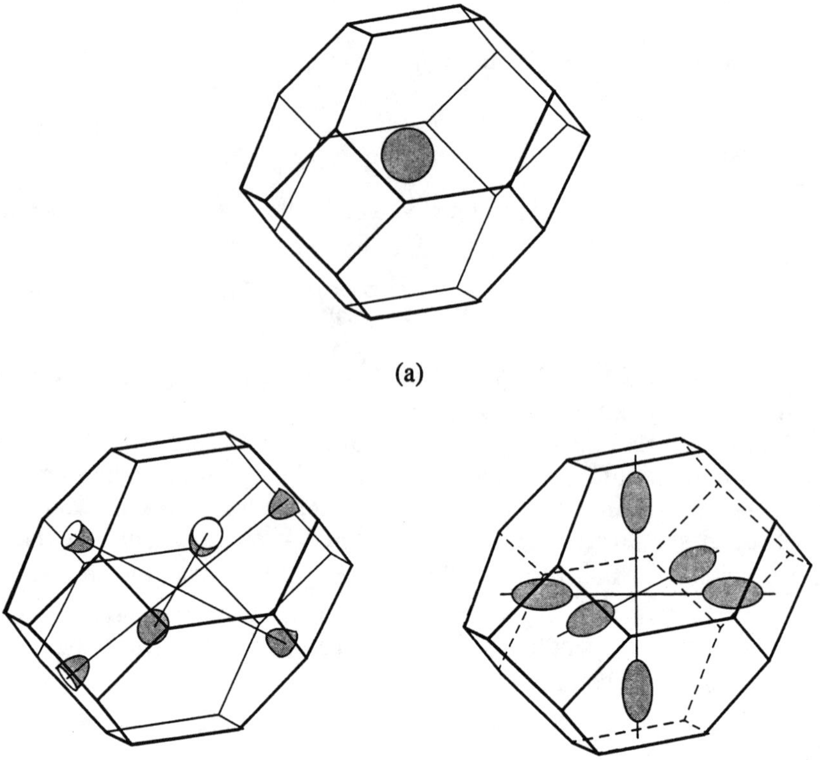

(a) (b) (c)

Figure 4.1.2 Surfaces of constant energy. (a) GaAs: The surface of the constant energy is a sphere at the center of the Brillouin zone. (b) Ge: There are eight half-ellipsoids of revolution along the <111> axes. The Brillouin zone boundaries are at the centers of the ellipsoids. (c) Si: There are six ellipsoids of revolution along the <100> axes. The centers of the ellipsoids are at three quarters of the distance from the Brillouin zone center to the Brillouin zone boundaries.

a 3-fold rotation axis C_3, which makes the energy ellipsoid to be an ellipsoid of revolution about the C_3 axis. The other two principal axes of the ellipsoid are arbitrary axes taken in the plane that is transverse to the C_3-axis. The electron energy from Eq.(4.1.18) takes the form

$$E_j(\vec{p}) = \frac{(p_x - p_{0x})^2 + (p_y - p_{0y})^2}{2m_1} + \frac{(p_z - p_{0z})^2}{2m_3}. \tag{4.1.19}$$

For an ellipsoid of revolution two principal values of the matrix (4.1.13) are equal $m_1^{-1} = m_2^{-1}$, and the third m_3^{-1} is different.

There are eight equivalent L-points of the same symmetry in the Brillouin zone of the cubic Ge crystal. Symmetry requires that the electron energy is a minimum at each point equivalent to $\vec{p}_0$. There are eight equivalent extremum positions in the Brillouin zone located on the body diagonals of the cube. There are therefore eight minima of the electron energy. Because the extremum of the energy occurs at the point L on the boundary of Brillouin zone, one half of each ellipsoid is inside the 1st Brillouin zone. Each ellipsoid does not have the symmetry of a cube; but the system of eight half-ellipsoids has the symmetry of a cube. The energy spectrum of Ge is shown in Fig.4.1.2b. The corresponding components of the effective mass tensor are

$$m_1 = m_2 = 0.082m_0,$$
$$m_3 = 1.59m_0, \tag{4.1.20}$$

where m_0 is the mass of a free electron.

In the case of silicon (Si) with O_h-symmetry, an extremum of the energy is located on each of the three 4-fold rotation axes of a cube C_4 which connect the midpoints of opposite cube faces. There are three equivalent axes and six equivalent positions of the energy minima in the Brillouin zone. Near each minimum, the surface of constant energy is an ellipsoid of revolution. It does not possess the symmetry of a cube. But, the six ellipsoids together have the total symmetry of a cube. The energy spectrum of silicon is shown in Fig.4.1.2c. The corresponding effective masses of Si are equal to

$$m_1 = m_2 = 0.98m_0,$$
$$m_3 = 0.19m_0\,. \tag{4.1.21}$$

Semiconductors with band structures of this type are often called ***many valley semiconductors***, and the ellipsoids near these extrema are called ***valleys***.

Equation (4.1.16) allows estimating the effective mass using dimension arguments. The matrix elements entering Eq.(4.1.16) have the dimensionality of $< u_{j0} \mid -i\hbar\nabla_\alpha \mid u_{j'0} > \approx \hbar/a$, where a is the lattice constant. The energy band separation in the denominator is of the order of the energy gap $E_j(0) - E_{j'}(0) \approx E_g$. Substituting these values into Eq.(4.1.16), one finds

$$\frac{1}{m^*} = \frac{1}{m_0} + \frac{1}{m_0^2 E_g}\frac{\hbar^2}{a^2} \quad \text{or} \quad \frac{m_0}{m^*} = 1 + \frac{\hbar^2}{a^2 m_0 E_g}.$$

One can see from this result the correlation of the effective mass with the lattice constant a and the semiconductor energy gap E_g.

The small parameter of the perturbation approximation in effective mass calculations is the ratio of the average electron energy in the conduction band to the energy gap

$$\frac{< E >}{E_g} \ll 1. \tag{4.1.22}$$

In the case of a nondegenerate electron gas, the average energy is of the order of $< E > \approx k_B T$. For a degenerate electron gas $< E > \approx E_F$, where E_F is the Fermi energy. It follows from the inequality (4.1.22) that the effective mass approximation works well near minima of the electron band.

4.2 EXPERIMENTAL DETERMINATION OF THE EFFECTIVE MASS

We have considered in Section 4.1 several different types of semiconductor band structures. These bands structures are verified by direct experimental measurements, some examples of which are considered below.

1. When the minimum of the electron energy occurs at the Γ point of the Brillouin zone, we have a spherically symmetric function $E_j(\vec{p})$ with a scalar isotropic effective mass m^*, which is defined by Eq.(4.1.16). This is the case for InSb and GaAs and some other materials. Semiconductors of this type are called ***direct gap semiconductors***. This name has come from optical measurements of the energy gap E_g. The absorption coefficient of these materials has the typical threshold dependence on the frequency of the incident light ω, shown in Fig.4.2.1. Starting from $\hbar\omega = E_g$ the electrons of the valence band receive enough energy from the light to be excited into the conduction band, see Fig.4.2.2. The energy conservation law for this process has the form

$$E_2 - E_1 = \hbar\omega \geq E_g, \tag{4.2.1}$$

where E_1 and E_2 are the energies of the initial and final states of the electron. The quasimomentum conservation law requires the quasimomentum transfered to the excited electron to be equal to the momentum of the exciting light $\hbar\vec{k}$:

$$\vec{p}_2 - \vec{p}_1 = \hbar\vec{k}. \tag{4.2.2}$$

The average value of the quasiwave vector of an electron participating in an optical process in the crystal has the order of magnitude

$$< u_{j0} | -i\nabla_\alpha | u_{j0} > \approx \frac{p}{\hbar} \approx \frac{1}{a} \approx 10^{10}\, m^{-1}.$$

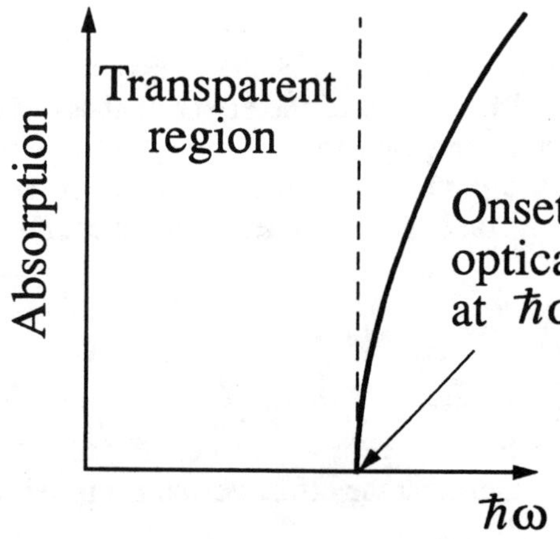

Figure 4.2.1 Optical absorption as a function of photon energy in pure insulators at absolute zero.

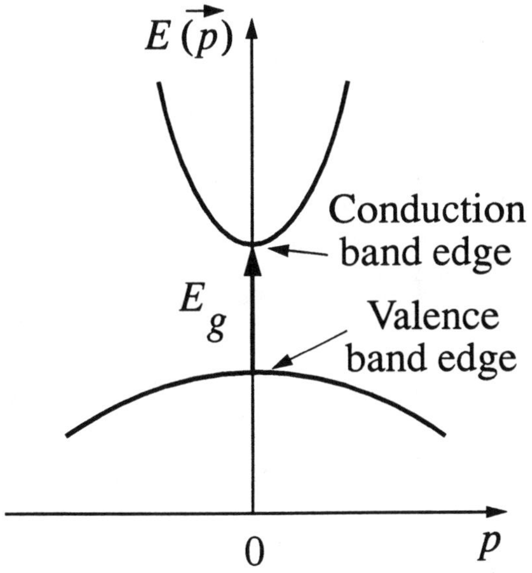

Figure 4.2.2 A direct optical transition is vertical with no change of $\vec{p}$. The threshold frequency ω_g is determined by the energy gap $h\omega_g = E_g$.

The light wave vector $k \approx \omega/c \approx 10^7\, m^{-1}$. The dimensionless product ka is very small, $ka \ll 1$. Neglecting $\hbar k \ll p_{1,2}$, we have from Eq.(4.2.2) that

$$\vec{p}_1 = \vec{p}_2. \tag{4.2.3}$$

This means that the electron makes a vertical transition from the valence to conduction band which is called a ***direct transition***. The position of the threshold of the absorption coefficient at $\hbar\omega = E_g$, is shown in Fig.4.2.3. It allows measuring the energy gap E_g.

Many-valley semiconductors, such as Ge or Si, have energy minima at some symmetry points $\vec{p}_0$ of the 1st Brillouin zone. In Ge, the minimum of the energy occurs at the L –point of the Brillouin zone (<111>-direction) and in Si, it occurs near the X –point of the Brillouin zone (<100>-direction). These band structures have been shown schematically in Fig.4.2.4. Experimental measurements show that there is no sharp threshold of the absorption coefficient, see Fig.4.2.5. This occurs because the band edges of the conduction and valence bands are separated in $\vec{p}$ –space by the quasimomentum $\vec{p}_A$ shown in Fig.4.2.4. The conservation laws of energy and momentum require in this case the participation of a third particle in the process of absorption. For example, this third particle could be an impurity atom or a lattice vibration. The absorption begins at the light frequency

$$\hbar\omega = E_2 - E_1 + \hbar\Omega(\vec{q}), \tag{4.2.4a}$$

where $\hbar\Omega(\vec{q})$ is the energy of the third particle. Absorption starts at some energy which is smaller than E_g. The momentum conservation law requires the third particle to have the momentum $\hbar\vec{q}$

$$\vec{p}_2 - \vec{p}_1 + \hbar\vec{q} = \hbar\vec{k}. \tag{4.2.4b}$$

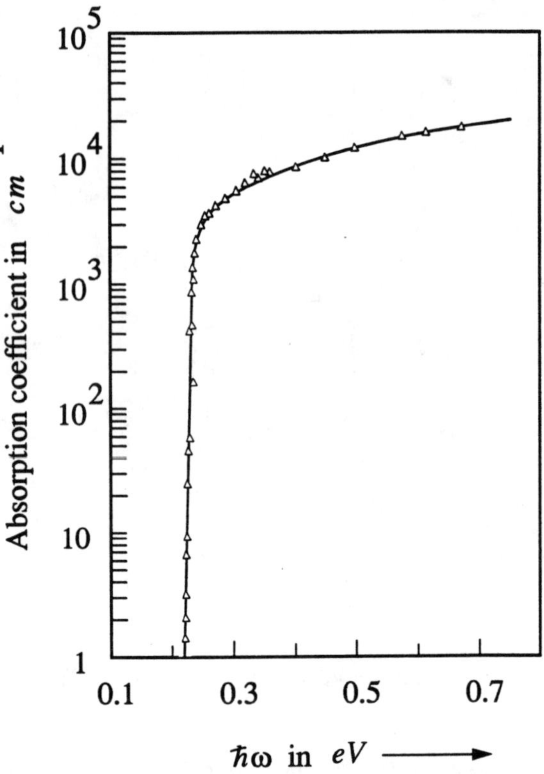

Figure 4.2.3 Optical absorption in pure indium antimonide, InSb. This is a direct gap material, because both the conduction and valence band edges are at the center of the Brillouin zone (After C. Kittel "Introduction to Solid State Physics", John Wiley & Sons, N.Y., 1986).

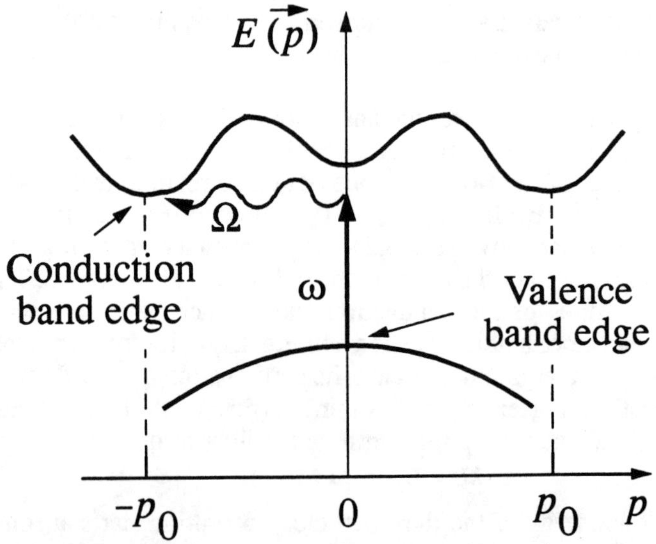

Figure 4.2.4 An indirect transition involves both a photon and a phonon because the band edges of the conduction and valence bands are separated by vector $\vec{p}_0$ in p-space. A phonon with energy $\hbar\Omega$ participates in the process.

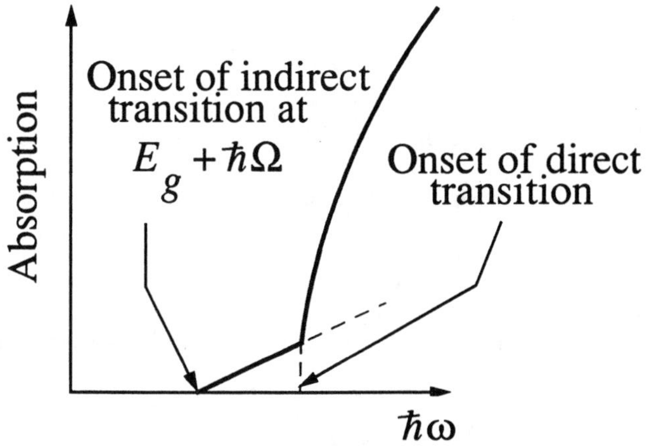

Figure 4.2.5 Optical absorption in an indirect gap material. There is no sharp threshold.

Neglecting the light momentum $\hbar\vec{k}$ as before, one finds

$$\vec{p}_2 - \vec{p}_1 = -\vec{p}_A = \pm\hbar\vec{q}. \tag{4.2.5}$$

The energy gap E_g is not well defined in this case.

2. Complete and reliable information about the band structure of semiconductors is obtained by the technique of ***cyclotron resonance***. The structure of energy bands and the values of effective masses in Ge and Si have been obtained this way. It has been shown in Section 2.5 that an electron in an external magnetic field moves on the surface of constant energy in the plane transverse to the magnetic field $\vec{B}$, see Fig.2.5.4. When a magnetic field $\vec{B}$ is applied in the z-direction, $\vec{B} = (0,0,B)$, the equations of motion (2.5.17) after substitution of Eqs.(2.5.13) and (4.1.17) take the form

$$\frac{dv_x}{dt} = -\omega_c v_y, \tag{4.2.6}$$

$$\frac{dv_y}{dt} = +\omega_c v_x, \tag{4.2.7}$$

$$\frac{dv_z}{dt} = 0. \tag{4.2.8}$$

Here the frequency

$$\omega_c = \frac{eB}{m^*} \tag{4.2.9}$$

is called the ***cyclotron frequency***.

Taking the time derivative of Eq.(4.2.6) and substituting dv_y/dt from Eq.(4.2.7), one finds

$$\frac{d^2 v_x}{dt^2} = -\omega_c^2 v_x. \tag{4.2.10}$$

The solution of this equation and of a similar equation for v_y has the form

$$\frac{dx}{dt} = v_x = v_0 \cos \omega_c t,$$

$$\frac{dy}{dt} = v_y = v_0 \sin \omega_c t.$$

Here v_0 is the initial value of the electron velocity. Integrating these equations, one obtains finally

$$x = \frac{v_0}{\omega_c} \sin \omega_c t,$$

$$y = \frac{v_0}{\omega_0} \cos \omega_c t.$$

The electron is accelerated in its helical orbit about the axis of the static magnetic field

$$r^2 = x^2 + y^2 = \frac{v_0^2}{\omega_c^2}.$$

It is seen from this equation that an electron in a magnetic field B behaves as an oscillator with angular cyclotron frequency ω_c.

Applying an extra *rf* electric field $\mathcal{E} = \mathcal{E}_0 e^{-i\omega t}$ in a direction perpendicular to $\vec{B}$, one can reach the resonance condition $\omega = \omega_c$ by tuning the frequency. Then the resonant absorption energy from the *rf* field occurs. The absorption coefficient has a sharp peak at the frequency $\omega = \omega_c$. The position of the peak allows finding the effective mass by direct measurement. The arrangement of the fields in a cyclotron resonance experiment is shown in Fig.4.2.6. The effective mass equals

$$m^* = \frac{eB}{\omega_c}. \tag{4.2.11}$$

Another possibility consists of varying the magnetic field at a fixed *rf* frequency ω. The maximum of absorption occurs at

$$B_c = \frac{m^*}{e} \omega_c, \tag{4.2.12}$$

when $\omega_c = \omega$. In the case of many-valley semiconductors (Ge, Si) where, according to Eq.(4.1.19), the inverse effective mass is anisotropic, the resonant frequency depends on the angle between $\vec{B}$ and the axes of the ellipsoids. In the case that the static magnetic field makes an angle θ with the longitudinal axis of an ellipsoid, the resonant cyclotron frequency equals

$$\omega_c^2 = (eB)^2 \left[\frac{\cos^2\theta}{m_1^2} + \frac{\sin^2\theta}{m_1 m_3} \right], \tag{4.2.13}$$

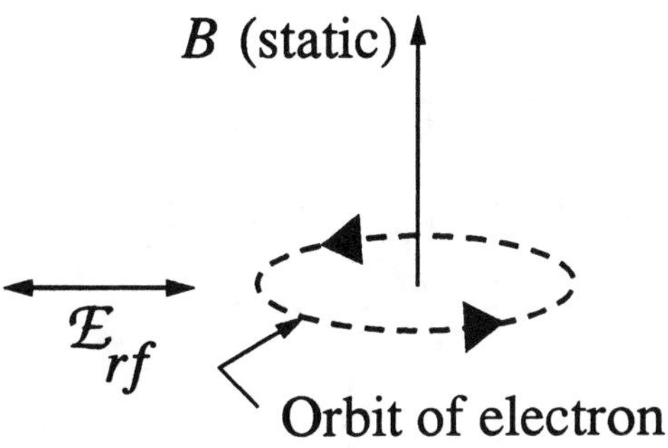

Figure 4.2.6 The arrangement of the fields in the cyclotron resonance experiment.

where $m_1 = m_2$ and m_3 are the transverse and longitudinal masses from Eq.(4.1.19). The position of the resonance peak in this case depends on the angle θ. Experimental curves for Ge at $4°K$ are shown in Fig.4.2.7. Since the eight minima of the electron energy occur in Ge at the eight L-points on boundaries of the 1-st Brillouin zone, there are only eight half-ellipsoids inside the 1st Brillouin zone.

When the magnetic field $\vec{B}$ is taken along the [001]-direction (along the edge of the cube), there is only one peak in absorption. This means that the longitudinal axes of all eight half-ellipsoids are situated symmetrically with respect to the direction of the magnetic field, i.e. they are along the four [111] diagonals of the cube.

If the magnetic field is parallel to the [111]-direction, there are two maxima in absorption. This means that there are two inequivalent groups of half-ellipsoids with respect to the magnetic field $\vec{B}$. One group of two half-ellipsoids has their longitudinal axes parallel to $\vec{B}$, and the second group of six half-ellipsoids have equivalent positions in the Brillouin zone with respect to $\vec{B}$. Two different groups of half-ellipsoids also exist in the case of $\vec{B}$ parallel to the [110]-direction. The longitudinal and transverse masses in Ge found by cyclotron resonance have been given by Eq.(4.1.20).

The effective mass approximation is a combined experimental and theoretical approach. The energy gap E_g and the reciprocal effective mass tensor $m_{\alpha\beta}^{-1}$ are found from experiment, but quantum mechanics and symmetry are used in order to obtain preliminary information about possible locations of energy minima in the Brillouin zone.

4.3 THE EFFECTIVE MASS APPROXIMATION: DEGENERATE ENERGY BANDS

Experimental evidence along with theoretical symmetry analysis and calculations show that the valence band of the important semiconductor materials (Ge,

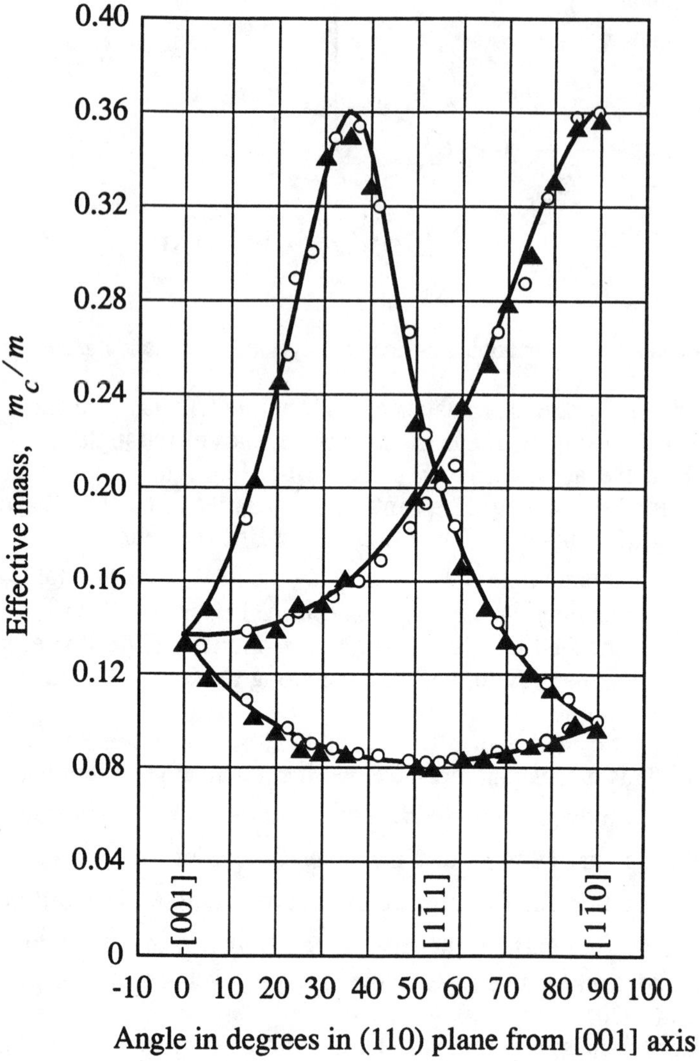

Figure 4.2.7 Effective cyclotron mass of an electron in Ge at 4 K as a function of the angle Θ between the direction of a static magnetic field $\vec{B}$ in the (110)-plane and the longitudinal axis of ellipsoids (After C. Kittel "Introduction to Solid State Physics", J.Wiley & Sons, N.Y., 1986).

Si, III-V- compounds) at the Γ–point has the symmetry Γ'_{25} indicated in Fig.4.3.1 where the band structure of these materials is shown. Table 3.4.3 shows that three wave functions $u_{10}^{(0)}(\vec{r})$, $u_{20}^{(0)}(\vec{r})$, and $u_{30}^{(0)}(\vec{r})$ correspond to the symmetry Γ'_{25} of the energy $E^{(0)}(0)$ at the top of the valence band

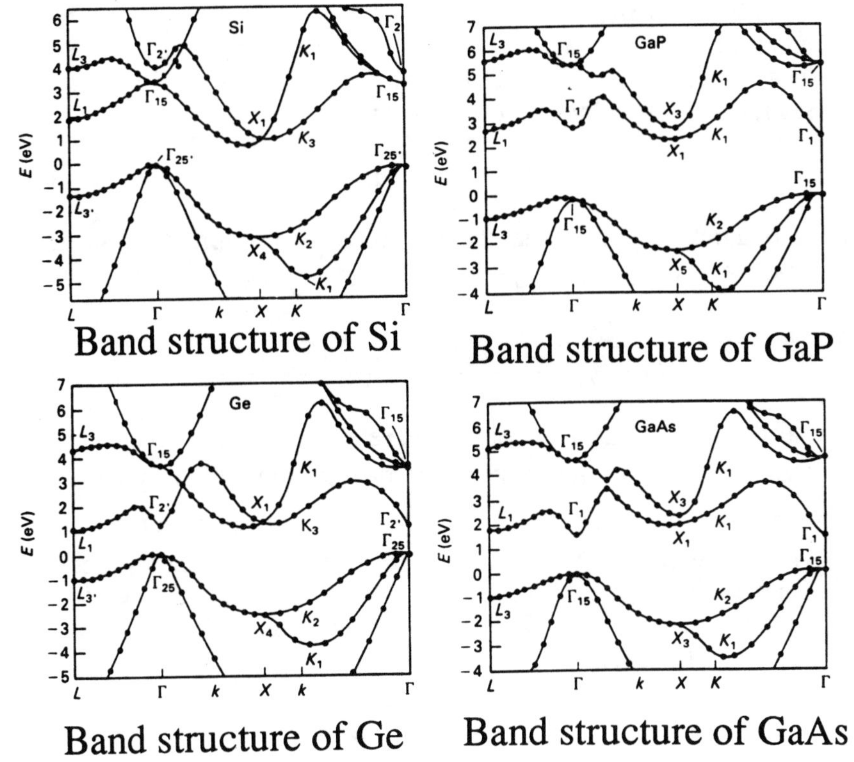

Figure 4.3.1 Band structures of important semiconductor materials: Si, Ge, GaP, and GaAs.

$$
\begin{aligned}
u_{10}^{(0)}(\vec{r}) &= f(r)yz, \\
u_{20}^{(0)}(\vec{r}) &= f(r)zx, \qquad (4.3.1)\\
u_{30}^{(0)}(\vec{r}) &= f(r)xy.
\end{aligned}
$$

A three-fold band degeneracy therefore occurs.

In order to find $E(\vec{p})$ for degenerate bands, one should find the corrections due to the small perturbation (4.1.6) to the energy $E^{(0)}(0)$ which satisfies the zero-order Schrödinger equation

$$\left[-\frac{\hbar^2}{2m_0}\nabla^2 + U(\vec{r})\right]u^{(0)}(\vec{r}) = E^{(0)}(0)u^{(0)}(\vec{r}). \tag{4.3.2a}$$

The perturbation approximation in case of degenerate bands requires the construction of linear combinations of the wave functions of the degenerate level and the wave functions of the other bands

$$u^{(0)} = \sum_s C_s u_{s0}^{(0)} + \sum_j C_j u_{j0}^{(0)} \tag{4.3.2b}$$

where $s = 1,2,3$ labels the degenerate bands and j labels the other bands. Substituting Eq.(4.3.2b) into the Schrödinger equation (4.1.4), multiplying by each function in turn, and integrating over the volume of the crystal results in a system of three linear equations for the expansion coefficients C_s and C_j. These equations are consistent, if the determinant of the system vanishes

$$\left\| \left[E^{(0)}(0) - E(\vec{p}) + \frac{\vec{p}^2}{2m_0} \right] \delta_{ss'} - \sum_{\alpha,\beta} p_\alpha p_\beta D_{\alpha\beta}^{ss'} \right\| = 0. \tag{4.3.3}$$

Here the second rank tensor $D_{\alpha\beta}^{ss'}$ has the dimensions of a reciprocal mass. It is the generalization of an effective mass for the case of degenerate bands, and has the form

$$D_{\alpha\beta}^{ss'} = \frac{\hbar^2}{m_0^2} \sum_j \frac{< u_{s0}^{(0)} \mid \nabla_\alpha \mid u_{j0}^{(0)} >< u_{j0}^{(0)} \mid \nabla_\beta \mid u_{s'0}^{(0)} >}{E_j^{(0)}(0) - E^{(0)}(0)}. \tag{4.3.4}$$

Since it is a coefficient in the expansion of the electron energy in the components of $\vec{p}$, $D_{\alpha\beta}^{ss'}$ should be invariant under the symmetry operations of the crystal. We shall see below which matrix elements of $D_{\alpha\beta}^{ss'}$ vanish. It is convenient to simplify Eq.(4.3.4) for the matrix D for the application of symmetry arguments. The energy separation in the denominator of Eq.(4.3.4) is a scalar quantity that does not change under the symmetry operations. We replace the denominator in Eq.(4.3.4) by the order of magnitude estimate

$$E_j^{(0)} - E^{(0)}(0) \approx E_g.$$

Then, $D_{\alpha\beta}^{ss'}$ takes the form

$$D_{\alpha\beta}^{ss'} = \frac{\hbar^2}{m_0^2} \frac{1}{E_g} \sum_j < u_{s0}^{(0)} \mid \nabla_\alpha \mid u_{j0}^{(0)} >< u_{j0}^{(0)} \mid \nabla_\beta \mid u_{s'0}^{(0)} > . \tag{4.3.5}$$

The product of matrix elements in Eq.(4.3.5) behaves under symmetry operations as the quantity

$$\sum_j < u_{s0}^{(0)} \mid \nabla_\alpha \mid u_{j0}^{(0)} >< u_{j0}^{(0)} \mid \nabla_\beta \mid u_{s'0}^{(0)} > = < u_{s0}^{(0)} \mid \nabla_\alpha \nabla_\beta \mid u_{s'0}^{(0)} > . \tag{4.3.6}$$

Substituting Eq.(4.3.6) into Eq.(4.3.5), one can find $D_{ss'}^{\alpha\beta}$ in a form that allows easy application of symmetry operations

$$D_{\alpha\beta}^{ss'} = \frac{\hbar^2}{m_0^2} \frac{1}{E_g} < u_{s0}^{(0)} \mid \nabla_\alpha \nabla_\beta \mid u_{s'0}^{(0)} > . \tag{4.3.7}$$

We consider first the diagonal element of D with $s = s' = 1$

$$D_{\alpha\beta}^{11} = \frac{\hbar^2}{m_0^2} \frac{1}{E_g} < u_{10}^{(0)} \mid \nabla_\alpha \nabla_\beta \mid u_{10}^{(0)} > .$$

Substituting the wave function $u_{10}^{(0)}$ from Eq.(4.3.1) results in

$$D_{\alpha\beta}^{11} = \frac{\hbar^2}{m_0^2} \frac{1}{E_g} < f(r)yz \mid \nabla_\alpha \nabla_\beta \mid f(r)yz > . \tag{4.3.8}$$

We shall show first that $D^{11}_{\alpha\beta}$ for $\alpha \neq \beta$ vanishes. Let us take $\alpha = x$ and $\beta = z$. Then D^{11}_{xz} takes the form

$$D^{11}_{xz} = \frac{\hbar^2}{m_0^2}\frac{1}{E_g} < f(r)yz \mid \frac{\partial}{\partial x}\frac{\partial}{\partial z} \mid f(r)yz > . \tag{4.3.9}$$

Equation (4.3.9) shows that the matrix D transforms under the symmetry operations from Table 3.4.1 like the product $\left[yz\frac{\partial}{\partial x}\frac{\partial}{\partial z}yz\right]$. For example, when the operation $S_2(x\overline{yz})$ is applied, we obtain

$$S_2(x\overline{yz})\left[yz\frac{\partial}{\partial x}\frac{\partial}{\partial z}yz\right] = -\left[yz\frac{\partial}{\partial x}\frac{\partial}{\partial z}yz\right].$$

On the other hand, since it is a contribution to the electron energy, D should be invariant under symmetry operations of the crystal. Therefore,

$$\left[yz\frac{\partial}{\partial x}\frac{\partial}{\partial z}yz\right] = -\left[yz\frac{\partial}{\partial x}\frac{\partial}{\partial z}yz\right],$$

whence it follows that $\left[yz\frac{\partial}{\partial x}\frac{\partial}{\partial z}yz\right] = 0$. This means that the matrix D does not vanish only when it is diagonal in α and β

$$D^{ss'}_{\alpha\beta} = D^{ss'}\delta_{\alpha\beta}. \tag{4.3.10}$$

When $\alpha = \beta = y$ is put in $D^{11}_{\alpha\beta}$ given by Eq.(4.3.8), we see that D transforms like $\left[yz\frac{\partial}{\partial y}\frac{\partial}{\partial y}yz\right]$. There are no operations S in the Table 3.4.1 or operations IS that transform $\left[yz\frac{\partial}{\partial y}\frac{\partial}{\partial y}yz\right]$ into $-\left[yz\frac{\partial}{\partial y}\frac{\partial}{\partial y}yz\right]$. Under symmetry operations D^{11}_{yy} goes into itself and does not vanish. Similar arguments for $\alpha = \beta = z$ show that we have two equal diagonal elements:

$$D^{11}_{yy} = D^{11}_{zz} = M. \tag{4.3.11}$$

Another example is the matrix element of $D^{11}_{\alpha\beta}$ with $\alpha = \beta = x$. This matrix element transforms as $\left[yz\frac{\partial}{\partial x}\frac{\partial}{\partial x}yz\right]$. For example, applying the operation $S_{13}(\overline{x}y\overline{z})$ from the Table 3.4.1, one can find that

$$S_{13}(\overline{x}z\overline{y})\left[yz\frac{\partial}{\partial x}\frac{\partial}{\partial x}yz\right] = \left[zy\frac{\partial}{\partial x}\frac{\partial}{\partial x}zy\right].$$

This result means that D^{zz}_{11} transforms into itself and does not vanish, but it has a value that is different from M in Eq.(4.3.11). The accepted notation for this matrix element is L:

$$D^{11}_{xx} = L. \tag{4.3.12}$$

Combining all the information from Eqs.(4.3.10) - (4.3.12), we find the general form of $D^{11}_{\alpha\beta}$,

$$D^{11}_{\alpha\beta} = \delta_{\alpha\beta}[L\delta_{\alpha x} + M(\delta_{\alpha y} + \delta_{\alpha z})], \tag{4.3.13a}$$

where $\delta_{\alpha\beta}$ is the Kronecker delta. Taking into account the equivalence of the coordinate axes in a cubic crystal, we can find the other diagonal matrix elements of $D_{\alpha\beta}^{ss'}$

$$D_{\alpha\beta}^{22} = \delta_{\alpha\beta}[L\delta_{\alpha y} + M(\delta_{\alpha x} + \delta_{\alpha z})], \tag{4.3.13b}$$

$$D_{\alpha\beta}^{33} = \delta_{\alpha\beta}[L\delta_{\alpha z} + M(\delta_{\alpha x} + \delta_{\alpha y}]. \tag{4.3.13c}$$

Similar applications of the symmetry operations of a cube lead to the following results for the nondiagonal matrix elements with $s \neq s'$:

$$D_{\alpha\beta}^{12} = D_{\alpha\beta}^{21} = N\delta_{\alpha x}\delta_{\beta y}, \quad D_{\alpha\beta}^{13} = D_{\alpha\beta}^{31} = N\delta_{\alpha x}\delta_{\beta z}, \quad D_{\alpha\beta}^{23} = D_{\alpha\beta}^{32} = N\delta_{\alpha y}\delta_{\beta z}. \tag{4.3.14}$$

Substitution of the matrix elements Eqs.(4.4.13), (4.4.14) in the determinant Eq.(4.3.3) results in

$$\begin{vmatrix} M(p_y^2+p_z^2)+L^2p_x^2-E' & Np_xp_y & Np_xp_z \\ Np_yp_x & M(p_x^2+p_z^2)+Lp_y^2-E' & Np_yp_z \\ Np_zp_x & Np_zp_y & M(p_x^2+p_y^2)+Lp_z^2-E' \end{vmatrix} = 0 \tag{4.3.15}$$

where $E' = E^{(0)}(0) + \vec{p}^2/2m_0 - E(\vec{p})$.

The solution of the characteristic equation (4.3.15) allows finding $E'(\vec{p})$ for an arbitrary direction of $\vec{p}$ in the Brillouin zone. To simplify the calculations, we solve the characteristic equation for two directions with different symmetry in the 1st Brillouin zone.

a. [100]-direction: $p_y = p_z = 0$; $p_x = p$. The characteristic equation (4.4.15) takes the form

$$\begin{vmatrix} Lp^2-E' & 0 & 0 \\ 0 & Mp^2-E' & 0 \\ 0 & 0 & Mp^2-E' \end{vmatrix} = 0$$

Calculating the determinant, one obtains the equation

$$(Lp^2 - E')(Mp^2 - E')^2 = 0.$$

This equation has three solutions

$$\begin{aligned} E'_{1,2} &= Mp^2, \\ E'_3 &= Lp^2. \end{aligned} \tag{4.3.16}$$

We see from these results that M and L are components of the reciprocal mass tensor. Two bands, $E'_{1,2}$, remain degenerate, see Fig.4.3.2.

b. [110]-direction: $p_z = 0, p_x = p_y = p/\sqrt{2}$. The characteristic determinant (4.3.15) equals

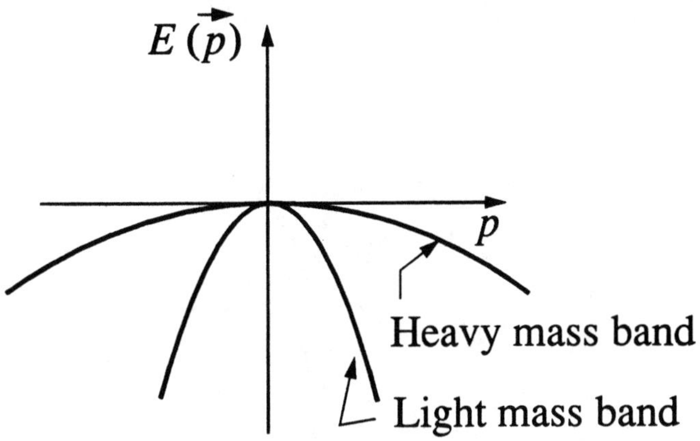

Figure 4.3.2 The structure of the valence band of a typical semiconductor.

$$\begin{vmatrix} Mp_y^2+Lp_x^2-E' & Np_xp_y & 0 \\ Np_xp_y & Mp_x^2+Lp_y^2-E' & 0 \\ 0 & 0 & M(p_x^2+p_y^2)-E' \end{vmatrix} = 0.$$

Calculating the determinant, one obtains the equations

$$Mp^2-E'=0,$$

$$\left((M+L)\frac{p^2}{2}-E'\right)^2-\left(N\frac{p^2}{2}\right)^2=0.$$

Solving these equations leads to

$$E'_1=Mp^2,$$

$$E'_{2,3}=(M+L\pm N)\frac{p^2}{2}. \qquad (4.3.17)$$

Here M, $(M+L+N)$, and $(M+L-N)$ are components of the reciprocal effective mass tensor. All three bands are nondegenerate in this direction. The corresponding surface of constant energy $E'_2=Constant$ is shown in Fig.4.3.3. Cyclotron resonance experiments yield values of N, L, and M, which are the values of the effective masses.

Equations (4.3.16) and (4.3.17) give the structure of the valence band, neglecting the angular momentum of the electron and the spin-orbit coupling, which lead to qualitative changes in the band structure given by Eqs.(4.3.16) and (4.3.17).

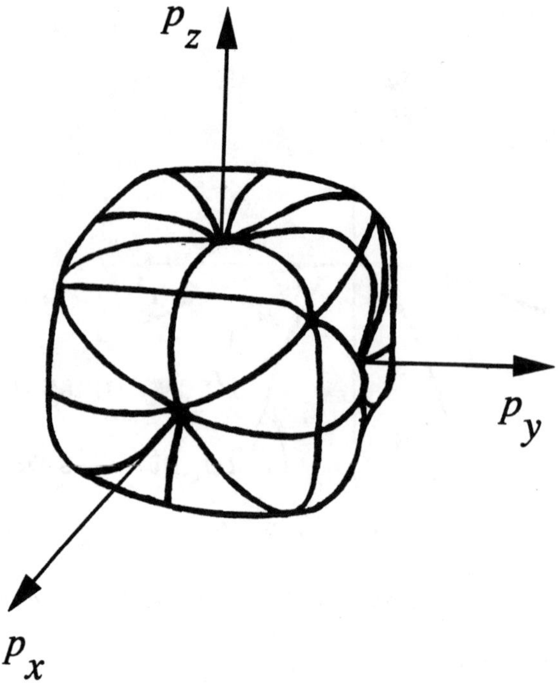

Figure 4.3.3 Sketch of warped constant energy surfaces for the valence band of a semiconductor.

4.4 EFFECTS OF THE SPIN-ORBIT INTERACTION ON THE BAND STRUCTURE OF SEMICONDUCTORS

The formation of the band structure of real semiconductors is considerably influenced by the spin-orbit interaction in the atoms that make up the crystal.

The origin of the spin-orbit interaction is the following. It has been shown by Dirac that the electron spin is a purely relativistic property of an electron. Therefore, the spin-orbit interaction is a small relativistic interaction. Nevertheless, this interaction leads in heavy atoms to a considerable splitting of some atomic levels. For example, the p-state of an electron is split into two states with total angular momenta $j = 3/2$ and $j = 1/2$. Similar splitting of degenerate bands also occurs in semiconductors. This splitting means the generation of new energy gaps, and the physical properties of semiconductors are defined by gaps in the band structure of these materials.

An electron in a crystal moves in an electric field $\vec{\mathcal{E}}$ connected with the periodic potential of the crystal lattice $U(\vec{r})$ by

$$\vec{\mathcal{E}} = -\frac{1}{(-e)}\nabla U, \tag{4.4.1}$$

where the charge of an electron is negative, $-e$, with e being the absolute value of the electron charge. In the coordinate system connected with the moving electron, there is a magnetic field

$$\vec{B} = \vec{\mathcal{E}} \times \vec{v}, \tag{4.4.2}$$

where $\vec{v}$ is the velocity of the electron. The magnetic field $\vec{B}$ acts on the magnetic moment of the electron $\vec{M}$, see Section 1.14, Eq.(1.14.1). This interaction is the origin of the spin-orbit interaction

$$\hat{H}_{s.o.} = -\vec{M} \cdot \vec{B}. \tag{4.4.3}$$

Pauli's theory of electron spin shows that it is convenient to represent spin through Pauli matrices which are defined by the following relation

$$\hat{\vec{S}} = \frac{\hbar}{2} \hat{\vec{\sigma}},$$

where the vector operator $\hat{\vec{\sigma}} = (\hat{\sigma}_x, \hat{\sigma}_y, \hat{\sigma}_z)$ has the matrix components

$$\hat{\sigma}_x = \begin{pmatrix} 0 & 1 \\ 1 & 0 \end{pmatrix}; \quad \hat{\sigma}_y = \begin{pmatrix} 0 & -i \\ i & 0 \end{pmatrix}; \quad \hat{\sigma}_z = \begin{pmatrix} 1 & 0 \\ 0 & -1 \end{pmatrix}. \tag{4.4.4}$$

These matrices operate on the spin wave functions

$$\uparrow = \begin{pmatrix} 1 \\ 0 \end{pmatrix}, \quad \downarrow = \begin{pmatrix} 0 \\ 1 \end{pmatrix}. \tag{4.4.5}$$

The action of the Pauli matrices Eq.(4.4.4) on the spin wave functions $\uparrow$ and $\downarrow$, is defined by the usual rules of matrix multiplication, with the following results

$$\begin{aligned} \hat{\sigma}_x \uparrow &= \begin{pmatrix} 0 & 1 \\ 1 & 0 \end{pmatrix} \begin{pmatrix} 1 \\ 0 \end{pmatrix} = \begin{pmatrix} 0 \\ 1 \end{pmatrix} = \downarrow, \\ \hat{\sigma}_y \uparrow &= \begin{pmatrix} 0 & -i \\ i & 0 \end{pmatrix} \begin{pmatrix} 1 \\ 0 \end{pmatrix} = i \begin{pmatrix} 0 \\ 1 \end{pmatrix} = i \downarrow, \\ \hat{\sigma}_z \uparrow &= \begin{pmatrix} 1 & 0 \\ 0 & -1 \end{pmatrix} \begin{pmatrix} 1 \\ 0 \end{pmatrix} = \begin{pmatrix} 1 \\ 0 \end{pmatrix} = \uparrow \end{aligned} \tag{4.4.6}$$

We see that $\hat{\sigma}_x$ and $\hat{\sigma}_y$ flip the electron spin, but $\hat{\sigma}_z$ leaves the spin unchanged.

The magnetic moment of the electron $\vec{M}$ is connected with $\hat{\vec{\sigma}}$ according to

$$\vec{M} = -\frac{e\hbar}{2m_0} \hat{\vec{\sigma}}. \tag{4.4.7}$$

The substitution of $\vec{M}$ from Eq.(4.4.7) into Eq.(4.4.3) results in

$$\hat{H}_{s.o.} = \frac{e\hbar}{2m_0} \hat{\vec{\sigma}} \cdot \vec{B}. \tag{4.4.8}$$

Combining $\vec{B}$ from Eqs.(4.4.2) and (4.4.1) with Eq.(4.4.8) results in the Hamiltonian of the spin-orbit interaction

$$\hat{H}_{s.o.} = -\frac{i\hbar^2}{2m_0^2}\hat{\vec{\sigma}} \cdot (\nabla U(\vec{r}) \times \nabla). \tag{4.4.9}$$

Our semiclassical derivation of the spin-orbit interaction results in equation (4.4.9) which is twice as large as the real spin-orbit interaction that follows from Dirac's relativistic theory of the electron.

The time-independent Schrödinger equation for an electron in the periodic potential of the lattice which takes into account the spin-orbit interaction has the form

$$\left[-\frac{\hbar^2}{2m_0}\nabla^2 + U(\vec{r}) - \frac{i\hbar^2}{4m_0^2}\hat{\vec{\sigma}} \cdot (\nabla U(\vec{r}) \times \nabla)\right]\psi(\vec{r}) = E\psi(\vec{r}), \tag{4.4.10}$$

where an extra factor 1/2 has been introduced into the spin-orbit interaction to be consistent with the result of Dirac's relativistic theory of the electron.

We apply Eq.(4.4.10) to correct the band structure at the Γ point of the Brillouin zone, using as before the example of Ge with O_h symmetry. When the spin-orbit interaction is taken into account the modulating Bloch wave functions from Eq.(4.3.1), $u_{j0}(\vec{r})$, are given an extra spin quantum number: $s_z = +1/2$ or $s_z = -1/2$. Because $\hat{H}^{(0)}$ is independent of spin, there is therefore a total of six wave functions of the zeroth approximation belonging to the same degenerate energy level $E^{(0)}(0)$

$$\begin{aligned} u_{10}^{(0)} &= f(r)\, yz \uparrow, & u_{40}^{(0)} &= f(r)\, yz \downarrow, \\ u_{20}^{(0)} &= f(r)\, zx \uparrow, & u_{50}^{(0)} &= f(r)\, zx \downarrow, \\ u_{30}^{(0)} &= f(r)\, xy \uparrow, & u_{60}^{(0)} &= f(r)\, xy \downarrow. \end{aligned} \tag{4.4.11}$$

In case of six-fold degeneracy the wave function of the zeroth approximation is taken in the form of a linear combination of the functions (4.4.11)

$$u^{(0)} = \sum_{s=1}^{6} C_s u_{s0}^{(0)}. \tag{4.4.12}$$

The substitution of the wave function (4.4.12) in the Schrödinger equation (4.4.10) results in a system of six linear equations for the expansion coefficients C_s. The equations are consistent when their (6 x 6) characteristic determinant vanishes. A symmetry analysis of the matrix elements that is similar to the one carried out in Section 4.3 results in the splitting off of one of three degenerate bands and in the creation of a new energy gap Δ which separates the split-off band from the others.

The origin of this splitting is clearly seen from the genesis of the band structure starting from the energy levels of the atoms that make up the crystal. The valence band of Ge originates from the p-state of the outer valence electron of a Ge atom. In a p-state (orbital quantum number $l = 1$), the total angular momentum of the electron $\hat{\vec{J}} = \hat{\vec{L}} + \hat{\vec{S}}$ has the eigenvalues $j = 3/2, 1/2$, see Eq.(1.14.11) and Fig.4.4.1a. The corresponding energy level of the atom is degenerate, see Fig.4.4.1a. When the spin-orbit interaction is taken into account, the energy level of the atom splits into two levels separated by the energy Δ. One of these levels corresponds to a total angular momentum $j = 3/2$, ($j_z = \pm 3/2, \pm 1/2$), and the second has $j = 1/2$, ($j_z = \pm 1/2$),

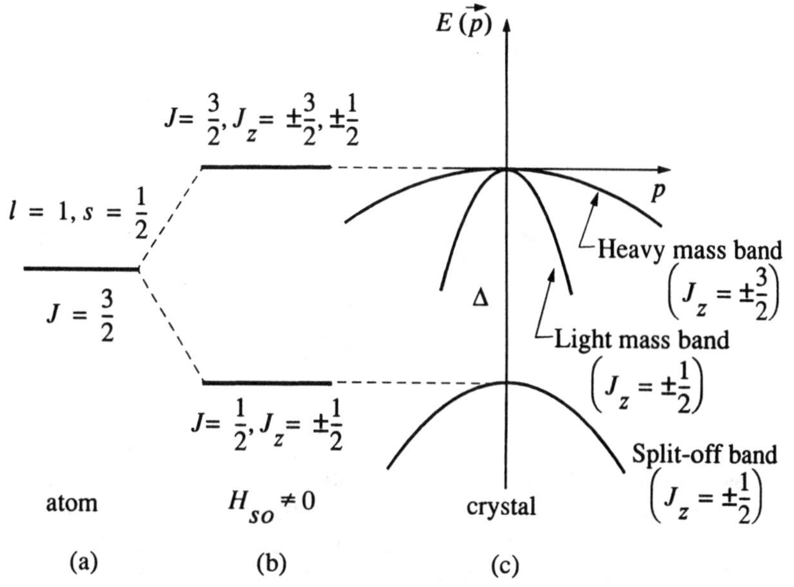

Figure 4.4.1 Genesis of the valence band structure. (a) p-state of an isolated atom, (b) p-state of an isolated atom with spin-orbit interaction taken into account, (c) valence band of a typical semiconductor.

see Fig.4.4.1b. When the crystal is made up of these atoms, electron acquire quasimomentum, and energy bands emerge from each level of the atom, see Fig.4.4.1c.

If the separation of the bands, Δ, due to the spin-orbit interaction is large compared with the kinetic energy of the electrons, the upper four-fold degenerate band and spin-orbit-split band can be considered separately.

The four-fold degenerate electron band at the Γ–point corresponds to the value of the total angular momentum $j = 3/2$, ($j_z = \pm 3/2, \pm 1/2$). Considerations of dimensionality and symmetry allowed Luttinger to write down the Hamiltonian of the four-fold degenerate electron bands in isotropic and quadratic in the quasimomentum $\vec{p}$ approximation

$$\hat{H}_L = -\frac{1}{2m_0}\left[\left(\gamma_1 + \frac{5}{2}\gamma_2\right)\hat{\vec{p}}^{\,2} - 2\gamma_2\left(\hat{\vec{p}} \cdot \hat{\vec{J}}\right)^2\right], \tag{4.4.13}$$

where γ_1 and γ_2 are called Luttinger parameters. Equation (4.4.13) represents a scalar function, energy, expressed in terms of the squares of the vectors $\hat{\vec{p}}$ and $\hat{\vec{J}}$. Taking $\hat{\vec{p}}$ in the z-direction and considering this direction as the quantization axis for angular momentum, one should substitute the eigenvalues $j_z = \pm 3/2, \pm 1/2$ in the Schrödinger equation $\hat{H}_L\psi = E\psi$ to find two values of the electron valence band energy

$$E_{3/2}(\vec{p}) = -\frac{p_z^2}{2m_0}(\gamma_1 - 2\gamma_2) \quad \text{for} \quad j_z = \pm\frac{3}{2}, \tag{4.4.14a}$$

$$E_{1/2}(\vec{p}) = -\frac{p_z^2}{2m_0}(\gamma_1 + 2\gamma_2) \quad \text{for} \quad j_z = \pm\frac{1}{2}. \tag{4.4.15b}$$

We see that in a given direction, the energy can be described in terms of negative effective masses of an electron in the valence band

$$m_{heavy} = -\frac{m_0}{\gamma_1 - 2\gamma_2}, \tag{4.4.15a}$$

$$m_{light} = -\frac{m_0}{\gamma_1 + 2\gamma_2}, \tag{4.4.15b}$$

where $|m_{heavy}|>|m_{light}|$. Since the reciprocal effective mass is the second derivative of the energy with respect to the components of the quasimomentum $\vec{p}$, see Eq.(4.1.3), the negative values of effective mass mean that the valence band has a maximum at the Γ–point. The correct behavior of the valence band has been shown in Fig.4.4.1c.

The two-fold degenerate spin-orbit-split band corresponds to $j = 1/2$, ($j_z = \pm 1/2$. It has a simple isotropic parabolic dependence of the electron energy on quasimomentum

$$E_{split}(\vec{p}) = -\Delta - \frac{p^2}{2m^*}. \tag{4.4.16}$$

Here Δ is the band gap that results from the spin-orbit interaction. This band also has a maximum at the Γ–point of the Brillouin zone. It should be noted that the fundamental quantum principle of time reversal symmetry does not remove the degeneracy between the Bloch state $\psi_{\vec{p}}(\vec{r})$ and its complex conjugate $\psi^*_{\vec{p}}(\vec{r})$ which belongs to the state in which both the quasimomentum and the angular momentum of the electron are reversed. This degeneracy is called ***Kramers degeneracy.***

4.5 KANE'S MODEL

The heavier the atom, the stronger is the spin-orbit interaction in atoms and the larger is the band gap Δ from Eq.(4.4.16). Thus, spin-orbit interaction effects are less prominent in Si (atomic number $Z_{Si} = 14$) than in Ge ($Z_{Ge} = 32$). In InSb the spin-orbit interaction becomes even more important ($Z_{In} = 49, Z_{Sb} = 51$). The energy gap Δ increases in InSb up to a values that exceeds the energy gap E_g between the conduction and valence bands

$$\Delta_{InSb} = 0.9\,eV, \qquad E_g = 0.17\,eV.$$

This means that for materials of this type the conduction band should also be taken into account in band structure calculations along with the triply degenerate valence band. The corresponding model which takes into account these four closely situated bands and neglects all the other bands is known as Kane's model.

The valence band of InSb at the Γ–point is triply degenerate as it is in Ge, but the symmetry of the corresponding wave functions is of Γ_{15}-type. This means, according to Table 3.4.2, that the wave functions transform under symmetry operations as components of the radius-vector $\vec{r} = (x, y, z)$. The wave function of the conduction band has spherical symmetry Γ_1. There are altogether four wave functions of the zeroth approximation.

To take into account the spin-orbit interaction, each wave function should be given one of two possible spin quantum numbers s_z. The total set of zero-approximation wave functions used in Kane's model is the following

$$
\begin{aligned}
u_{10}^{(0)} &= f(r)\uparrow, & u_{50}^{(0)} &= f(r)\downarrow, \\
u_{20}^{(0)} &= f(r)x\uparrow, & u_{60}^{(0)} &= f(r)x\downarrow, \\
u_{30}^{(0)} &= f(r)y\uparrow, & u_{70}^{(0)} &= f(r)y\downarrow, \\
u_{40}^{(0)} &= f(r)z\uparrow, & u_{80}^{(0)} &= f(r)z\downarrow .
\end{aligned}
\tag{4.5.1}
$$

A linear combination of these functions should be substituted in the time-independent Schrödinger equation (4.4.10), and a system of eight linear equations results from it. The secular equation now becomes an (8 x 8) determinant. A symmetry analysis of this determinant leads to four doubly degenerate energy bands

$$E'_c(\vec{p}) = \frac{E_g}{2} + \left[\left(\frac{E_g}{2}\right)^2 + \frac{2}{3}(Pp)^2\right]^{1/2}, \tag{4.5.2a}$$

$$E'_{v_1}(\vec{p}) = 0, \tag{4.5.2b}$$

$$E'_{v_2}(\vec{p}) = \frac{E_g}{2} - \left[\left(\frac{E_g}{2}\right)^2 + \frac{2}{3}(Pp)^2\right]^{1/2}, \tag{4.5.2c}$$

$$E'_{v_3}(\vec{p}) = -\Delta - \frac{(Pp)^2}{3(\Delta + E_g)}. \tag{4.5.2d}$$

Here P and Δ are standard notations for constant nonzero matrix elements in Kane's model.

Equation (4.5.2a) gives $E_c(\vec{p})$ for the conduction band. The most impressive result here is that the p-dependence of $E_c(\vec{p})$ is ***nonparabolic***. Three other energy bands Eqs.(4.5.2b,c,d) represent the electron band structure of the valence band. In our simple approximation, the energy of the heavy mass band $E'_{v_1}(\vec{p})$ has no dependence on $\vec{p}$. Some other energy bands should be taken into consideration in order to find the negative curvature of the heavy mass valence band. The light mass band $E'_{v_2}(\vec{p})$ also has a considerable nonparabolicity. The fourth band $E'_{v_3}(\vec{p})$ is the spin-orbit-split band. The gap Δ is a direct measure of the spin-orbit interaction strength. It is well measured in optical experiments.

The simplified view of the electronic band structure corresponding to Kane's model is shown in Fig.4.5.1. The nonparabolicity of the electronic band structure in materials with $\Delta \geq E_g$ has been observed experimentally as a strong dependence of the effective mass in InSb on the electron concentration. The reciprocal effective mass $\frac{\partial^2 E(p)}{\partial p_\alpha \partial p_\beta}$ is the constant curvature for a parabolic dependence of $E(\vec{p})$ on $\vec{p}$. In case of a nonparabolic dependence of $E(\vec{p})$ on $\vec{p}$, for each value of the energy the curvature is different and the effective mass is different as well.

The interaction of electronic states of the conduction band and of the valence band considered in Kane's model yields an adequate description of the band structure of the so-called ***direct gap nonparabolic semiconductors*** such as GaAs, InSb and InP.

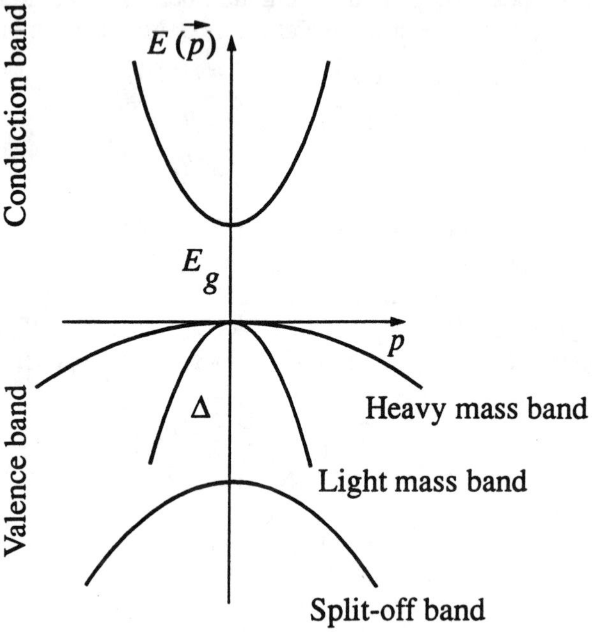

Figure 4.5.1 Four-band Kane model (a simplified view of the electronic band structure for a direct gap semiconductor).

4.6 ELECTRON BAND STRUCTURE OF SEMICONDUCTOR QUANTUM NANOSTRUCTURES

The discovery of new crystal growth techniques such as MBE, LPE, and MOCVD has opened the way to the growth of semiconductors atomic layer by atomic layer. Molecular Beam Epitaxy (MBE) uses a molecular beam of chemical elements which impinge on a substrate placed in a high vacuum. Liquid Phase Epitaxy (LPE) uses the cooling of a saturated solution containing the chemical components in contact with a substrate. Metalloorganic Chemical Vapor Deposition

Epitaxy (LPE) uses the cooling of a saturated solution containing the chemical components in contact with a substrate. Metalloorganic Chemical Vapor Deposition (MOCVD) is based on the gas flow in a reactor. The composition and doping of the growing layers are governed by the arrival rates of the molecules and by the temperature of the substrate. These technologies allow to fabricate semiconductor structures with characteristic size an order of magnitude or more smaller than the de Broglie wavelength of an electron in the crystal. In a typical semiconductor, the de Broglie wavelength is about 50-100 *nm*. Spatial size quantization occurs in thin semiconductor nanostructures with thicknesses of the order of the electron de Broglie wave length. The band structure of bulk materials is no longer valid. There are three types of structures which are of considerable interest: these are quantum wells, quantum wires and quantum dots. These objects demonstrate a number of new phenomena which are of fundamental value in basic physics. They have also found some applications in quantum well lasers, field effect transistors, and nonlinear optical elements.

a. Quantum well. A quantum well is an ultrathin layer of a semiconductor whose bandgap is smaller than the bandgap of the surrounding semiconductors, see Fig.4.6.1a,b. An example is a thin GaAs layer sandwiched between layers of $Al_xGa_{1-x}As$.

The potential energy of a quantum well structure has a sandwich form for both the conduction and the valence bands. The motion of an electron in both the conduction and valence bands is confined in the material with the smaller bandgap in the direction that is transverse to the plane of the thin layer, say the z-direction in Fig.4.6.1a. At the same time electrons move as free particles in the two other directions (x – and y –directions).

The wave function and energy of an electron in a thin layer can be obtained from the time-independent Schrödinger equation with potential energy $U(z)$. The problem is solved by the separation of variables

$$\psi(\vec{r}) = e^{\frac{i}{\hbar}(p_x x + p_y y)}\chi(z), \tag{4.6.1}$$

$$E = E_c + \frac{p_x^2 + p_y^2}{2m^*} + E_{n_z}. \tag{4.6.2}$$

Here the exponential factor in the wave function (4.6.1) corresponds to the free motion of an electron in the x – and y –directions, and the function $\chi(z)$ represents the localization of the electron in the z –direction in a one-dimensional potential well. The second term in the energy (4.6.2) is the energy of free motion, and E_{n_z} is the quantized energy in the quantum well.

In the simplest approximation of a potential well with infinite walls, considered in Section 1.12, the size quantization of p_z occurs

$$\frac{p_z}{\hbar} = \frac{\pi}{d_3} n_z, \tag{4.6.3}$$

where d_3 is the thickness of the layer. The wave function of an electron in the conduction band has the form

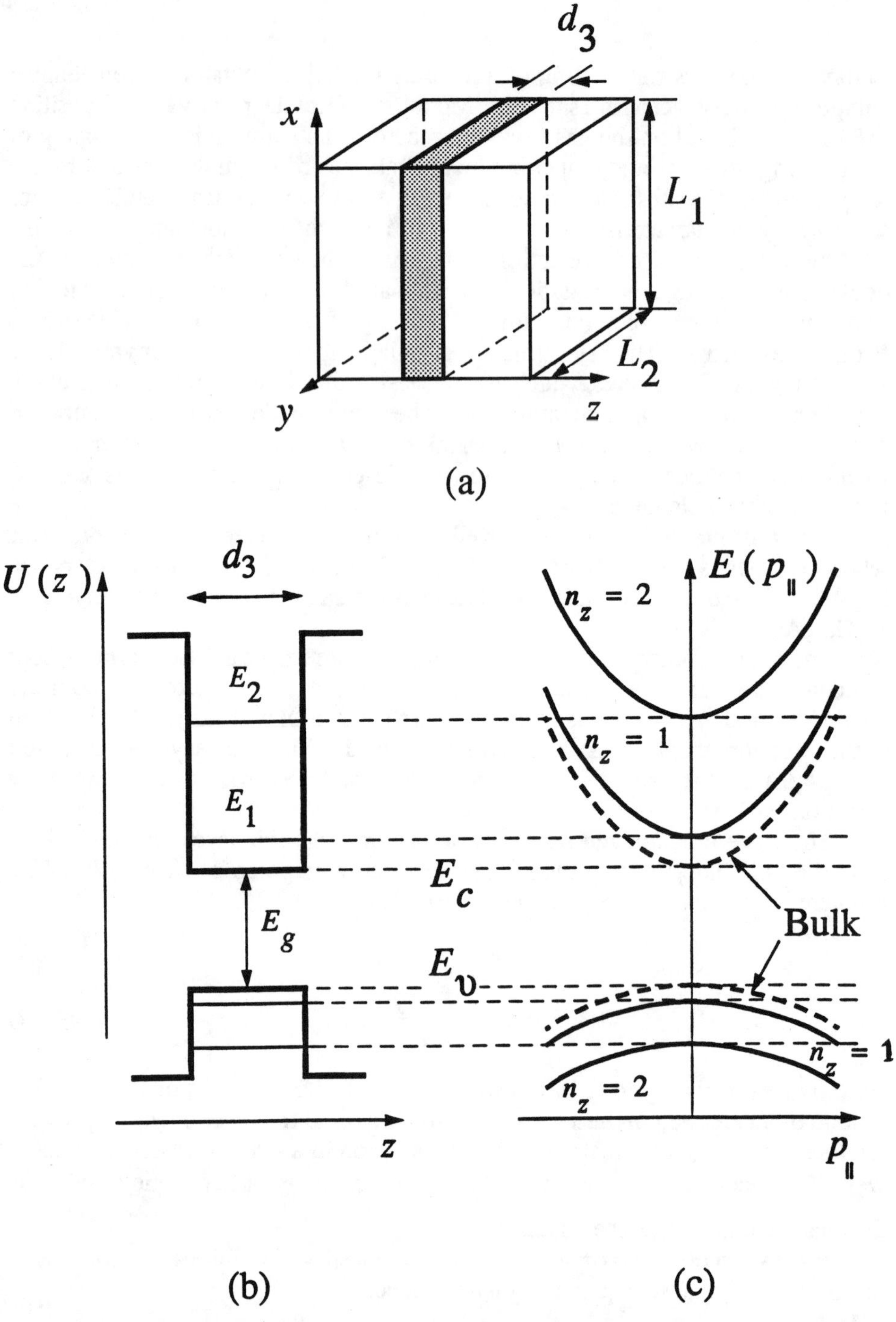

Figure 4.6.1 (a) Quantum well structure, where a thin region of a narrow gap semiconductor (GaAs) is sandwiched between layers of a wide band semiconductor (AlGaAs). (b) The potential barriers for quantum well. (c) The electronic band structure of quantum well.

$$\chi(z) = A \sin\left(\frac{\pi}{d_3} n_z z\right). \tag{4.6.4}$$

Here $A = \sqrt{(2/d_3)}$ is the normalization constant. The corresponding quantized energy in the quantum well equals

$$E_{n_z} = \frac{\hbar^2}{2m^*}\left(\frac{\pi}{d_3}\right)^2 n_z^2 \quad \text{and } n_z = 1,2,3,\ldots \tag{4.6.5}$$

It is seen from Eqs.(4.6.2) and (4.6.5) that each quantum number n_z is associated with a two-dimensional energy subband with a parabolic dependence of the energy on the quasimomentum $\vec{p}_{\parallel} = (p_x, p_y)$. The lowest energy corresponds to $n_z = 1$.

Figure 4.6.1c shows the parabolic dependence of the energy E on the two-dimensional quasimomentum $\vec{p}_{\parallel} = (p_x, p_y)$ for several possible values of the discrete quantum number n_z. The function $E(\vec{p})$ for the case of a bulk semiconductor is shown by the dashed line.

In order to obtain solutions of the Schrödinger equation for a potential well of ***finite*** height the interface boundary conditions for $\chi(z)$ and for its derivative $d\chi(z)/dz$ are needed. These boundary conditions are the continuity of $\chi(z)$ and of $(1/m^*)\partial\chi(z)/\partial z$. The second boundary condition differs from the boundary condition Eq.(1.10.12b) by the factor $1/m_n$ which is necessary for conservation of the particle current through the interface. The problem of a potential well of finite height leads to a transcendental equation for the energy levels, which requires numerical solution.

Calculations of the energy in the electron valence band are more complicated, because of the degeneracy of the valence band. The simplest classification of energy levels is obtained at the Γ point $p_x = p_y = 0$ of the two-dimensional 1st Brillouin zone. The Luttinger Hamiltonian (4.4.13) of an electron in the valence band at the Γ–point is

$$\hat{H}_L = -\frac{1}{2m_0}\left\{\left(\gamma_1 + \frac{5}{2}\gamma_2\right)\hat{p}_z^2 - 2\gamma_2\left(\hat{p}_z \cdot \hat{J}_z\right)^2\right\}. \tag{4.6.6}$$

In the case of an infinite potential well, the electron wave functions of the valence band should vanish on the boundaries, see Eq.(1.12.10). This condition imposes the same limitations on the possible values of the quasimomentum, as was already discussed in considering Eq.(4.6.3),

$$\frac{p_z}{\hbar} = \frac{\pi}{d_3} n_z, \tag{4.6.7}$$

where the quantum number n_z as before takes integer values. The quantization axis of the electron angular momentum is taken in the direction of p_z. Then, the z-component of the total angular momentum is quantized according to Eq.(1.14.11) and has the values $j_z = \pm 3/2, \pm 1/2$. Substituting the allowed values of p_z from Eq.(4.6.6) and the eigenvalues of J_z in the Schrödinger equation

$$\hat{H}_L \psi = E\psi,$$

we find the following values of the electron energy at the Γ–point

$$E_1 = -\frac{1}{2m_0}(\gamma_1 - 2\gamma_2)\left(\frac{\pi}{d_3}n_z\right)^2 = -\frac{1}{2m_{heavy}}\left(\frac{\pi}{d_3}n_z\right)^2, \tag{4.6.8a}$$

$$E_2 = -\frac{1}{2m_0}(\gamma_1 + 2\gamma_2)\left(\frac{\pi}{d_3}n_z\right)^2 = -\frac{1}{2m_{light}}\left(\frac{\pi}{d_3}n_z\right)^2. \tag{4.6.8b}$$

To compare the electron valence band structure of a quantum well with the valence band structure of the bulk material, we display results of band structure calculations for a GaAs - $Ga_xAl_{1-x}As$ quantum well with width $d_3 = 150\,\mathring{A}$ in the approximation of the Luttinger Hamiltonian (4.6.6) see Fig.4.6.2.

The energy spectrum in the quadratic approximation is shown by the dashed lines. The dashed curve (1) corresponds to the light mass band and size quantization quantum number $n_z = 1$, the dashed curve (2) is the heavy mass band and $n_z = 1$, and the dashed curve (3) is again the heavy mass band with $n_z = 2$. We see that the four-fold degenerate bands of the bulk material, shown in Fig.4.4.1, are split into separate bands with heavy and light masses. This happens due to the reduction of

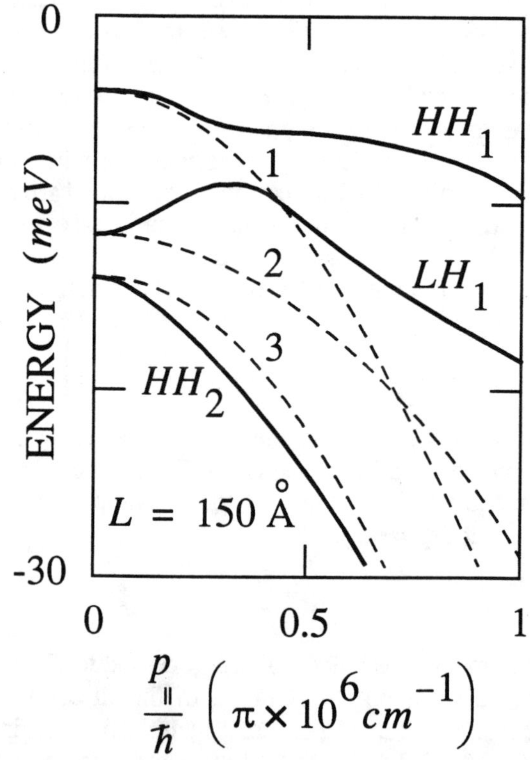

Figure 4.6.2 In-plane dispersion relations of the valence subbands of a GaAs - $Ga_{0.7}Al_{0.3}As$ quantum well ($L = 150\,\mathring{A}$). The dashed lines are the subband dispersions obtained in the diagonal approximation (after G. Bastard "Wave Mechanics Applied to Semiconductor Heterostructures". John Wiley & Sons, N. Y., 1988)

the electron symmetry in a quantum well with respect to its symmetry in the bulk material. In bulk material with the symmetry of a cube or tetrahedron, the three orthogonal x-, y-, z- directions are equivalent. In a quantum well, the equivalency remains only in the x- and y- directions. In the third direction, the size quantization expressed by Eq.(4.6.7) occurs.

There is also the crossing of the curves (1) and (2). It is known in quantum mechanics that when energy levels of the same symmetry cross, a strong repulsive interaction of the levels occurs, which results in the band structure shown by the solid lines in Fig.4.6.2. We see that a strong nonparabolicity of the bands appears, and even a reversal of the sign of the effective mass in curve (2) at the Γ–point. The energy of this valence band has a minimum at the Γ–point, and the effective mass of the electron becomes positive, as in the conduction band.

The study of quantum well band structures is developing very rapidly at the present time, both experimentally and theoretically.

b. Quantum wire. A semiconductor structure that has the form of a thin wire of rectangular cross section, surrounded by material with a wider band gap, is called a quantum wire if the cross section of these wires is of the order of, or smaller, than, the electron de Broglie wavelength, see Fig.4.6.3.

The wire is a two-dimensional potential well which confines the electron in two directions, say, z and y. An electron moves as a free particle in the third, x, direction, which is along the axis of the wire. In the approximation of a potential well with infinite walls, the electron wave function and energy are obtained by solving the Schrödinger equation with potential energy $U(y)+U(z)$, which is zero inside the quantum wire and infinite outside it. The procedure of separation of variables works in this case in the form

$$\psi(\vec{r}) = e^{\frac{i}{\hbar}p_x x}\chi(y)\eta(z), \tag{4.6.9}$$

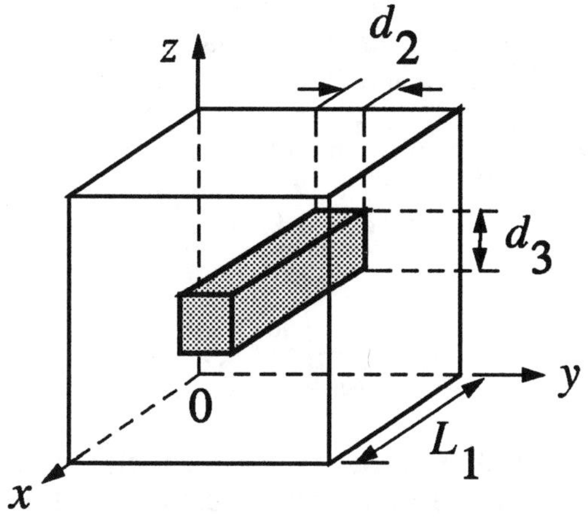

Figure 4.6.3 Quasi-one-dimensional quantum wire.

and

$$E = E_c + \frac{p_x^2}{2m_n} + E_{n_y} + E_{n_z}. \tag{4.6.10}$$

Here the exponential factor in the ψ–function represents the free motion of the electron along the x–direction, and the functions $\chi(y)$ and $\eta(z)$ describe the confinement of the electron inside the two-dimensional potential well with dimensions d_2 and d_3 in y– and z– directions, respectively. These functions and the corresponding discrete energies are found in the same way as this was done in Section 1.12,

$$\psi(\vec{r}) = A e^{\frac{i}{\hbar} p_x x} \sin\left(\frac{\pi}{d_2} n_y y\right) \sin\left(\frac{\pi}{d_3} n_z z\right), \tag{4.6.11a}$$

where the normalization constant is $A = \sqrt{4/d_2 d_3}$. The corresponding energy equals

$$E = E_c + \frac{p_x^2}{2m^*} + E_{n_y} + E_{n_z}, \tag{4.6.11b}$$

where

$$E_{n_y} = \frac{\hbar^2}{2m^*}\left(\frac{\pi}{d_2}\right)^2 n_y^2, \quad E_{n_z} = \frac{\hbar^2}{2m^*}\left(\frac{\pi}{d_3}\right)^2 n_z^2, \tag{4.6.12}$$

with n_y, n_z = 1, 2, 3,..

It follows from Eqs.(4.6.11) and (4.6.12) that each pair of quantum numbers (n_y, n_z) is associated with a one-dimensional energy subband with a parabolic dependence on the quasimomentum p_x.

c. Quantum dot. In a quantum dot, an electron is confined by the walls of a three-dimensional potential well with dimensions d_1, d_2, d_3. The corresponding wave function of the electron and the energy are obtained by a generalization of the problem of a one-dimensional potential well:

$$\psi(\vec{r}) = \left[\frac{8}{d_1 d_2 d_3}\right]^{1/2} \sin\left(\frac{\pi}{d_1} n_x x\right) \sin\left(\frac{\pi}{d_2} n_y y\right) \sin\left(\frac{\pi}{d_3} n_z z\right), \tag{4.6.13}$$

$$E = E_c + E_{n_x} + E_{n_y} + E_{n_z}, \tag{4.6.14}$$

where

$$E_{n_x} = \frac{\hbar^2}{2m^*}\left(\frac{\pi}{d_1}\right)^2 n_x^2; \quad E_{n_y} = \frac{\hbar^2}{2m^*}\left(\frac{\pi}{d_2}\right)^2 n_y^2; \quad E_{n_z} = \frac{\hbar^2}{2m^*}\left(\frac{\pi}{d_3}\right)^2 n_z^2, \tag{4.6.15}$$

and n_x, n_y, n_z = 1, 2, 3,.. The allowed energy levels are discrete and well separated. Quantum dots are often called artificial atoms, but it should be remembered that they consist of thousands of atoms.

4.7 CONCEPT OF A HOLE

It was shown in Chapter 3 that at the point Γ of the Brillouin zone, the electron energy $E(\vec{p})$ can have a maximum. The corresponding reciprocal effective mass in the simplest case of isotropic energy bands $1/m^* = \partial^2 E(\vec{p})/\partial p^2$ is negative. This occurs, for example, in the valence band considered in Sections 4.3 and 4.5. If the origin of the energy of electron is taken at the top of the valence band, the electron energy in the valence band is

$$E(\vec{p}_e) = E_v - \frac{\vec{p}_e^{\,2}}{2m^*}, \tag{4.7.1}$$

where $\vec{p}_e$ is the electron quasimomentum.

It is shown below that the electrons in a totally filled valence band contribute neither to the electric current density nor to the energy current density. Since the quasimomentum $\vec{p}_e$ takes equal number of positive and negative values in the band, see Eq.(2.4.17), the total quasimomentum of the filled band vanishes: $\sum \vec{p}_e = 0$.

The electric current density $\vec{j}_e$ is equal to the sum of all electron velocities $\vec{v}(\vec{p}_e)$ in the filled states multiplied by the negative elementary charge of an electron

$$\vec{j}_e = (-e)2 \sum_{\substack{p_e \\ (filled)}} \vec{v}(\vec{p}_e). \tag{4.7.2}$$

The factor of 2 here arises from the two values of the electron spin in each state. Separating contributions from positive and negative values of $\vec{p}_e$, one obtains

$$\vec{j}_e = (-e)2\left[\sum_{\vec{p}_e>0} \vec{v}(\vec{p}_e) + \sum_{\vec{p}_e<0} \vec{v}(\vec{p}_e)\right] = (-e)2\left[\sum_{\vec{p}_e>0} \vec{v}(\vec{p}_e) + \sum_{\vec{p}_e>0} \vec{v}(-\vec{p}_e)\right]. \tag{4.7.3}$$

It follows from Eqs.(2.6.10) and (2.5.12) that

$$\vec{v}(-\vec{p}_e) = \frac{\partial E(-\vec{p}_e)}{\partial(-\vec{p}_e)} = -\frac{\partial E(\vec{p}_e)}{\partial \vec{p}_e} = -\vec{v}(\vec{p}_e). \tag{4.7.4}$$

Combining Eq.(4.7.4) with Eq.(4.7.3) results in

$$\vec{j}_e = (-e)2\left[\sum_{\vec{p}_e>0} \vec{v}(\vec{p}_e) - \sum_{\vec{p}_e>0} \vec{v}(\vec{p}_e)\right] \equiv 0. \tag{4.7.5}$$

Equation (4.7.5) means that there is no electric current in a filled valence band. In a similar way, the energy current density $\vec{Q}$ in a filled band vanishes

$$\vec{Q} = \sum_{p_e s_z} E(\vec{p}_e)\vec{v}(\vec{p}_e) \equiv 0. \tag{4.7.6}$$

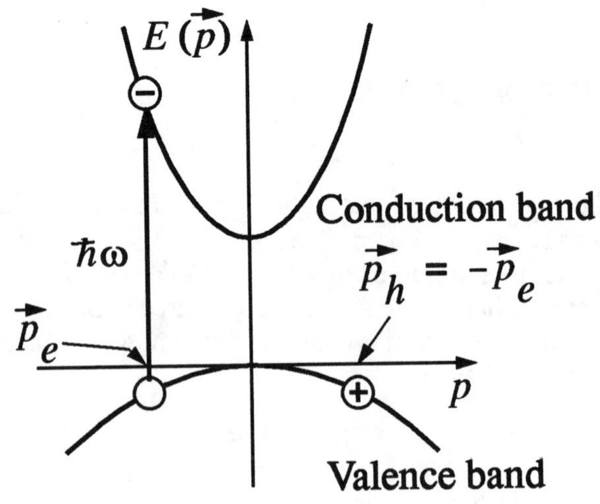

Figure 4.7.1 Absorption of $\hbar\omega$ takes an electron from the filled valence band to the conduction band. If $\vec{p}_e$ is the momentum of an electron in the initial state, it becomes the wavevector of the electron in the conduction band. The total wavevector of the valence band after absorption is $-\vec{p}_e$. This vector is ascribed to a single hole in the valence band: $\vec{p}_h = -\vec{p}_e$.

The circumstances are different when there are some missing electron states in the valence band. If one electron with quasimomentum $\vec{p}_e$ is excited from the valence band to the conduction band, the valence band has a total quasimomentum $(-\vec{p}_e)$. We can considerably simplify the consideration of an almost filled band by introducing the concept of a "hole". ***The vacant orbitals of the valence band are called "holes"***. The total quasimomentum of the valence band with a missing electron $(-\vec{p}_e)$ is ascribed to the hole, see Fig.4.7.1. The quasimomentum of the hole $\vec{p}_h$ is given by

$$\vec{p}_h = -\vec{p}_e\,. \tag{4.7.7}$$

The vacant orbital, that is the hole, contributes to the electric current density in the valence band. If the state with quasimomentum $\vec{p}\,'_e$ and the spin $s_z = \uparrow$ is not occupied by an electron, and the state with the same quasimomentum $\vec{p}\,'_e$ but with the opposite spin $s_z = \downarrow$ is filled, the current density takes the form

$$\vec{j}_e = (-e)\{2\sum_{\substack{\vec{p}_e \neq \vec{p}_e{}' \\ (filled)}} \vec{v}(\vec{p}_e) + \vec{v}(\vec{p}_e{}')\}.$$

Here the first term represents the contribution of the occupied states in the valence band, and the second term is the contribution of the electron with the unpaired spin. Completing the sum in the first term with $\vec{v}(\vec{p}\,'_e)$ and subtracting $\vec{v}(\vec{p}\,'_e)$, one obtains

$$\vec{j}_e = (-e)\{2\sum_{\substack{p_e \\ (total)}} \vec{v}(\vec{p}_e) - \vec{v}(\vec{p}_e{}')\}. \tag{4.7.8}$$

The first term in Eq.(4.7.8) vanishes according to Eq.(4.7.5). Finally, we find the current density

$$\vec{j}_e = (+e)\vec{v}(\vec{p}_e{}'), \tag{4.7.9}$$

which means that the electric current density in the band with a missing electron equals the current of a positively charged (+e) particle corresponding to the vacant $\vec{p}'_e$ state and moving with velocity $\vec{v}(\vec{p}'_e)$. This positively charged particle is the hole.

The analysis of the energy current density allows obtaining the relation between the energy of the electron and the energy of the hole. In the valence band with one missing electron, the energy current density is

$$\vec{Q} = 2 \sum_{\substack{\vec{p}_e \neq \vec{p}_e' \\ (filled)}} E(\vec{p}_e)\vec{v}(\vec{p}_e) + E(\vec{p}_e')\vec{v}(\vec{p}_e') =$$

$$2 \sum_{\substack{\vec{p}_e \\ (total)}} E(\vec{p}_e)\vec{v}(\vec{p}_e) - E(\vec{p}_e')\vec{v}(\vec{p}_e'). \tag{4.7.10}$$

The first term on the right hand side of Eq.(4.7.10) is zero according to Eq.(4.7.6). One finds

$$\vec{Q} = -E(\vec{p}_e')\vec{v}(\vec{p}_e'). \tag{4.7.11}$$

We see that the energy current density from the valence band with one vacant state is equal in magnitude and is opposite in direction to the energy current density of a single particle with the energy $-E(\vec{p}'_e)$ and velocity $\vec{v}(\vec{p}'_e)$. The energy of the vacant state is attributed to the hole,

$$E(\vec{p}_h) = -E(\vec{p}_e). \tag{4.7.12}$$

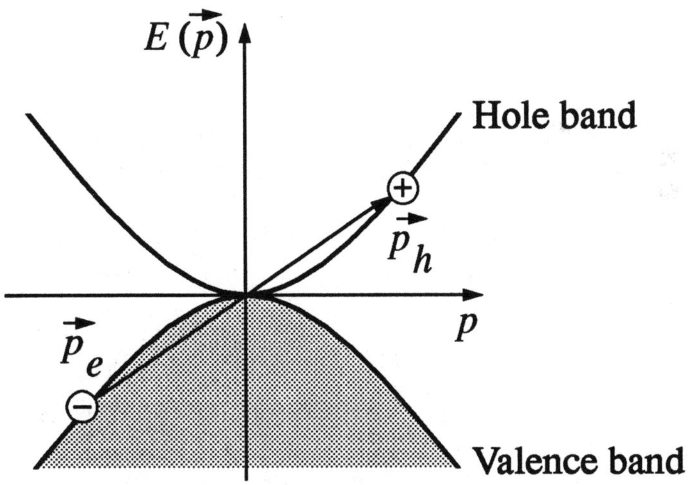

Figure 4.7.2 The upper half of the figure shows the hole band that simulates the dynamics of a hole. It is constructed by the inversion of the valence band through the origin.

We see from Eq.(4.7.12) that the energy of the hole has a minimum at the top of the valence band, and the corresponding curvature is positive, see Fig.4.7.2. This means that the effective mass of the hole is positive. The concept of a hole gives an alternative description of the electron valence band with a missing electron.

According to the rule (4.7.12), the Luttinger Hamiltonian (4.4.13) for an electron in the valence band becomes positive for the hole. The corresponding negative effective masses of electrons from Eq.(4.4.16) are positive effective masses of heavy and light holes.

Combining Eqs.(4.7.7) and (4.7.12), we find the velocity of the hole is equal to the velocity of the valence electron

$$\vec{v}_h = \vec{v}(\vec{p}_h) = \frac{dE(\vec{p}_h)}{d\vec{p}_h} = \frac{d[-E(\vec{p}_e)]}{d(-\vec{p}_e)} = \frac{dE(\vec{p}_e)}{d\vec{p}_e} = \vec{v}(\vec{p}_e). \tag{4.7.13}$$

Equation (4.7.3) states that the velocity of the hole equals the velocity of the electron in the valence band. The substitution of $\vec{v}(\vec{p}_e)$ from Eq.(4.7.13) into Eq.(4.7.9) and the use of Eq.(4.7.12) result in the final result for the hole current density

$$\vec{j}_h = (+e)\vec{v}(\vec{p}_h'). \tag{4.7.14}$$

The same substitution in Eq.(4.7.11) leads to

$$\vec{Q} = E(\vec{p}_h')v(\vec{p}_h'). \tag{4.7.15}$$

States with missing electrons are near the top of the valence band. Since the origin of the electron energy is taken at the top of the valence band, the hole energy equals

$$E(\vec{p}_h) = E_v - \frac{p_e^2}{2m_e} = E_v + \frac{p_h^2}{2m_h}. \tag{4.7.16}$$

In considering transport phenomena in semiconductors or optical properties of semiconductors, the concept of holes is very profitable, since it is very convenient to speak about a gas of electrons in the conduction band and a gas of holes in the hole band.

The equation of motion for the hole is easily obtained from the equation of motion for the electron Eq.(2.5.17) after substitution of $\vec{p}_e$ from Eq.(4.7.7) and $\vec{v}(\vec{p}_e)$ from Eq.(4.7.13):

$$\frac{d\vec{p}_h}{dt} = (+e)\,[\vec{\mathcal{E}} + (\vec{v}_h \times \vec{B})]. \tag{4.7.17}$$

It is seen from Eq.(4.7.17) that the equation of motion for a hole is the equation of a particle with a positive charge.

The direction of motion of an electron in the conduction band and a hole in the valence band in an external field $\vec{\mathcal{E}}$ is shown in Fig.4.7.3. The hole and the electron velocities are in opposite directions. Their electric currents are in the direction of the electric field. In cyclotron resonance experiments discussed in Section 4.2, the direction of circulation is opposite for electrons and holes.

The similarity in the description of electrons and holes is not complete.

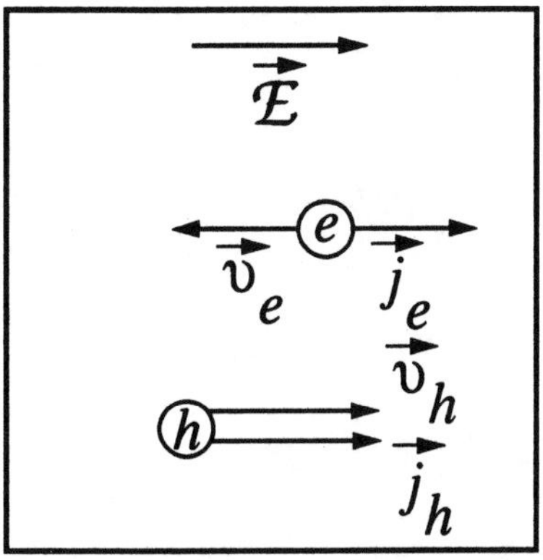

Figure 4.7.3 The direction of motion of an electron in the conduction band and a hole in the valence band in external electric field $\vec{\mathcal{E}}$.

Electrons are real particles that can exist in vacuum; but holes represent vacant electron orbitals in the valence band. They do not exist in vacuum and are often called ***quasiparticles.***

BIBLIOGRAPHY

1. J.F. Nye *Physical Properties of Crystals, their Representation by Tensors and Matrices* (Clarendon Press, Oxford, 1957).
2. G.L. Bir, G.E. Pikus *Symmetry and Strain Induced Effects in Semiconductors* (John Wiley & Sons, N.Y., 1974).
3. G. Bastard *Wave Mechanics Applied to Semiconductor Heterostructures* (John Wiley & Sons, N.Y., 1988).

CHAPTER 5

Lattice Vibrations

5.1 ATOMIC VIBRATIONS IN A THREE-DIMENSIONAL CRYSTAL LATTICE

Atoms of a crystal lattice are not at rest. They perform thermal vibrations, which consist of small displacements of all atoms from their equilibrium positions.

Let $\vec{u}_{ls}$ be the displacement of the s-th atom with mass M_{ls} in the l-th primitive cell of the crystal lattice. The conjugate momentum of the atom is $\vec{p}_{ls} = M_s \dot{\vec{u}}_{ls}$. The total kinetic energy of the crystal lattice vibrations has the form

$$T = \sum_{ls} \frac{\vec{p}_{ls}^{\,2}}{2M_s}. \tag{5.1.1}$$

The total potential energy of the crystal Φ is a function of the instantaneous positions of all atoms. It can be expanded in powers of the displacements u_{ls}^{α}, $\alpha = x, y, z$, of the atoms from their equilibrium sites

$$\Phi(\{u_{ls}^{\alpha}\}) = \Phi_0 + \sum_{ls\alpha} \left(\frac{\partial \Phi}{\partial u_{ls}^{\alpha}} \right)_0 u_{ls}^{\alpha} + \frac{1}{2} \sum_{ls\alpha, l's'\beta} \left(\frac{\partial^2 \Phi}{\partial u_{ls}^{\alpha} \partial u_{l's'}^{\beta}} \right)_0 u_{ls}^{\alpha} u_{l's'}^{\beta}. \tag{5.1.2}$$

Here Φ_0 is the first, constant, term in the expansion (5.1.2). The linear term in the expansion is zero, because the derivative is evaluated in the equilibrium configuration, in which there is no force $F_{ls}^{\alpha} = -\left(\frac{\partial \Phi}{\partial u_{ls}^{\alpha}}\right)_0 = 0$ acting on an atom. The coefficients

$$\Phi^{\alpha\beta}(ls;l's') \equiv \left(\frac{\partial^2 \Phi}{\partial u_{ls}^{\alpha} \partial u_{l's'}}\right)_0$$

are called ***atomic force constants.*** The approximation which retains terms of up to the second order in the atomic displacements in the expansion Eq.(5.1.2) is called the ***harmonic approximation***. The neglected terms of higher order in $\vec{u}_{ls}$ are called ***anharmonic***. They are assumed to be small. The Hamiltonian of the atomic vibrations is

$$H = T + \Phi . \tag{5.1.3}$$

The force constants satisfy some general symmetry relations that follow from the symmetry of the potential energy Φ.

a. Since $\Phi^{\alpha\beta}(ls;l's')$ are coefficients in the quadratic expansion (5.1.2), they are symmetric in the indices $ls\alpha$ and $l's'\beta$,

$$\Phi^{\alpha\beta}(ls;l's') = \Phi^{\beta\alpha}(l's';ls). \tag{5.1.4}$$

The relations (5.1.4) reduce the number of independent force constants $\Phi^{\alpha\beta}(ls;l's')$.

b. There are also some restrictions imposed on $\Phi^{\alpha\beta}(ls;l's')$ by the invariance of the potential energy against a rigid body displacement of the crystal as a whole. A rigid body displacement means that all atoms of the crystal lattice are shifted through the same displacement v^{α} in a certain direction

$$u_{ls}^{\alpha} = v^{\alpha}. \tag{5.1.5}$$

The potential energy Eq.(5.1.2) should be the same under the displacement defined by Eq.(5.1.5), that is $\Phi(\{u^{\alpha}_{ls}+v^{\alpha}\}) = \Phi(\{u^{\alpha}_{ls}\})$. This means that the contribution to the potential energy Eq.(5.1.2) from the displacement (5.1.5) should vanish:

$$\sum_{ls\alpha, l's'\beta} \Phi^{\alpha\beta}(ls;l's') v^{\alpha} v^{\beta} \equiv \sum_{\alpha\beta} v^{\alpha} v^{\beta} \sum_{ls, l's'} \Phi^{\alpha\beta}(ls;l's') = 0.$$

Since v^{α} is an arbitrary displacement, one obtains

$$\sum_{ls, l's'} \Phi^{\alpha\beta}(ls;l's') = 0. \tag{5.1.6}$$

c. The force in the α-direction acting on the atom ls in the lattice is equal to

$$F_{ls}^{\alpha} = -\frac{\partial \Phi}{\partial u_{ls}^{\alpha}} = -\sum_{l's'\beta} \Phi^{\alpha\beta}(ls;l's') u_{l's'}^{\beta} .$$

This force is also invariant with respect to the rigid body displacement Eq.(5.1.5). This means that

$$\sum_{l's'} \Phi^{\alpha\beta}(ls;l's') = 0, \tag{5.1.7}$$

which is a stronger condition on the atomic force constant than Eq.(5.1.6). This condition is called the ***condition of equilibrium***.

d. The lattice potential energy Φ does not change under a displacement of the crystal through any translation vector $\vec{R}_l$ from Eq.(2.2.1). This means that the atomic force constants $\Phi^{\alpha\beta}(ls;l's')$ depend only on the difference of the primitive cell coordinates $l - l' \equiv \vec{R}_l - \vec{R}_{l'}$, that is

$$\Phi^{\alpha\beta}(ls;l's') = \Phi^{\alpha\beta}(l - l';ss'). \tag{5.1.8}$$

e. There are also some extra restrictions on the atomic force constants that follow from the space symmetry of each crystal. If S is a symmetry operation of the crystal, the following relation should hold

$$S\Phi = \Phi(S^{-1}\{u_{ls}^{\alpha}\}) = \Phi. \tag{5.1.9}$$

From the Hamiltonian $H = T + \Phi$ one can obtain the classical equations of motion for the crystal lattice following the usual procedure of classical mechanics, $M_s\ddot{u}_{ls}^{\alpha} = -\partial H/\partial u_{ls}^{\alpha}$,

$$M_s\ddot{u}_{ls}^{\alpha} = -\sum_{l's'\beta} \Phi^{\alpha\beta}(ls;l's')u_{l's'}^{\beta}. \tag{5.1.10}$$

The right hand side of Eq.(5.1.10) is an elastic force which is proportional to the atomic displacement $\vec{u}_{ls}$. Equation (5.1.10) is a linear system of $3Nr$ equations for finding $3Nr$ atomic displacements $\vec{u}_{ls}$, where N is the number of primitive cells in the crystal and r is the number of atoms in the primitive cell. $3Nr$ is the total number of freedom degrees of the crystal lattice.

We now turn to the solution of Eq.(5.1.10) for a crystal with simple cubic symmetry with a single atom in the primitive cell. The displacement of the atom from the equilibrium position is now $\vec{u}_l$. It is labeled only by the index of the primitive cell l. The equations of motion (5.1.10) take the form

$$M\ddot{u}_l^{\alpha} = -\sum_{l'\beta} \Phi^{\alpha\beta}(l - l')u_{l'}^{\beta}. \tag{5.1.11}$$

Cubic symmetry with three orthogonal three-fold axes C_3 makes the second rank force constant tensor $\Phi^{\alpha\beta}(l - l')$ diagonal, see Appendix 1,

$$\Phi^{\alpha\beta}(l - l') = \Phi(l - l')\delta_{\alpha\beta}. \tag{5.1.12}$$

The condition (5.1.7), which expresses the invariance of the crystal lattice potential energy against a rigid body displacement reduces to

$$\sum_{l'} \Phi(l - l') = 0. \tag{5.1.13}$$

Taking into account Eq.(5.1.13), one obtains from Eq.(5.1.11) the equations of the motion in the form

$$M\ddot{\vec{u}}_l = \sum_{l'} \Phi(l - l')(\vec{u}_{l'} - \vec{u}_l). \tag{5.1.14}$$

We seek the stationary solution of this equation in the form of a traveling wave

$$\vec{u}_l = \vec{u}(\vec{q})e^{i\vec{q}\cdot\vec{R}_l - i\omega t}. \tag{5.1.15}$$

Here $\vec{q}$ is the ***quasiwave vector*** of the lattice vibration, ω is the frequency of vibration and the amplitude of the vibration $\vec{u}(\vec{q})$ is independent of l. The substitution of $\vec{u}_l$ from Eq.(5.1.15) in Eq.(5.1.14) results in

$$M\omega^2\vec{u}(\vec{q}) = \sum_{l'} \Phi(l-l')\left[e^{i\vec{q}\cdot(\vec{R}_{l'}-\vec{R}_l)} - 1\right]\vec{u}(\vec{q}). \tag{5.1.16}$$

For each value of $\vec{q}$ this is a system of three equations for the three components of the vector $\vec{u}(\vec{q})$. Equation (5.1.16) has nonvanishing solutions when the frequency of the lattice vibration is

$$\omega^2 = \frac{1}{M}\sum_{l'} \Phi(l-l')\left[e^{i\vec{q}\cdot(\vec{R}_{l'}-\vec{R}_l)} - 1\right].$$

In an infinite lattice, the indices l and $(l-l')$ take the same set of values. Therefore, the sum over l' can be changed to the sum over $(l-l')$

$$\omega^2 = \frac{1}{M}\sum_{l} \Phi(l)\left[e^{-i\vec{q}\cdot\vec{R}_l} - 1\right]. \tag{5.1.17}$$

Equation (5.1.17) relates the possible frequencies of the lattice vibrations to the quasiwave vector $\vec{q}$. A relation of this type is called the a ***dispersion law.***

In the simple cubic lattice each atom is at a center of inversion. This means that $\Phi(l)$ is an even function of l

$$\Phi(l) = \Phi(-l). \tag{5.1.18}$$

This relation allows to represent $\omega^2(\vec{q})$ as a real quantity

$$\omega^2(\vec{q}) = \frac{1}{M}\sum_{l} \Phi(l)(\cos\vec{q}\cdot\vec{R}_l - 1). \tag{5.1.19}$$

It is seen from Eq.(5.1.19) that dispersion law defines the frequency $\omega^2(\vec{q})$ to be a periodic function of $\vec{q}$ with the periodicity of the reciprocal lattice,

$$\omega^2(\vec{q}) = \omega^2(\vec{q}+\vec{K}) \tag{5.1.20}$$

Therefore, the quasiwave vector $\vec{q}$ in the crystal is not a single-valued quantity, because $\vec{q}$ and $\vec{q}+\vec{K}$ are equivalent. This equality is a direct result of the translational symmetry of the crystal lattice. The dispersion law Eq.(5.1.19) should be plotted in reciprocal space, within the primitive cell of the reciprocal lattice, that is within the 1st Brillouin zone

$$-\frac{\pi}{a} \le q_i \le \frac{\pi}{a}, i = x, y, z. \tag{5.1.21}$$

If we replace $\vec{q}$ by $-\vec{q}$ in Eq.(5.1.19), we see that $\omega^2(-\vec{q})$ is an even function of $\vec{q}$

$$\omega^2(\vec{q}) = \omega^2(-\vec{q}). \tag{5.1.22}$$

This equality is a result of the time reversal symmetry.

The allowed values of the quasiwave vector $\vec{q}$ are defined for a crystal with edges $G_1\vec{a}_1$, $G_2\vec{a}_2$, $G_3\vec{a}_3$ by periodic boundary conditions imposed on the atomic displacements given by Eq.(5.1.15). Similar boundary conditions have been discussed for electron wave functions in Section 2.4. The allowed values of the quasiwave vector $\vec{q}$ are

$$q_i = \frac{2\pi}{G_i a_i} m_i\,, \tag{5.1.23}$$

where $i = 1,2,3$.

We consider now the solution for longwavelength vibrations with small values of $\vec{q}$ (long wavelength limit)

$$qa \ll 1. \tag{5.1.24}$$

Using the small parameter defined by Eq.(5.1.24) we can expand the cosine in Eq.(5.1.19) to obtain

$$\omega^2(\vec{q}) = \frac{1}{2M}\sum_l (\vec{q}\cdot\vec{R}_l)^2 \Phi(l). \tag{5.1.25}$$

One can define the unit vector $\vec{n}$ in the direction of the quasiwave vector $\vec{q}$

$$\vec{n} = \frac{\vec{q}}{q}. \tag{5.1.26}$$

The dispersion law Eq.(5.1.25) takes the following form in terms of $\vec{n}$:

$$\omega^2(\vec{q}) = \frac{1}{2M} q^2 \sum_l (\vec{n}\cdot\vec{R}_l)^2 \Phi(l) = s^2(\vec{n}) q^2, \tag{5.1.27}$$

where

$$s^2(\vec{n}) = \frac{1}{2M}\sum_l (\vec{n}\cdot\vec{R}_l)^2 \Phi(l). \tag{5.1.28}$$

For small values of $\vec{q}$, we obtain a linear dependence of the frequency of lattice vibrations on the modulus of the quasiwave vector $\vec{q}$

$$\omega = s(\vec{n}) q\,, \tag{5.1.29}$$

which is known for anisotropic continuous media. It represents the frequency of a ***sound wave*** or ***acoustic wave*** with the phase velocity $s(\vec{n})$. The result is clear, since small values of q from Eq.(5.1.24) correspond to the large wavelengths $\lambda = 2\pi/q$. The inequality (5.1.24) is equivalent to the requirement $\lambda \gg a$ that defines the transition from discrete crystal lattice dynamics to the mechanics of continuous media. Atoms of the crystal lattice move in an acoustic mode of vibration as it is shown in Fig.5.1.1a.

In the longwavelength limit $q \to 0$ the frequency Eq.(5.1.29) vanishes. Because the equations of motion (5.1.16) are a three-dimensional system of equations for finding the vector amplitudes $\vec{u}(\vec{q})$, there are three acoustic waves with vanishing frequencies in the longwavelength limit. The three acoustic waves differ in their velocities and polarizations. In an isotropic continuum, one of the vibrations

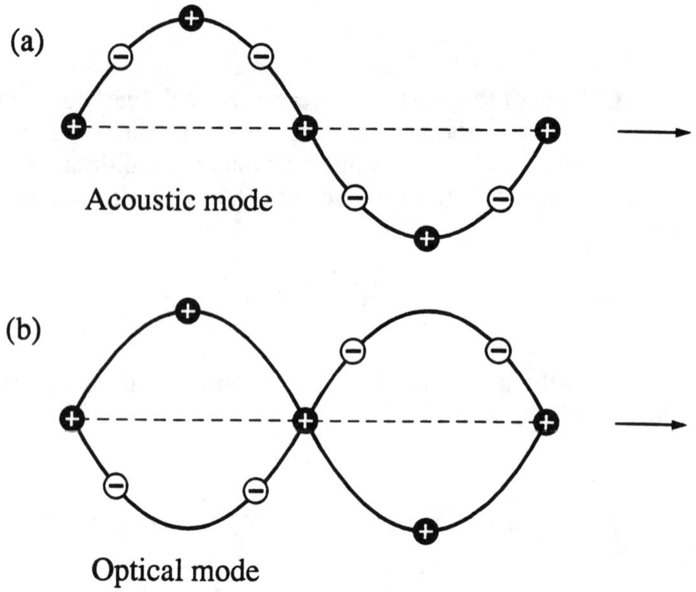

Figure 5.1.1 (a) Transverse acoustic displacements, and (b) transverse optical displacements in a diatomic linear chain.

is a longitudinal acoustic wave. This means that the atomic displacements in this vibration are parallel to the direction of the wave propagation $\vec{n}$. The dependence of frequency on the quasiwave vector in this case is

$$\omega_{LA} = s_L q,$$

where s_L is the logitudinal sound velocity.

Two other vibrations have the same velocity and transverse polarization. This means that the atoms move in the plane normal to $\vec{n}$. The frequencies of the two transverse waves are equal to

$$\omega_{TA} = s_T q.$$

It is shown in the mechanics of continuous media that the longitudinal wave has a higher velocity than the transverse wave

$$s_L > s_T.$$

The schematic dispersion law for acoustic waves is shown in Fig 5.1.2 in the 1st Brillouin zone. The corresponding surfaces of the constant frequency in an isotropic medium are spheres of different radii defined by s_L and s_T, see Fig.5.1.3a. In real crystals, the surfaces of the constant frequency are anisotropic as it is shown in Fig.5.1.3b.

The frequency Eq.(5.1.29) has been obtained with the help of the equilibrium condition Eq.(5.1.13) which follows in turn from the invariance of the crystal lattice against three rigid body displacements. This means that the existence of ***three zero frequencies*** in the longwavelength limit is a ***direct consequence of rigid body displacement*** invariance.

For an arbitrary value of q $(aq{\sim}1)$, the dispersion law is defined by the

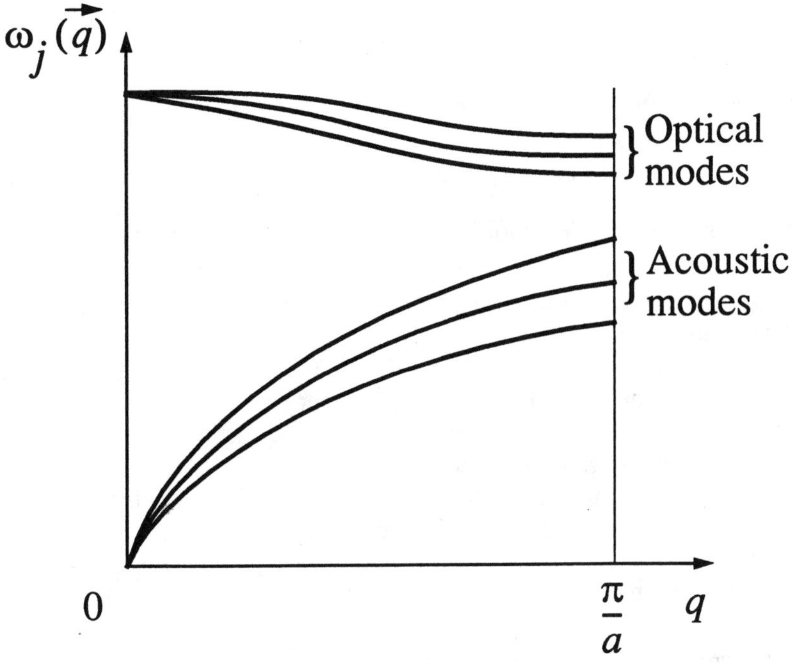

Figure 5.1.2 Schematic plot of $\omega_j(q)$ for a diatomic lattice for an arbitrary direction of $\vec{q}$ in the Brillouin zone.

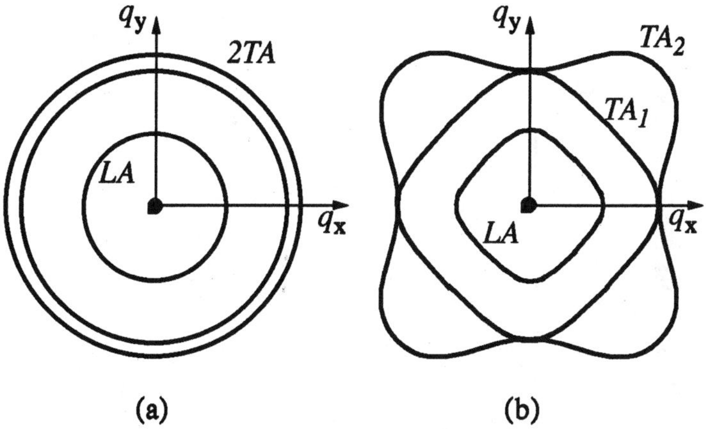

Figure 5.1.3 Constant frequency surfaces (a) for an isotropic medium, and (b) for a real crystal.

atomic force constants $\Phi(l)$ and by Eq.(5.1.19). Since the crystal lattice has a finite number of degrees of freedom, $3Nr$, the possible frequencies of vibrations lie within a band of finite width. In order of magnitude the maximum frequency equals

$$\omega_{max} \sim \frac{s}{a} \sim 10^3 s^{-1}, \qquad (5.1.30)$$

where $a \sim 10^{-8}$ $cm = 10^{-10}$ m and $s \sim 10^3$ m/s.

Near the maximum frequency ω_{max}, another expansion of Eq.(5.1.19) is possible

$$\omega^2 = \omega_{max}^2 - \sum_{\alpha\beta} \gamma_{\alpha\beta}(q_\alpha - q_{max\,\alpha})(q_\beta - q_{max\,\beta}), \quad (5.1.31)$$

where q_{max} is defined by the condition $\omega(q_{max}) = \omega_{max}$ and $\gamma_{\alpha\beta}$ is a positive definite second rank tensor. There is no linear term $(\vec{q} - \vec{q}_{max})$ in the expansion (5.1.31), because ω_{max} is by definition the maximum frequency. The dispersion law Eq.(5.1.31) is quadratic one.

In semiconductor crystals such as Si, Ge, GaAs, there are two atoms in a primitive cell. This means that there are altogether $3r$=6 modes in the vibrational spectrum of the crystal. Three of these modes are acoustic or sound waves. The other three are called ***optical vibrations.*** Atoms of the crystal lattice move in an optical mode of vibration as it is shown in Fig.5.1.1b. The physical properties of optical vibrations are discussed below in Section 5.2. The schematic dispersion law for optical modes for a covalent crystal is shown in Fig.5.1.2. Figure 5.1.4 shows the phonon dispersion curves in the [111] direction of Ge at $T = 0K$, for which

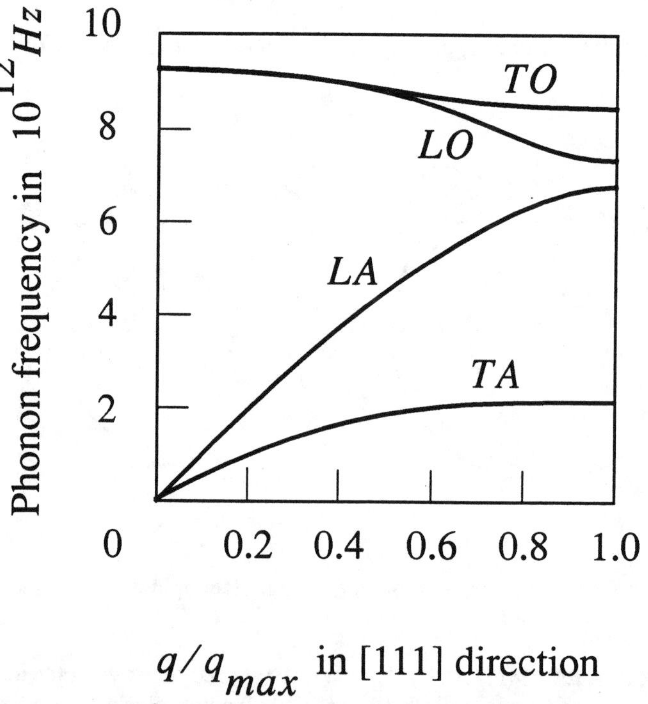

Figure 5.1.4 The phonon dispersion relation in the [111] direction in germanium at 80 K, $q_{max} = \frac{2\pi}{a}\left(\frac{1}{2}, \frac{1}{2}, \frac{1}{2}\right)$ (After C. Kittel "Introduction to Solid State Physics" John Wiley & Sons, N.Y., 1986).

$q_{max} = \frac{2\pi}{a}\left(\frac{1}{2}, \frac{1}{2}, \frac{1}{2}\right)$. There is a single *LA* mode and two degenerate *TA* modes. The others are optical modes.

When a superlattice is fabricated from two different semiconductor crystals, see Fig.2.8.1, the vibrational spectrum of the superlattice appears to be different from the vibrational spectrum of the starting crystals. This happens because the macroscopic period of the superlattice is superimposed on the translational period of the crystal, which results in the reduction of the 1st Brillouin zone.

It is instructive to consider the transition from the monoatomic one-dimensional chain shown in Fig.5.1.5a to the chain with doubled periodicity shown in Fig.5.1.5b. The monoatomic chain has the single acoustic mode shown in Fig.5.1.6a in the 1st Brillouin zone for values of the wave vector from $-\pi/a$ up to $+\pi/a$. In the case of a diatomic lattice, the number of vibration modes is doubled. If the primitive translation of the monoatomic lattice is a, the primitive translation of the doubled lattice is $2a$. In going from the lattice in Fig.5.1.5a to the lattice in Fig.5.1.5b, we are doubling the length of the primitive cell. The size of the Brillouin zone is halved, correspondingly. The acoustic branch of Fig.5.1.6a in the halved 1st Brillouin zone from $-\pi/(2a)$ up to $+\pi/(2a)$ creates an optical branch by the shift of parts *BC* and *B'C'* of the acoustic branch through the minimum reciprocal lattice vector of the lattice with the doubled primitive cell $K = \pi/(2a)$. We see in Fig.5.1.6b that the reduction of the 1st Brillouin zone results in a new optical branch.

In a superlattice, the 1st Brillouin zone is reduced to a small part of the 1st Brillouin zone of a monoatomic lattice. The new Brillouin zone extends from $-\pi/d$ up to $+\pi/d$, where $d \gg a$ is the period of the superlattice. The acoustic branch gives

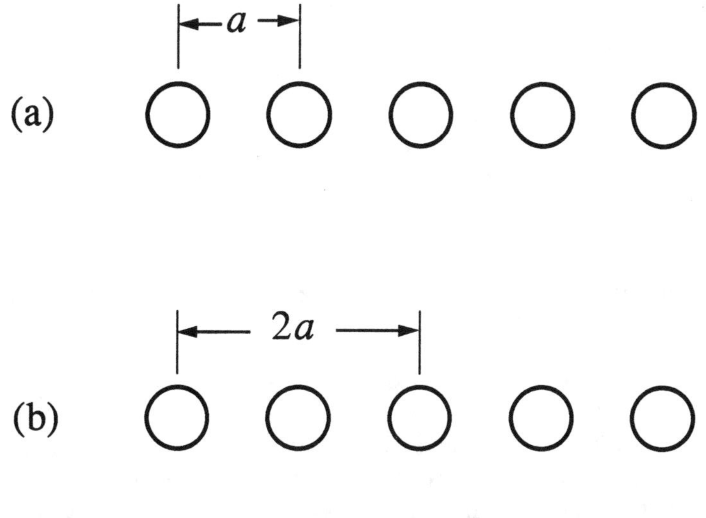

Figure 5.1.5 One-dimensional (a) monatomic linear chain, and (b) monatomic linear chain, treated as a diatomic chain.

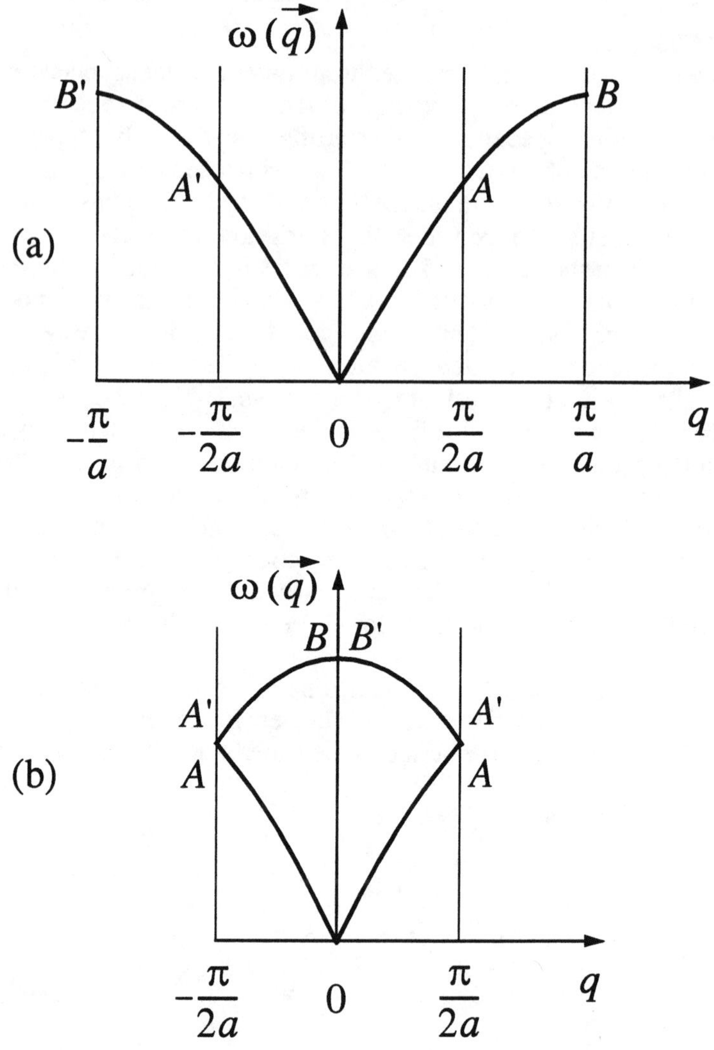

Figure 5.1.6 Vibration frequencies of (a) monatomic linear chain and (b) monatomic linear chain, treated as a diatomic chain.

rise to the family of new optical modes shown in Fig.5.1.7. These new optical modes are called ***folded modes***.

The interfaces between different layers of the superlattice can create some extra modes of vibration which exist inside the separate potential wells of the superlattice. These modes are similar to standing waves, and are called ***confined modes***.

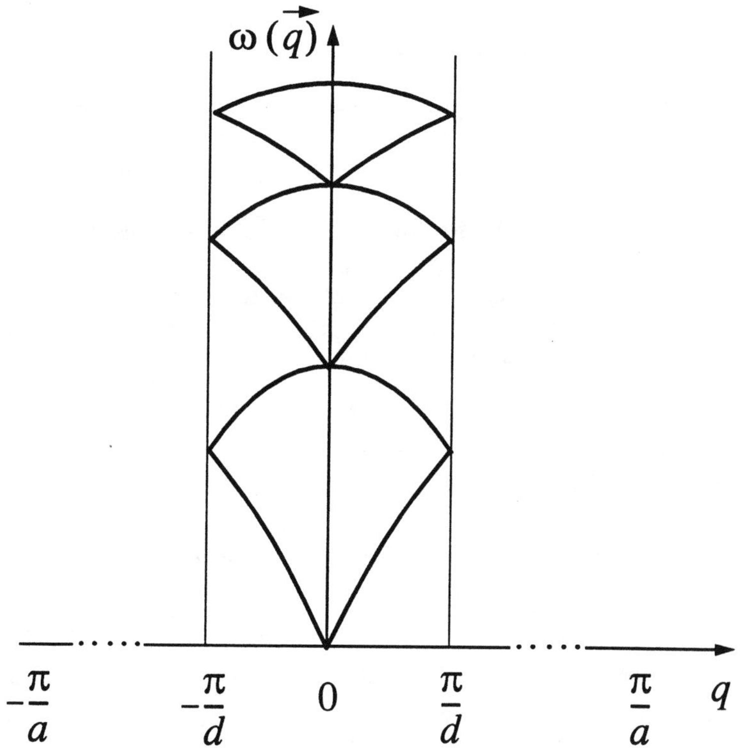

Figure 5.1.7 Folded vibrational modes in quantum well.

5.2 LATTICE DYNAMICS OF IONIC CRYSTALS

In most of the semiconductors compounds $A^{III}B^{V}$, $A^{II}B^{VI}$ the covalent bond in crystals is completed by some percent of the ionic bond. These crystals do not have a center of inversion, and lattice vibrations are performed by ions of opposite sign. The Coulomb interaction of charged ions should be taken into account in the lattice dynamics of these crystals, along with the elastic forces considered in Section 5.1.

A direct attempt to include the Coulomb interaction of charged ions in the definition of the force constants leads to the result:

$$\Phi^{\alpha\beta}(ls;l's') = -\frac{\partial^2}{\partial r^{\alpha}\partial r^{\beta}}\phi_{ss'}(r)\,|_{r=|\vec{R}_{ls}-\vec{R}_{l's'}|}\,. \tag{5.2.1}$$

Here $\phi_{ss'}$ is the pair Coulomb interaction of atoms in the lattice

$$\phi_{ss'} = -\frac{e_s^* e_{s'}^*}{4\pi\varepsilon_0\,|\,r\,|}\,, \tag{5.2.2}$$

where $e_s^*, e_{s'}^*$ are effective charges of the two types of ions in the crystal. The substitution of Eq.(5.2.2) into Eq.(5.2.1), and then into $\sum_l \Phi^{\alpha\beta}(l)$ given by Eq.(5.1.10), leads to an infinite value of $\sum_l \Phi^{\alpha\beta}(l)$ (a logarithmic singularity of the type of $\int d^3r/r^3$). The Coulomb potential decreases so slowly that the usual expansion in Eq.(5.1.2) is not valid. Let us discuss a more rigorous approach.

In each primitive cell of a III-V or II-VI crystal, one of the ions has an electric charge $Z > 4$ and the other has an effective charge $Z < 4$. When performing the optical vibration, these ions oscillate in opposite directions in each primitive cell creating dipoles and the polarization $\vec{P}$ of the medium. Therefore, an internal electric field $\vec{\mathcal{E}}$ is present in an ionic crystal.

We shall consider the long wavelength vibrations of an ionic crystal with two atoms in the primitive cell. The internal electric field $\vec{\mathcal{E}}$ created by the charged ions will be taken into consideration as an extra dynamical variable along with the displacements $\vec{u}_{ls}$. In the long wavelength limit ($q \to 0$), we should take into account the static macroscopic field $\vec{\mathcal{E}}$ which interacts with the macroscopic dipole moment of the crystal $\vec{M}$. This means that in the potential energy of the crystal, we should take into account the extra energy of the crystal dipole moment $\vec{M}$ in the electric field $\vec{\mathcal{E}}$

$$\Phi = \Phi_0 + \sum_{ls\alpha, l's'\beta} \Phi^{\alpha\beta}(ls;l's')u_{ls}^{\alpha}u_{l's'}^{\beta} - \sum_{\mu} \mathcal{E}_{\mu}M_{\mu}\,, \tag{5.2.3}$$

where $\mu = x, y, z$. The dipole moment $\vec{M}(\{u_{ls}^{\alpha}\})$ depends on the atomic displacements. Expanding $\vec{M}$ in u_{ls}^{α} up to terms linear in u_{ls}^{α}, we find in linear approximation

$$M_{\mu} = M_{\mu}^{(0)} + \sum_{ls\alpha} M_{\mu\alpha}(ls)u_{ls}^{\alpha}\,. \tag{5.2.4}$$

The expansion coefficients $M_{\mu\alpha}(ls)$, which are elements of a second rank tensor, are called the ***first-order dipole moments***. $M_{\mu\alpha}(ls)$ has the dimensions of charge, and has the physical meaning of the charge of the (ls)–th atom.

We shall consider semiconductor crystals with the high symmetry of a cube or tetrahedron O_h, T_d, which does not allow them to have an electric dipole moment in equilibrium, when $\vec{u}_{ls} = 0$. Therefore, the term $M_{\mu}^{(0)}$ vanishes in Eq.(5.2.4). The substitution of Eq.(5.2.4) into the potential energy (5.2.3) results in the following total potential energy

$$\Phi = \Phi^{(0)} + \sum_{ls\alpha, l's'\beta} \Phi^{\alpha\beta}(ls;l's')u_{ls}^{\alpha}u_{ls}^{\beta} + \sum_{\mu, ls\alpha} M_{\mu\alpha}(ls)u_{ls}^{\alpha}\mathcal{E}_{\mu}\,. \tag{5.2.5}$$

The potential energy Eq.(5.2.5) should satisfy the translational symmetry of the crystal. It follows from this requirement that $M_{\mu\alpha}(ls)$ is independent of l. The invariance of the crystal potential energy against a rigid body displacement results in a condition that is similar to Eq.(5.1.7),

$$\sum_s M_{\mu\alpha}(s) = 0. \tag{5.2.6}$$

The Hamiltonian of the crystal consists of the kinetic energy, Eq.(5.1.1), and the potential energy, Eq.(5.2.5): $H = T + \Phi$. Since the total set of dynamical variables now consists of the atomic displacements $\{u_{ls}^{\alpha}\}$ and the internal electric field $\vec{\mathcal{E}}$, the equations of motion have the form

$$\dot{p}_{ls}^{\alpha} = -\frac{\partial H}{\partial u_{ls}^{\alpha}}, \tag{5.2.7a}$$

and the polarization of the medium is

$$\vec{P} = -\frac{\partial H}{\partial \vec{\mathcal{E}}}. \tag{5.2.7b}$$

Substituting H into Eqs.(5.2.7a,b) results in the equations of motion

$$M_s \ddot{u}_{ls}^{\alpha} = -\sum_{l's'\beta} \Phi^{\alpha\beta}(ls;l's')u_{l's'}^{\beta} + \sum_{\mu} M_{\mu\alpha}(s)\mathcal{E}_{\mu}.$$

The first term on the right side of this equation is the elastic force which has been considered before. The second term results from the long range forces and depends on the electric field $\vec{\mathcal{E}}$. We are seeking for solutions of these equations with a harmonic time dependence $u_{ls}^{\alpha} \sim e^{-i\omega t}$, see Eq.(5.1.10). Then, the time-independent equations of motion take the form

$$M_s \omega^2 u_{ls}^{\alpha} = \sum_{l's'\beta} \Phi^{\alpha\beta}(ls;l's')u_{l's'}^{\beta} - \sum_{\mu} M_{\mu\alpha}(s)\mathcal{E}_{\mu}. \tag{5.2.8}$$

We consider for simplicity a cubic ionic crystal with two atoms in the primitive cell, see Fig 5.2.1. The index s takes only two values: $s = +,-$. Cubic symmetry

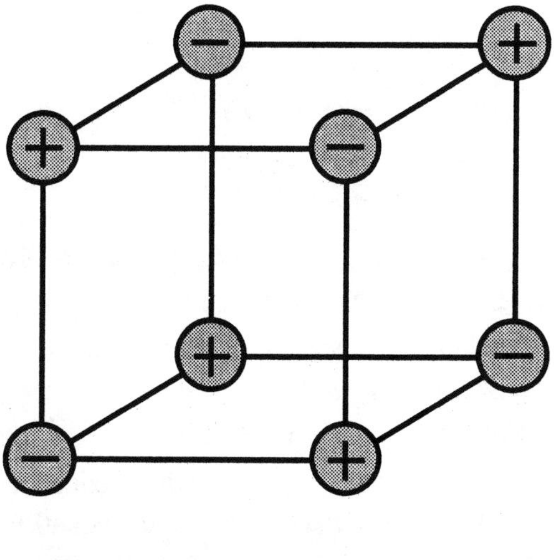

Figure 5.2.1 Ionic crystal of NaCl type.

makes the force constant matrix to be diagonal: $\Phi^{\alpha\beta}(ls;l's') = \Phi(ls;l's')\delta_{\alpha\beta}$. In the simplest approximation, we assume that the short range force constants $\Phi(ls;l's')$ connect nearest neighbor ions only. There remain only two force constants $\Phi(++) = \Phi(--)$ and $\Phi(+-) = \Phi(-+)$ in this case.

Generally speaking, the quantity $M_{\mu\alpha}$ is a second rank tensor. In crystals of cubic and tetrahedral symmetry the corresponding matrix is diagonal, that is

$$M_{\mu\alpha}(s) = e_s^* \delta_{\mu\alpha}, \tag{5.2.9}$$

where e_s^* is the effective charge of the ion of s-type in the primitive cell, and the $\delta_{\mu\alpha}$ is the Kronecker delta. Substituting Eq.(5.2.9) in Eq.(5.2.6), and representing the sum over s in explicit form, one obtains

$$e_+^* = -e_-^*. \tag{5.2.10}$$

This is the neutrality condition for the primitive cell.

We are seeking solutions of the equations of lattice vibrations in the long-wavelength limit ($q \rightarrow 0$). In this limit, the atomic displacements u_{ls}^{α} and the force constants $\Phi(ls;l's')$ do not depend on the primitive cell indices l and l'

$$\vec{u}_{ls} = \vec{u}_s,$$

and

$$\Phi(ls;l's') = \Phi(s;s').$$

The two nonzero force constants are related by the equilibrium condition (5.1.7):

$$\Phi(++) = \Phi(--) = -\Phi(+-) = -\Phi(-+). \tag{5.2.11}$$

Substituting Eqs.(5.2.9), (5.2.10), and (5.2.11) in Eq.(5.2.8) we find the equations of motion of the negative and positive ions in a very simple form

$$\omega^2 M_+ \vec{u}_+ = \Phi(\vec{u}_+ - \vec{u}_-) - e^* \vec{\mathcal{E}}, \tag{5.2.12a}$$

$$\omega^2 M_- \vec{u}_- = \Phi(u_- - u_+) + e^* \vec{\mathcal{E}}, \tag{5.2.12b}$$

$$\vec{P} = \frac{N}{2V} e^* (u_+ - u_-). \tag{5.2.12c}$$

Here N is the total number of atoms in the lattice. The long-range Coulomb forces are represented by the internal electrostatic electric field $\vec{\mathcal{E}}$. Because the condition (5.2.10) has already been taken into account, the field $\vec{\mathcal{E}}$ has an opposite effect on the two different ions in the primitive cell.

In Eq.(5.2.12), we have four unknown quantities $\vec{u}_+$, $\vec{u}_-$, $\vec{\mathcal{E}}$, and $\vec{P}$. To close the system of equations, we need one more equation which connects $\vec{\mathcal{E}}$ and $\vec{P}$. In the case of long wavelength lattice vibrations this can be the equation of electrostatics

$$\nabla \cdot \vec{\mathcal{E}} = -\nabla \cdot \vec{P}. \tag{5.2.12d}$$

Equations (5.2.12) have $3r = 6$ solutions $\omega_j(0)$. To find them it is convenient to consider separately the symmetric and antisymmetric combinations of displacements

$$u_s = \frac{u_+ + u_-}{2} \quad \text{and} \quad u_a = \frac{u_+ - u_-}{2}. \tag{5.2.13}$$

Multiplying Eqs.(5.2.12a,b) by M_- and M_+, respectively, and subtracting the second equation from the first, we obtain the equations for antisymmetric vibrations.

$$\omega^2 \frac{M_+ M_-}{M_+ + M_-} \vec{u}_a = \Phi \vec{u}_a - e^* \vec{\mathcal{E}}, \tag{5.2.14a}$$

$$\vec{P} = \frac{1}{2v_0} e^* \vec{u}_a, \tag{5.2.14b}$$

where $v_0 = V/N$ is the volume of the primitive cell.

The antisymmetric mode is connected with the electrostatic field $\vec{\mathcal{E}}$, since two atoms moving against each other in the electric field, $\vec{\mathcal{E}}$, produce a dipole and a macroscopic polarization $\vec{P}$. Substituting $\vec{P}$ from Eq.(5.2.14b) in Eq.(5.2.12d), one obtains

$$\nabla \cdot \vec{\mathcal{E}} + \frac{1}{2v_0} e^* \nabla \cdot \vec{u}_a = 0. \tag{5.2.15}$$

It is known from vector algebra that any vector can be represented as a sum of longitudinal and transverse components

$$\vec{u} = \vec{u}_{\parallel} + \vec{u}_{\perp}, \tag{5.2.16}$$

where $\nabla \cdot \vec{u}_{\perp} = 0$ and $\nabla \times \vec{u}_{\parallel} = 0$. Therefore, since $\nabla \cdot \vec{u}_a = \nabla \cdot \vec{\mathcal{E}}_{\parallel}$,

$$\nabla \cdot \left(\vec{\mathcal{E}} + \frac{1}{2v_0} e^* \vec{u}_{a\parallel} \right) = 0, \tag{5.2.17a}$$

and, since the electrostatic field satisfies the equation $\nabla \times \vec{\mathcal{E}} = 0$, we have

$$\nabla \times \left(\vec{\mathcal{E}} + \frac{1}{2v_0} e^* \vec{u}_{a\parallel} \right) = 0. \tag{5.2.17b}$$

Because the divergence and the curl of the vector vanish, the vector itself must vanish. This means that

$$\vec{\mathcal{E}} = -\frac{1}{2v_0} e^* \vec{u}_{a\parallel} = \frac{1}{2v_0} e^* \frac{(u_+ - u_-)_{\parallel}}{2}. \tag{5.2.18}$$

It is seen that the internal electric field $\vec{\mathcal{E}}$ is defined by the relative displacement of the ions in the primitive cell.

The substitution of Eq.(5.2.18) into Eq.(5.2.14a) results in the longitudinal equation

$$\omega^2\mu\vec{u}_{\parallel} = \Phi\vec{u}_{\parallel} + \frac{1}{2v_0}(e^*)^2\vec{u}_{\parallel}, \tag{5.2.19}$$

and two transverse equations

$$\omega^2\mu\vec{u}_{\perp} = \Phi\vec{u}_{\perp}, \tag{5.2.20}$$

where $\mu = M_+M_-/(M_+ + M_-)$ is the reduced mass of the primitive cell. We see that in the longitudinal equation there is an extra force that originates from the internal electric field $\vec{\mathcal{E}}$. The corresponding frequencies of vibrations are the longitudinal optical (LO) and transverse optical (TO) mode frequencies respectively,

$$\omega_{TO}^2 = \frac{\Phi}{\mu}, \tag{5.2.21a}$$

$$\omega_{LO}^2 = \frac{\Phi}{\mu} + \frac{1}{2\mu v_0}(e^*)^2 = \omega_{TO}^2 + \Omega^2, \tag{5.2.21b}$$

where

$$\Omega^2 = \frac{(e^*)^2}{2v_0\mu} \tag{5.2.22}$$

is the plasma frequency corresponding to the reduced mass μ of the two atoms in the primitive cell.

The quantity Eq.(5.2.22) characterizes the splitting of the longitudinal and transverse optical frequencies. It is called the $LO-TO$ -splitting caused by the internal electrostatic field $\vec{\mathcal{E}}$. For transverse vibrations there is no electrostatic field $\vec{\mathcal{E}}$, and the corresponding frequency ω^2_{TO} is defined by the elastic force constant Φ.

Equations (5.2.12) also have solutions of acoustic type. The displacements in acoustic modes satisfy the condition (5.1.24), which takes the form $\vec{u}_+ = \vec{u}_-$ for our diatomic lattice. It is seen from Eq.(5.2.15) that the internal electric field vanishes in this case, and therefore does not affect the equations of motion for the acoustic modes. Multiplying Eqs.(5.2.12a) and (5.2.12b) by M_- and M_+, respectively, adding the resulting equations, and applying condition Eq.(5.1.24) results in the equations for the symmetric combination given by Eq.(5.2.11):

$$\omega^2\vec{u}_s = 0. \tag{5.2.23}$$

Corresponding to the three components of $\vec{u}_s$, we have three acoustic frequencies $\omega_j(\vec{q})$, which vanish in the limit $q \to 0$

$$\omega_j(0) = 0, \quad j = 1,2,3. \tag{5.2.24}$$

These are the acoustic modes of diatomic ionic crystal.

Equation (5.2.14a) allows us to find the dielectric permittivity of the crystal $\varepsilon(\omega)$. Solving this equation with respect to $\vec{u}_a$, one finds

$$\vec{u}_a = -\frac{e^*\vec{\mathcal{E}}}{\omega^2\mu - \Phi}. \tag{5.2.25}$$

Substituting Eqs.(5.2.25) and (5.2.21a) into Eq.(5.2.14b) results in

$$\vec{P} = \frac{1}{2v_0}\frac{(e^*)^2}{\mu}\frac{1}{\omega_{TO}^2 - \omega^2}\vec{\mathcal{E}}. \tag{5.2.26}$$

The dielectric permittivity is defined by the equation $\vec{D} = \varepsilon_\infty\vec{\mathcal{E}} + P$, where ε_∞ is the high frequency electronic contribution to ε. Taking $\vec{P}$ from Eq.(5.2.25), we find

$$\vec{D} = \varepsilon\vec{\mathcal{E}} = \varepsilon_\infty\vec{\mathcal{E}} + \vec{P} = \varepsilon_\infty\left(1 + \frac{(e^*)^2}{\varepsilon_\infty 2v_0\mu}\frac{1}{\omega_{TO}^2 - \omega^2}\right)\vec{\mathcal{E}}.$$

The dielectric permittivity therefore equals

$$\varepsilon(\omega) = \varepsilon_\infty\left(1 + \frac{\Omega^2}{\omega_{TO}^2 - \omega^2}\right) = \varepsilon_\infty\frac{\omega_{LO}^2 - \omega^2}{\omega_{TO}^2 - \omega^2}. \tag{5.2.27}$$

The frequency dependence of $\varepsilon(\omega)$ is shown in Fig. 5.2.2. It vanishes at $\omega^2 = \omega^2_{LO}$, and has a singularity at $\omega^2 = \omega^2_{TO}$.

Taking $\omega = 0$ in Eq.(5.2.26), we find very useful relation

$$\frac{\varepsilon(0)}{\varepsilon_\infty} = \frac{\omega_{LO}^2}{\omega_{TO}^2}. \tag{5.2.28}$$

This is the ***Lyddane-Sachs-Teller relation*** that shows the direct relation of the $LO-TO$-splitting to the high frequency (ε_∞) and static ($\varepsilon(0)$) dielectric permittivities. The $LO-TO$- splitting observed in the optical spectra of semiconductors is a

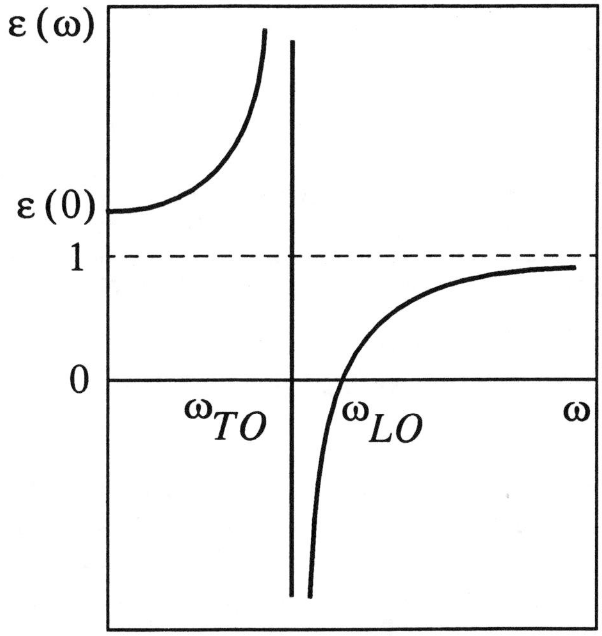

Figure 5.2.2 Plot of $\varepsilon(\omega)$ from Eq.5.2.24 for $\varepsilon(\infty) = 2$ and $\varepsilon(0) = 3$. $\varepsilon(\omega)$ is negative for $\omega_{TO} < \omega < \omega_{LO}$.

direct measure of the ionic effective charge and of the crystal ionicity. In the covalent crystals Ge or Si there is neither an effective charge nor an $LO-TO$-splitting.

It follows from Eq.(5.2.26) that in the interval $\omega_{TO} < \omega < \omega_{LO}$, where $\varepsilon(\omega)$ is negative, the crystal has an imaginary index of refraction n. The crystal reflects the incident light in this frequency region. The reflection spectrum of ionic materials has the very typical frequency dependence shown in Fig. 5.2.3. There is a wide band, corresponding to negative values of $\varepsilon(\omega)$, where crystals strongly reflect the incident light.

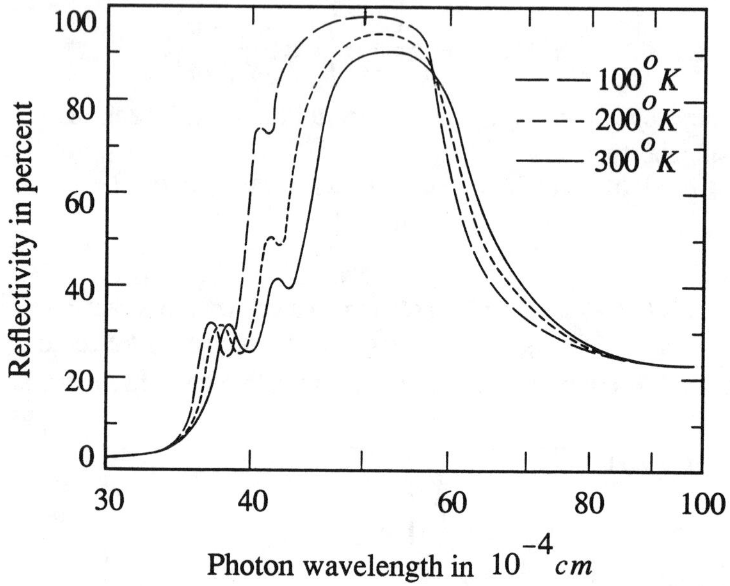

Figure 5.2.3 Reflectivity of a thick crystal of NaCl at several temperatures, versus the wavelength of light. The values of ω_{LO} and ω_{TO} at room temperature are $4.96\times 10^{11}\ s^{-1}$ and $3.09\times 10^{11}\ s^{-1}$ corresponding to a phonon wave length 38×10^{-4} cm and 61×10^{-4} cm. (After C. Kittel "Introduction to Solid State Physics", John Wiley & Sons, N.Y., 1986)

5.3 QUANTIZATION OF LATTICE VIBRATIONS

So far crystal lattice vibrations were considered using classical mechanics. The quantum mechanics approach is introduced in this section. The Hamiltonian of the lattice vibrations $H = T + \Phi$ is a quadratic form in p_{ls}^{α} and u_{ls}^{α} that can be diagonalized by the expansion of p_{ls}^{α} and u_{ls}^{α} in traveling plane waves. In certain new variables $p(\vec{q}j)$ and $s(\vec{q}j)$ the Hamiltonian becomes the sum of the energies of independent harmonic oscillators

$$H = \frac{1}{2}\sum_{\vec{q}j}[p^2(\vec{q}j) + \omega_j^2(\vec{q})s^2(\vec{q}j)]. \tag{5.3.1}$$

The oscillations described by the Hamiltonian (5.3.1) are called ***normal modes*** and $p(\vec{q}j)$ and $s(\vec{q}j)$ are called ***normal coordinates,*** $p(\vec{q}j) = \dot{s}(\vec{q}j)$. This representation of H allows its direct quantization. We must regard $\hat{s}(\vec{q}j)$ and $\hat{p}(\vec{q}j)$ as quantum-mechanical operators which obey the commutator relation discussed in Chapter 1, Section 1.8, namely:

$$\left[\hat{s}(\vec{q}j), \hat{p}(\vec{q}j')\right] = i\hbar\delta_{jj'}\delta_{qq'}. \tag{5.3.2}$$

It follows from Eq.(5.3.2) that

$$\hat{p}(\vec{q}j) = -i\hbar\frac{\partial}{\partial s(\vec{q}j)}. \tag{5.3.3}$$

The quantum-mechanical Hamiltonian takes the form of a sum over independent harmonic oscillator Hamiltonians

$$\hat{H} = \frac{1}{2}\sum_{\vec{q}j}\left[-\hbar^2\frac{\partial^2}{\partial s^2(\vec{q}j)} + \omega_j^2(\vec{q})s^2(\vec{q}j)\right]. \tag{5.3.4}$$

The corresponding Schrödinger equation has the form

$$\sum_{\vec{q}j}\left[-\frac{\hbar^2}{2}\frac{\partial^2}{\partial s^2(\vec{q}j)} + \frac{1}{2}\omega_j^2(\vec{q})s^2(\vec{q}j)\right]\Psi = E\Psi. \tag{5.3.5}$$

Since the Hamiltonian is the sum of Hamiltonians of individual oscillators, the problem (5.3.5) is solved by separation of variables

$$\Psi = \prod_{\vec{q}j}\psi_{n_j(\vec{q})}(s(\vec{q}j)),$$

$$E = \sum_{\vec{q}j}E_{n_j(\vec{q})}. \tag{5.3.6}$$

Here $n_j(\vec{q})$ is a quantum number that labels the energy level of the oscillator $(\vec{q}j)$. The Schrödinger equation for an individual mode is

$$\left\{-\frac{\hbar^2}{2}\frac{\partial^2}{\partial s^2(\vec{q}j)} + \frac{1}{2}\omega_j^2(\vec{q})s^2(\vec{q}j)\right\}\psi_{n_j(\vec{q})}(s(\vec{q}j)) = E_{n_j(q)}\psi_{n_j(\vec{q})}(s(\vec{q}j)). \tag{5.3.7}$$

This equation is solved exactly in quantum mechanics. The energy levels are

$$E_{n_j(\vec{q})} = \left[n_j(\vec{q}) + \frac{1}{2}\right]\hbar\omega_j(\vec{q}), \tag{5.3.8}$$

and the corresponding wave functions are equal to

$$\psi_{n_j}(s) = \left(\frac{\alpha_j}{\pi^{1/2}2^n n!}\right)^{1/2} e^{-\frac{1}{2}\alpha_j^2 s^2} H_n(\alpha_j s), \tag{5.3.9}$$

where $\alpha_j^2(\vec{q}) = \omega_j(\vec{q})/\hbar$ and $H_n(x)$ is the Hermite polynomial of n-th degree, n=0,1,2,3.... .

The total vibrational energy of the crystal equals

$$E_{\{n_j(\vec{q})\}} = \sum_{\vec{q}j}\left[n_j(\vec{q}) + \frac{1}{2}\right]\hbar\omega_j(\vec{q}). \tag{5.3.10}$$

The integer $n_j(\vec{q})$ is the number of quanta of vibrational energy $\hbar\omega_j(\vec{q})$ in the normal mode $(\vec{q}j)$. These quanta are called ***phonons.***

The average number of phonons in the mode $(\vec{q}j)$ is defined by a statistical average and is temperature dependent

$$< n(\omega_j(\vec{q})) >= \frac{\sum_{n_j(\vec{q})=0}^{\infty} n_j(\vec{q}) e^{-\frac{\left(n_j(\vec{q})+\frac{1}{2}\right)\hbar\omega_j(\vec{q})}{k_B T}}}{\sum_{n_j(\vec{q})=0}^{\infty} e^{-\frac{\left(n_j(\vec{q})+\frac{1}{2}\right)\hbar\omega_j(\vec{q})}{k_B T}}}. \tag{5.3.11}$$

Here <...> means statistical averaging. The sum can easily be evaluated as a power series, with the result that

$$< n(\omega_j(\vec{q})) > \; = \frac{1}{e^{\frac{\hbar\omega_j(\vec{q})}{k_B T}} - 1}. \tag{5.3.12}$$

We see that the equilibrium distribution of phonons in the normal mode $(\vec{q}j)$ is described by the ***Bose-Einstein distribution function***, Eq.(5.3.12), where the chemical potential of the phonons $\mu = 0$. This fact indicates that the average number of phonons is defined only by the temperature, moreover the creation and destruction of phonons equalize each other and does not require any external energy.

In the limit of low temperatures, $k_B T << \hbar\omega_j(\vec{q})$, the average number of phonons in the mode $(\vec{q}j)$ is

$$n(\omega_j(\vec{q})) = e^{-\frac{\hbar\omega_j(\vec{q})}{k_B T}}, \tag{5.3.13}$$

which means that there are few phonons in the mode $(\vec{q}j)$.

At high temperatures, $k_B T >> \hbar\omega_j(\vec{q})$,

$$n(\omega_j(\vec{q})) = \frac{k_B T}{\hbar\omega_j(\vec{q})}. \tag{5.3.14}$$

The average number of phonons in the mode $(\vec{q}j)$ is proportional to the temperature T.

BIBLIOGRAPHY

A.A. Maradudin, E. Montroll, J. Weiss, and I. Ipatova *Dynamics of the Crystal Lattice in Harmonic Approximation* (Academic Press, N.Y., 1971).

CHAPTER 6

Equilibrium Electrons and Holes in Semiconductors

6.1 INTRINSIC SEMICONDUCTORS

The most important feature of the electron band structure of pure semiconductor materials is the considerable energy gap ($\approx 1eV$) between the conduction and valence bands.

In thermal equilibrium, there is a certain distribution of electrons over the different quantum states. The Pauli exclusion principle, see Section 1.14, allows only double occupancy of each level (one electron for each spin). Therefore, in the absence of thermal excitations (at T = 0 K) electrons occupy the lowest energy levels. The valence band is completely filled with electrons and the conduction band is empty. The semiconductor behaves as an insulator.

At nonzero temperatures, thermal excitation causes electrons to jump from time to time across the energy gap from the valence to the conduction band, and to leave empty states (holes) in the valence band. Because the electrons and holes appear in pairs, the equilibrium concentration of electrons n_0 is equal to the equilibrium concentration of holes p_0

$$n_0 = p_0 \, . \tag{6.1.1}$$

Semiconductor materials, in which free carriers appear due to the direct excitation of electrons from the valence to the conduction band, are called ***intrinsic semiconductors***. Equation (6.1.1) is called the charge ***neutrality condition*** for an intrinsic semiconductor at equilibrium.

The simplest case of direct gap materials with spherical constant energy surfaces is considered first. In the effective mass approximation, the conduction band is described by a parabolic dependence of the electron energy on the quasi-momentum $\vec{p}$

$$E(\vec{p}) = E_c + \frac{\vec{p}^2}{2m_n}, \tag{6.1.2}$$

where m_n is the electron effective mass in the conduction band, and E_c is the energy at the bottom of the conduction band. The density of electron states is calculated as in the case of free electrons in Section 2.3. The only difference come from the substitution of the effective mass m_n for the mass of the free electron m_0. The density of states $D_c(E)$ in the conduction band of a three-dimensional crystal is

$$D_c(E) = \frac{1}{2\pi^2}\left(\frac{2m_n}{\hbar^2}\right)^{3/2} (E - E_c)^{1/2} \quad \text{for} \quad E \geq E_c . \tag{6.1.3}$$

The square root energy dependence here is a result of the quadratic energy-quasimomentum dependence Eq.(6.1.2).

How many orbitals are occupied by electrons depends on the statistics of the electrons. Because electrons have half-integer spin, they obey Fermi-Dirac statistics. The corresponding distribution function has been given in the case of free electrons in Eq.(2.3.13). Replacing the energy of free electron by the electron energy in the conduction band Eq.(6.1.2), one finds

$$f_e^{(0)}(E) = \frac{1}{e^{\frac{E(\vec{p})-\mu}{k_B T}} + 1}, \tag{6.1.4}$$

where μ is the ***chemical potential*** of an electron in the crystal. In the physics of semiconductors, μ is referred to as the ***Fermi level***. The function $f_e^{(0)}(E)$ is the probability that the electron state with energy E in the conduction band is occupied in equilibrium at some temperature T. The energy level E is empty with probability $[1 - f_e^{(0)}(E(\vec{p}))]$.

The concentration of electrons n_0 in the conduction band equals

$$n_0 = \frac{2}{V}\sum_{\vec{p}} f_e^{(0)}(E(\vec{p})), \tag{6.1.5}$$

where the sum is taken over all possible values of $\vec{p}$ in the Brillouin zone. The factor of "2" arises from the two possible orientations of the electron's spin in each band state. Because the spectrum of $\vec{p}$-values is very dense, one can use the Eq.(2.3.24) and replace the sum over $\vec{p}$ by an integral over $\vec{p}$. The electron concentration in this case equals

$$n_0 = \frac{2}{(2\pi\hbar)^3} \int_{E_c}^{\infty} f_e^{(0)}(E(\vec{p}))d\vec{p}. \tag{6.1.6}$$

In order to evaluate the integral over $\vec{p}$, we write Eq.(6.1.6) in terms of the density of states given by Eq.(6.1.3),

$$n_0 = \int_{E_c}^{\infty} D_c(E) f_e^{(0)}(E) dE. \tag{6.1.7}$$

Here the integrand is the product of the density of states and the probability of occupation of these states. Substituting Eq.(6.1.3) into Eq.(6.1.7), one obtains for the concentration of electrons in the conduction band

$$n_0 = \frac{1}{2\pi^2}\left(\frac{2m_n}{\hbar^2}\right)^{3/2} \int_{E_c}^{\infty} \frac{(E-E_c)^{1/2}}{e^{\frac{E-\mu}{k_BT}}+1} dE. \tag{6.1.8}$$

In principle, Eq.(6.1.8) defines μ as function of n_0 and T. In fact, the integral in Eq.(6.1.8) can not be expressed in terms of elementary functions, but it is well tabulated.

Similar considerations lead to the density of states for electrons in the valence band $D_v(E)$. In the approximation of isotropic nondegenerate bands, when $E(\vec{p})$ is given by Eq.(4.7.1), $E = E_v - p_v^2/2m_p$, m_p being the effective mass in the valence band, $D_v(E)$ equals

$$D_v(E) = \frac{1}{2\pi^2}\left(\frac{2m_p}{\hbar^2}\right)^{3/2} (E_v - E)^{1/2} \quad \text{for} \quad E \le E_v. \tag{6.1.9}$$

The concentration of positive holes is defined by the number of empty states in the valence band. The probability that a given quantum state with the energy E is empty is related to the electron distribution function by

$$f_p(E) = [1 - f_e(E)] = \frac{1}{e^{\frac{\mu-E}{k_BT}}+1}. \tag{6.1.10}$$

The concentration of holes p_0 follows from Eqs.(6.1.9) and (6.1.10),

$$p_0 = \int_{-\infty}^{E_v} D_v(E) f_p(E) dE = \frac{1}{2\pi^2}\left(\frac{2m_p}{\hbar^2}\right)^{3/2} \int_{-\infty}^{E_v} \frac{(E_v - E)^{1/2} dE}{e^{\frac{\mu-E}{k_BT}}+1}. \tag{6.1.11}$$

At high temperatures, $k_BT \gg |\mu(T)|$, when $e^{-\mu/k_BT} \gg 1$ for electrons, and $e^{\mu/k_BT} \gg 1$ for holes one can neglect the unity in the denominators of Eqs.(6.1.4) and (6.1.10). The Fermi-Dirac distribution functions reduce to classical Boltzman distributions. We have for nondegenerate electrons

$$f_e^{(0)}(E) = C_n e^{-\frac{E}{k_BT}}, \tag{6.1.12a}$$

and for nondegenerate holes

$$f_p^{(0)}(E) = C_p e^{\frac{E}{k_BT}}, \tag{6.1.12b}$$

where C_n and C_p are normalization constants. Neglecting unity with respect to the exponential in the denominator of Eq.(6.1.8) and integrating over E one obtains the electron concentration of nondegenerate gas in the form

$$n_0 = N_c e^{\frac{\mu - E_c}{k_B T}}, \tag{6.1.13a}$$

where the quantity

$$N_c = 2\left(\frac{2\pi m_n k_B T}{(2\pi\hbar)^2}\right)^{3/2} \tag{6.1.13b}$$

is called the ***effective density of states*** in the conduction band.

The concentration of a nondegenerate gas of holes is obtained in the same way,

$$p_0 = N_v e^{\frac{E_v - \mu}{k_B T}}, \tag{6.1.14a}$$

where the effective density of states is

$$N_v = 2\left(\frac{2\pi m_p k_B T}{(2\pi\hbar)^2}\right)^{3/2}. \tag{6.1.14b}$$

An order of magnitude estimate of the effective densities N_c and N_v at room temperature, where $T = 0.026\ eV$, for the case that $m_n = m_p = m_0$, yields

$$N_c = N_v \approx 2.5 \times 10^{17} cm^{-3}.$$

In nondegenerate semiconductors, the real concentrations n_0 and p_0 are considerably lower than N_c and N_v.

It follows from Eqs.(6.1.12) and (6.1.13) that for nondegenerate classical electrons the concentration dependence of μ at a given T has the form

$$\mu(T) = E_c + k_B T \ln\frac{n_0}{N_c}. \tag{6.1.15}$$

Inverting Eq.(6.1.14a) with respect of μ, one obtains the dependence of μ on the hole concentration

$$\mu(T) = E_v - k_B T \ln\frac{p_0}{N_v}. \tag{6.1.16}$$

The charge neutrality condition (the equilibrium condition) Eq.(6.1.1) enables us to express the Fermi level $\mu(T)$ in terms of the band structure parameters. Substituting Eqs.(6.1.13a) and (6.1.14a) in Eq.(6.1.1), and using Eqs.(6.1.13b), (6.1.14b), one obtains

$$\mu(T) = \frac{1}{2}(E_v + E_c) + \frac{3}{4}k_B T \ln\frac{m_p}{m_n}. \tag{6.1.17a}$$

Substituting $E_c - E_v = E_g$ in Eq.(6.1.17a) results in

$$\mu(T) = E_v + \frac{1}{2}E_g + \frac{3}{4}k_BT\ln\frac{m_p}{m_n}. \qquad (6.1.17b)$$

Since the factor $3/4\ln(m_p/m_n)$ is of the order of unity, the second term in Eq.(6.1.17b) is defined by the temperature T. At room temperature, the Fermi level of the intrinsic material is nearly midway between E_c and E_v, that is, near the center of the energy gap E_g. With increasing temperature, $\mu(T)$ shifts from the band edge of the heavier carriers to the band edge of the lighter carriers.

Multiplying together n_0 from Eq.(6.1.13a) and p_0 from Eq.(6.1.14a) results in very useful relation

$$n_0p_0 = N_cN_ve^{-\frac{E_g}{k_BT}}. \qquad (6.1.18)$$

We see that the product n_0p_0 is independent of the location of μ within the bandgap and of the carrier concentrations. The constancy of the product of the electron and hole concentrations is called the ***law of mass action***.

The condition of charge neutrality Eq.(6.1.1) in intrinsic semiconductor means that

$$n_0 = p_0 \equiv n_i, \qquad (6.1.19)$$

where n_i is the equilibrium concentration of electrons or holes in an intrinsic semiconductor. Combining Eqs.(6.1.18) and (6.1.19) leads to the relation

$$n_i = (N_cN_v)^{1/2}e^{-\frac{E_g}{2k_BT}}, \qquad (6.1.20)$$

which shows that the intrinsic concentration of electrons and holes increases with temperature T at an exponential rate. Typical temperature dependences of n_i for some semiconductors are shown in Fig.6.1.1. At $T = 300\,K$, $n_i = 1.5 \times 10^{10} cm^{-3}$ in Si, and $n_i = 1.8 \times 10^{10} cm^{-3}$ in GaAs.

The law of mass action may be written in the form

$$n_0p_0 = n_i^2. \qquad (6.1.21)$$

The values of n_i for different materials vary, since the bandgap energies and effective masses of materials are different.

It should be noted that the concentrations given by Eqs.(6.1.13) and (6.1.14), and the law of mass action, are obtained in the approximation of classical Boltzman statistics for the carriers, which is valid when the chemical potential μ (Fermi level) lies within the bandgap, but away from its edges by an energy of at least several times k_BT.

It follows from Eq.(2.1.18) for μ that, if m_n and m_p differ considerably, increasing the temperature shifts the Fermi level into the band of the lighter carriers. A further increase of the temperature results in a further shift of the Fermi level into the energy band. This leads to the ***degeneracy*** of the electron gas in the energy band. The condition for this degeneracy is

$$e^{\frac{E_c-\mu}{k_BT}} \ll 1. \qquad (6.1.22)$$

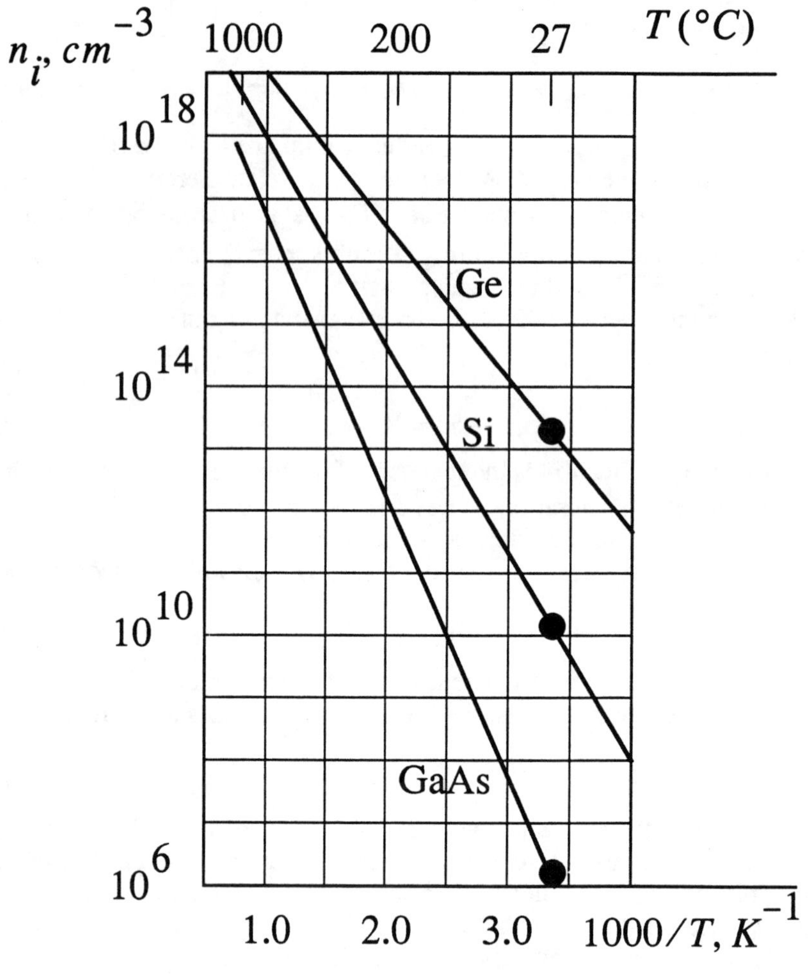

Figure 6.1.1 The intrinsic carrier concentrations in Ge, Si, and GaAs as functions of temperature.

Under this, the Fermi-Dirac distribution function $f_e^{(0)}(E)$ has the form of the step-function shown in Fig.2.3.5. Replacing the distribution function in Eq.(6.1.8) by unity, and taking for the upper limit of the integral $E = E_F$ (compare the definition of the Fermi energy E_F for the free electron gas in Section 2.3), one finds

$$n_0 = \frac{4}{3\pi^2} N_c \left(\frac{E_F^{(e)} - E_c}{k_B T} \right)^{3/2} \qquad \text{for} \quad E_F^{(e)} > E_c \,. \tag{6.1.23}$$

Inverting this equation leads to the concentration dependence of the Fermi energy

$$E_F^{(e)} - E_c = \frac{(3\pi^2)^{2/3} \hbar^2}{2m_n} n_0^{2/3} \,. \tag{6.1.24}$$

A similar procedure gives for holes

$$E_v - E_F^{(h)} = \frac{(3\pi^2)^{2/3}\hbar^2}{2m_p} p_0^{2/3}. \qquad (6.1.25)$$

6.2 DENSITY OF STATES EFFECTIVE MASS

Real semiconductors never have a simple isotropic dependence of the energy on the quasimomentum given by Eqs.(4.7.1) or (6.1.2). It was shown in Section 4.1 that in many-valley semiconductors the surface of constant energy consists of several ellipsoids of revolution (or valleys) centered at some symmetry points in the Brillouin zone. The electron energy in a single valley is given by

$$E(\vec{p}) = E_c + \frac{(p_x - p_{0x})^2}{2m_1} + \frac{(p_y - p_{0y})^2}{2m_1} + \frac{(p_z - p_{0z})^2}{2m_3}. \qquad (6.2.1)$$

The concentration of electrons in a single valley is given by Eq.(6.1.8), where the energy E takes the value Eq.(6.2.1). The following scaling in quasimomentum space,

$$p_x = p_x'\left(\frac{m_1}{m_d^*}\right)^{1/2}; \quad p_y = p_y'\left(\frac{m_1}{m_d^*}\right)^{1/2}; \quad p_z = p_z'\left(\frac{m_3}{m_d^*}\right)^{1/2}, \qquad (6.2.2)$$

with

$$m_d^* = (m_1 m_1 m_3)^{1/3}, \qquad (6.2.3)$$

reduces the energy Eq.(6.2.1) to the simpler form

$$E(\vec{p}') = E_c + \frac{(\vec{p}')^2}{2m_d^*}. \qquad (6.2.4)$$

Calculations of the electron density of states in the case of a spherical constant energy surface given by Eq.(6.1.3), see Section 2.3, leads to the density of states

$$D_c(E) = \frac{1}{2\pi^2}\left(\frac{2}{\hbar^2}\right)^{3/2} (m_1 m_1 m_3)^{1/2} (E - E_c)^{1/2} \quad \text{for} \quad E \geq E_c. \qquad (6.2.5)$$

If there are ν_c valleys, the density of states $D_c(E)$ is multiplied by ν_c. Comparing the result with Eq.(6.1.3), one can define the quantity

$$m_{eff} = \nu_c^{2/3}(m_1 m_1 m_3)^{1/3}, \qquad (6.2.6)$$

which is called the ***density of states effective mass***. By substituting this mass for the electron mass in the equations of Section 6.1, we can use the results of that section for many-valley semiconductors.

The energy bands of electrons in some materials are nonparabolic, and the energy bands of holes are degenerate. The surfaces of constant energy are warped, see Fig.4.3.3. Nevertheless, the density of states effective mass can be found for an arbitrary dependence of the energy $E(\vec{p})$ on the quazimomentum $\vec{p}$.

In order to do this, we consider two close surfaces of constant energy, shown in Fig.6.2.1a. The coordinate system is chosen such that the p_x- and p_y- directions are on the surface of constant energy, and the p_z- direction is normal to the surface. The increase of the energy, dE, at a given point E is defined by the gradient $\partial E(\vec{p})/\partial \vec{p}$,

$$dE = \frac{\partial E(\vec{p})}{\partial \vec{p}} \cdot d\vec{p}. \qquad (6.2.7)$$

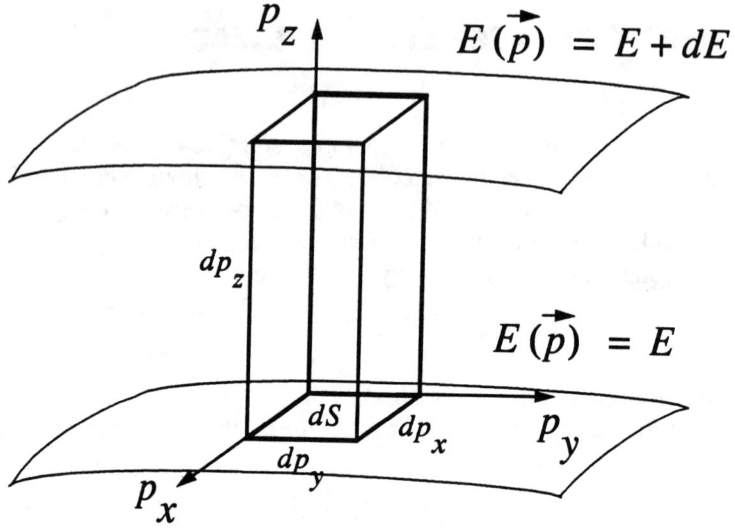

(a)

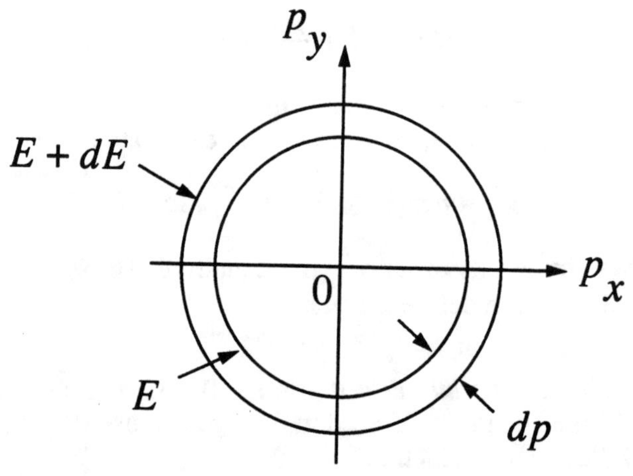

(b)

Figure 6.2.1 Calculation of density of quantum states in the band. (a) Two elements of the constant energy surfaces $E(\vec{p}) = E$ and $E(\vec{p}) = E + dE$. The volume element shown is $dp_x dp_y dp_z = dS dp_z$. (b) Surfaces of constant energy in $\vec{p}$–space.

Since dp_x and dp_y are taken on the surface of constant energy, and the gradient $\partial E(\vec{p})/\partial\vec{p}$ is directed along the normal to it, the cosine in the scalar product of Eq.(6.2.7) is equal to unity, and the scalar product in our reference system takes the form

$$dE = \left|\frac{dE(\vec{p})}{d\vec{p}}\right| dp_z\,,$$

or

$$dp_z = \frac{dE}{\left|\frac{\partial E(\vec{p})}{\partial\vec{p}}\right|}. \tag{6.2.8}$$

The elementary volume $d\vec{p} = dp_x dp_y dp_z$ in $\vec{p}$-space equals

$$d\vec{p} = dS\frac{dE}{\left|\frac{\partial E(\vec{p})}{\partial\vec{p}}\right|}, \tag{6.2.9}$$

where $dS = dp_x dp_y$ is the element of surface area. Substitution of Eq.(6.2.9) in Eq.(6.1.6) leads to

$$n_0 = \int f_e(E)\left[\frac{2}{(2\pi\hbar)^3}\oint_S \frac{dS}{\left|\frac{\partial E(\vec{p})}{\partial\vec{p}}\right|}\right] dE. \tag{6.2.10}$$

The comparison of Eq.(6.2.10) with Eq.(6.1.6) shows that the integration over $\vec{p}$-space is replaced in Eq.(6.2.10) by integration over the surface of constant energy $E(\vec{p}) = E$ and integration over the energy E, see Fig.6.2.1b. The quantity in the square brackets in Eq.(6.2.10) is the density of states. The quantity $|\,\partial E(\vec{p})/\partial\vec{p}\,|$ in the denominator of the density of states is the absolute value of the average group velocity of the electron in the conduction band.

A comparison of Eq.(6.2.10) and Eq.(6.1.8) yields the density of states effective mass in the case of the classical Boltzman statistics of the electrons

$$m_{eff}^{3/2} = \frac{1}{(2\pi K_B T)^{3/2}}\int e^{-\frac{E}{k_B T}}\left[\oint_S \frac{dS}{\left|\frac{\partial E(\vec{p})}{\partial\vec{p}}\right|}\right] dE. \tag{6.2.11}$$

The application of the concept of the density of states effective mass in the equations of Section 6.1 enables the results of that section to be applied to semiconductors with complex band structures.

When holes in the hole band are described by the Luttinger Hamiltonian, Eq.(4.6.3), and the energy takes the values Eq.(4.6.7a,b) corresponding to heavy m_{heavy} and light m_{light} masses, the density of states effective mass equals

$$m_{eff} = (m_{heavy}^{3/2} + m_{light}^{3/2})^{2/3}. \tag{6.2.12}$$

The mass m_{eff} is defined mainly by the mass m_{heavy}.

Another useful form of the density of states is obtained by using the rule of integration of the δ-function given in Section 1.16, Eq.(1.16.20). In the integral

$$I = \int \delta(E - E(\vec{p}))d\vec{p}, \tag{6.2.13}$$

one can replace the integration over $\vec{p}$-space by integration over the surface of constant energy $E(\vec{p}) = E = Constant$ and integration over the energy E. As in the derivation of Eq.(6.2.10), one obtains

$$I = \int dS \int \frac{\delta(E - E(\vec{p}))}{|\nabla_{\vec{p}} E(\vec{p})|} dE. \tag{6.2.14}$$

Evaluating the integral with the help of the δ-function results in

$$I = \int_S \frac{dS}{|\nabla_{\vec{p}} E(\vec{p})|_{E(\vec{p})=E}}. \tag{6.2.15}$$

We have here the density of states appearing in Eq.(6.2.10). Therefore, alternative forms of the densities of states of electrons

$$D_c(E) = \frac{2}{(2\pi\hbar)^3} \int \delta(E - E_c(\vec{p})) d\vec{p}, \tag{6.2.16}$$

and holes

$$D_v(E) = \frac{2}{(2\pi\hbar)^3} \int \delta(E - E_v(\vec{p})) d\vec{p}. \tag{6.2.17}$$

It is seen from these equations that $D_c(E) = 0$ for $E < E_c$, and $D_v(E) = 0$ for $E > E_v$.

6.3 DENSITY OF STATES OF SEMICONDUCTOR NANOSTRUCTURES

a. Quantum Well. Because the components of quasimomentum $\vec{p} = (p_x, p_y)$ take the discrete set of values given by Eq.(2.3.9), the area per state in two-dimensional $\vec{p}$-space is defined for a quantum well, in a manner similar to the way in which this was done in Section 2.3, by the element of area

$$\Delta p_x \Delta p_y = \left(\frac{2\pi\hbar}{L_1}\right)\left(\frac{2\pi\hbar}{L_2}\right) \Delta m_x \Delta m_y, \tag{6.3.1}$$

where Δm_x, Δm_y are equal to unity, and L_1 and L_2 are the dimensions of the crystal in the directions of free motion.

The number of states below the surface of constant energy $E = Constant$ is equal to the number of states inside the circle of radius $p = \sqrt{p_x^2 + p_y^2}$. The area of the circle is $\pi(p_x^2 + p_y^2)$. The total number of states N_{n_z} inside the circle for each value of n_z is

$$N_{n_z} = 2 \frac{\pi(p_x^2 + p_y^2)}{\left(\frac{2\pi\hbar}{L_1}\right)\left(\frac{2\pi\hbar}{L_2}\right)}. \tag{6.3.2}$$

The factor of "2" arises from the spin of electron.

Expressing $(p^2_x + p^2_y)$ in terms of the energy E given by Eq.(4.6.4), we find for each value of n_z

$$N_{n_z} = L_1 L_2 \frac{m_n}{\pi\hbar^2}\left(E - E_c - E_{n_z}\right), \tag{6.3.3}$$

where $L_1 L_2 = S$ is the cross section area of the quantum well. The density of states D_{n_z} of the quantum level n_z is defined by the number of states per unit interval of energy,

$$D_{n_z} = \frac{dN}{dE} = L_1 L_2 \frac{m_n}{\pi\hbar^2}. \tag{6.3.4}$$

We see from Eq.(6.3.4) that for a quantum well the density of states $D_{n_z}(E)$ has a constant value that does not depend on the energy E.

If we associate the two-dimensional subband which represents the kinetic energy of the in-plane motion of an electron with each bound state E_{n_z}, then the total density of states is equal to the sum of the densities for all values of n_z. It exhibits the staircase-like distribution function shown in Fig.6.3.1

$$D_{QW}(E) = \frac{dN_{QW}}{dE} = \frac{m_n L_1 L_2}{\pi\hbar^2} \sum_{n_z} \theta\left(E - E_c - E_{n_z}\right), \tag{6.3.5}$$

where $\theta(E)$ is the step function

$$\theta\left(E - E_c - E_{n_z}\right) = \begin{cases} 1 & \text{for} \quad E > E_c + E_{n_z} \\ 0 & \text{for} \quad E < E_c + E_{n_z} \end{cases}. \tag{6.3.6}$$

We see from Eq.(6.3.5) that $D_{QW} = 0$ when $E < E_c + E_{n_z}$. This result follows from the confinement associated with localization of the particle along the z- direction. D_{QW} also possesses jumps of finite height when the energy passes through the edge of a two-dimensional subband. This behavior differs from the smoothly varying dependence $D(E) \sim \sqrt{E}$ obtained in Section 6.1, Eq.(6.1.3), in the three-dimensional case of electrons in the conduction band. We note that the reduction of the dimensionality is accompanied by an increasingly singular behavior of the density of states.

Each step of the staircase corresponds to a different quantum number n_z and may be considered as a subband within the conduction band. The bottoms of these subbands have higher energies for higher quantum numbers n_z. The density of states of the valence band has a similar staircase behavior.

The main difference of these results from those for bulk semiconductor consists of the considerable density of states for the lowest allowed conduction-band energy level and for the highest allowed valence-band energy level. This property is very important for the optical properties of quantum wells.

b. Quantum wire. Each pair of quantum numbers n_y, n_z from Eqs.(4.6.10) and

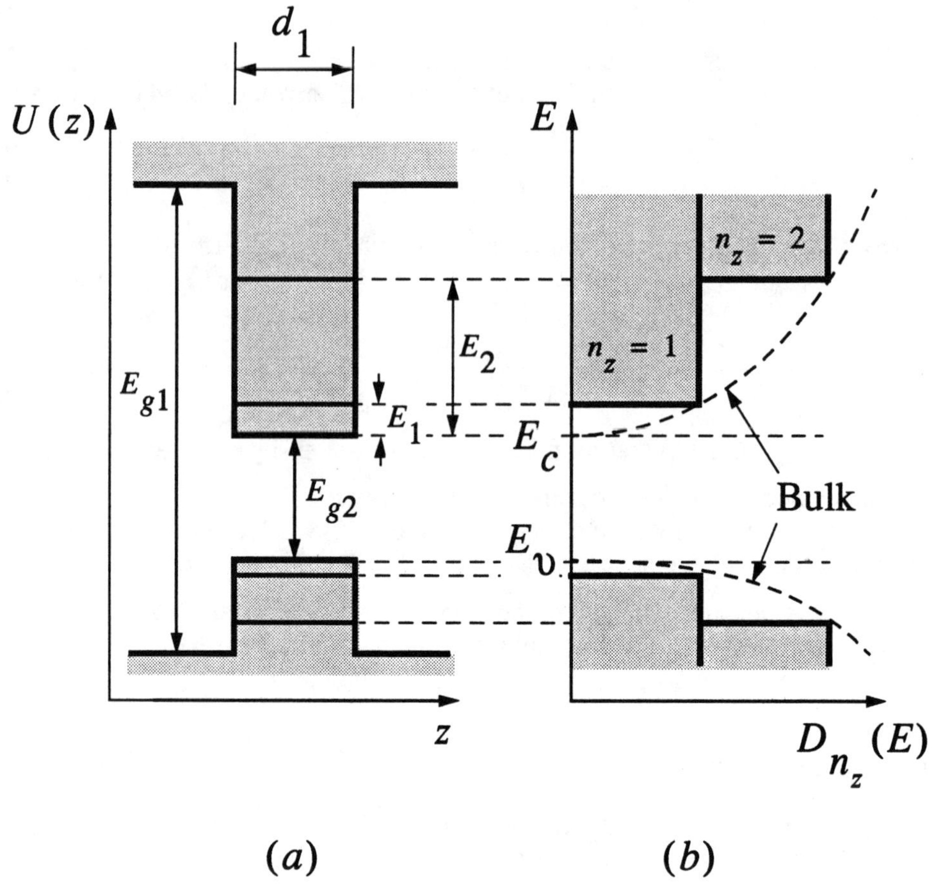

Figure 6.3.1 (a) The potential barrier for quantum well. (b) Density of states for a quantum-well structure (solid line), and for a bulk semiconductor (dashed line).

(4.6.11) is connected with an energy subband. The length per state of a quantum wire in one-dimensional p-space according to Eq.(2.3.9) is defined by the element of length

$$\Delta p_x = \left(\frac{2\pi\hbar}{L_1}\right)\Delta m_x\,, \tag{6.3.7}$$

where Δm_x is equal to unity, and L_1 is the length of the wire. The total number of electrons below the surface of the constant energy for each value of n_y, n_z is equal to the ratio of p_x to the length per state from Eq.(6.3.7),

$$N_{n_y n_z} = 2\frac{p_x}{(2\pi\hbar/L_1)}. \tag{6.3.8}$$

Substituting for p_x the result given by Eq.(4.6.10) we obtain

$$N_{n_y n_z} = L_1\frac{\sqrt{2m_n}}{\pi\hbar}\left(E - E_c - E_{n_y} - E_{n_z}\right)^{1/2}. \tag{6.3.9}$$

The density of states at the quantum level (n_y, n_z) equals

$$D_{n_y n_z} = L_1 \frac{\sqrt{m_n}}{\pi\hbar\sqrt{2}} \left(E - E_c - E_{n_y} - E_{n_z}\right)^{-1/2} \quad \text{for} \quad E > E_c + E_{n_x} + E_{n_y}, \qquad (6.3.10)$$

and is zero otherwise. This is a decreasing function of energy, as is shown in Fig.6.3.2. The energy subbands in quantum wires are narrower than those in a quantum well.

(a)

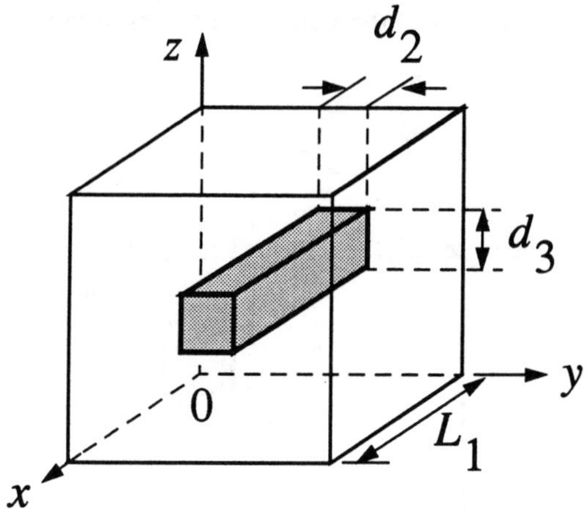

(b)

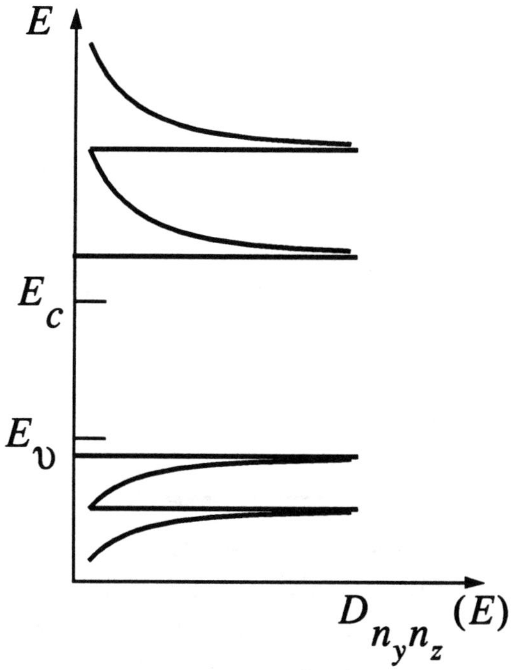

Figure 6.3.2 (a) The quantum wire. (b) The density of states in a quantum wire.

c. Quantum dot. It follows from the discrete structure of the electron energy spectrum of a quantum dot that the density of states has the form of a sequence of delta-functions at the allowed energies, see Fig.6.3.3,

$$D_{n_x,n_y,n_z} = \sum_{n_x,n_y,n_z} \delta\left(E - E_{n_x,n_y,n_z}\right). \tag{6.3.11}$$

Due to this discrete structure of the density of states, quantum dots are sometimes called artificial atoms.

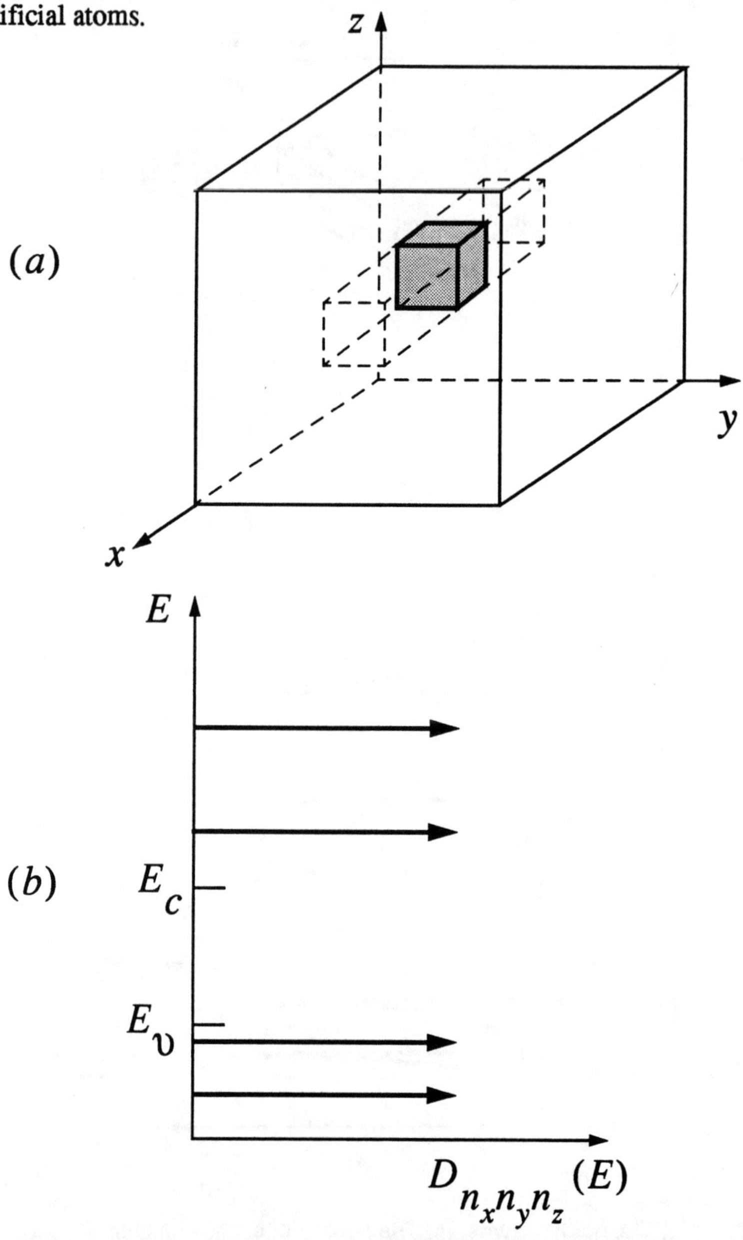

Figure 6.3.3 (a) The quantum dot. (b) The density of states in a quantum dot.

6.4 EXTRINSIC SEMICONDUCTORS

Adding small controlled amounts of impurities to a crystal affects its properties significantly. It is possible to obtain carrier concentrations several orders of magnitude larger than in intrinsic materials. Doped semiconductors are called ***extrinsic semiconductors***.

In this section we shall consider the effects of substitutional impurities, which replace some atoms of the host semiconductor. Ge and Si are elements of Group IV, and they form crystals of cubic symmetry. The Group V elements (e.g. phosphorus) have five valence electrons. When a Group V impurity is introduced into a host material of the type of Ge or Si, four electrons contribute to the covalent bonds with the host crystal. The fifth electron, being an excess electron, is more weakly bound to the impurity atom (Fig.6.4.1). As a result, its energy level E_D is below the bottom of the conduction band, $E_D < E_c$, see Fig.6.4.2.

To estimate the binding energy E_D of this electron, one can use the known energy of the hydrogen atom, Eq.(1.13.40), in which the dielectric permittivity of the medium, $\varepsilon = \varepsilon_0\varepsilon_r$, where ε_r is the relative permittivity, is substituted for the permittivity of vacuum ε_0. The electron effective mass m_n is also substituted for the mass m_0 of the free electron, and we obtain

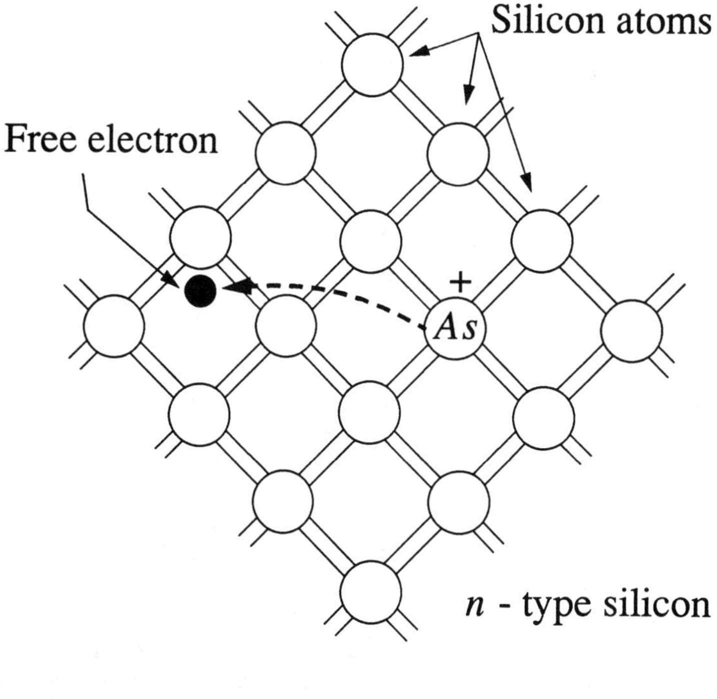

Figure 6.4.1 A substitutional arsenic impurity atom loses an electron becoming a positive ion (ionized donor). A free electron appears in the conduction band.

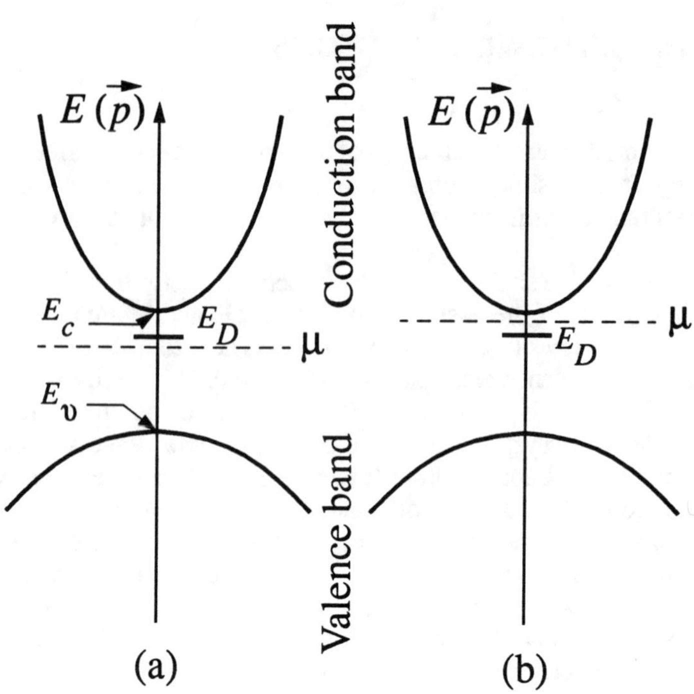

Figure 6.4.2 Fermi-level μ in an n-semiconductor. (a) $\mu < E_D$, ionized donors. (b) $\mu > E_D$, freezing out of donors.

$$E_D = -\frac{e^4 m_n}{2\hbar^2(4\pi\varepsilon_r\varepsilon_0)^2}. \tag{6.4.1}$$

The Bohr radius of the excess electron in the ground state equals

$$a_D = \frac{4\pi\varepsilon_r\varepsilon_0\hbar^2}{m_n e^2}, \tag{6.4.2}$$

where ($-e$) is the charge of an electron with e being positive. For a typical value of the effective mass of an electron in a semiconductor, $m_n = 0.1\, m_0$, one finds

$$E_D = (13.6\, eV)_{hydrogen}\frac{m_n}{m_0\varepsilon_r} = 6\, meV. \tag{6.4.3}$$

The considerable decrease of the binding energy in a semiconductor with respect to that for a hydrogen atom results in the increase of the Bohr radius by the factor $\varepsilon_r m_0/m_n$. Instead of $a_0 = 0.53\,\text{Å}$ in a hydrogen atom, one finds $a^{(Ge)} = 80\,\text{Å}$ in Ge, and $a^{(Si)} = 30\,\text{Å}$ in Si. These are large radii, which greatly exceed the lattice parameter of the crystal, a. The binding energy Eq.(6.4.3) in Ge and Si ($\approx 0.01\, eV$) is comparable with $k_B T$ at room temperature ($T = 0.026\, eV$). As a result, most impurity electrons at room temperature are thermally excited into the conduction band. This makes the thermal ionization of impurity atoms a very important source of current carriers.

Impurity atoms of this type are called ***donors.***

At low temperatures, the donor electron is bound to the impurity atom. When thermal excitation raises the donor electron into the conduction band, the ionized positively charged donor atom is left in the crystal. Materials in which the electrons in the conduction band are supplied by donor atoms are called ***n- type semiconductors***. The donor ***ionization energy*** equals $E_c - E_D$. Since $E_D << E_g$, these donors are called ***shallow impurities***.

According to Eq.(6.1.13a), an increase of the electron concentration leads to an increase of the chemical potential (Fermi level) μ. As a result, the Fermi level in n- type material lies above the middle of the bandgap, see Fig.6.4.3.

If an impurity atom from Group III (e.g. boron) is a substitutional impurity in a cubic semiconductor formed from atoms of Group IV in the periodic table, it does not have enough electrons to fill the tetrahedral bonds with surrounding atoms. To accommodate itself in the lattice, the impurity atom borrows an electron from the valence band and leaves a hole there. Impurity atoms of this type are called ***acceptors,*** see Fig.6.4.4. Because the net charge of an acceptor is negative, the energy of the electron trapped by an acceptor, E_A, is greater than that of a valence

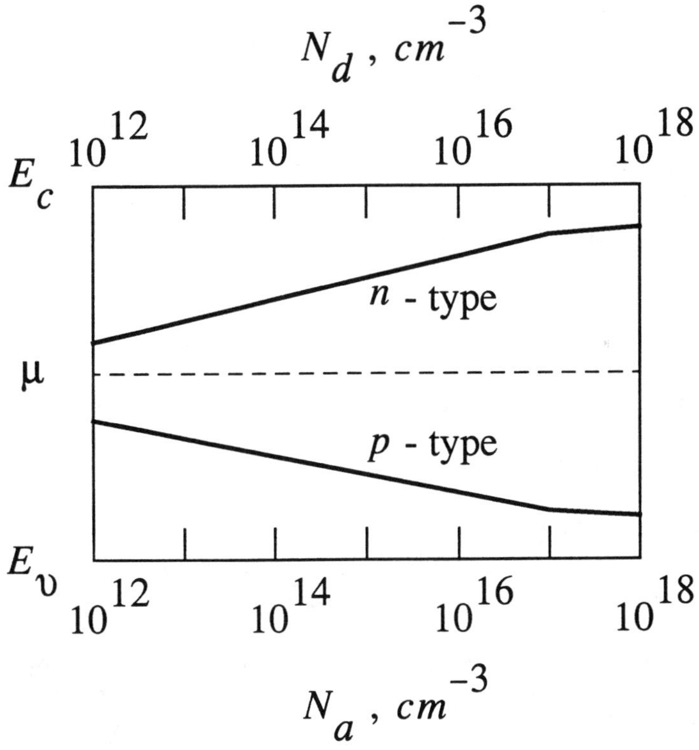

Figure 6.4.3 The qualitative impurity concentration dependence of the Fermi-level μ for n-type and p-type semiconductors.

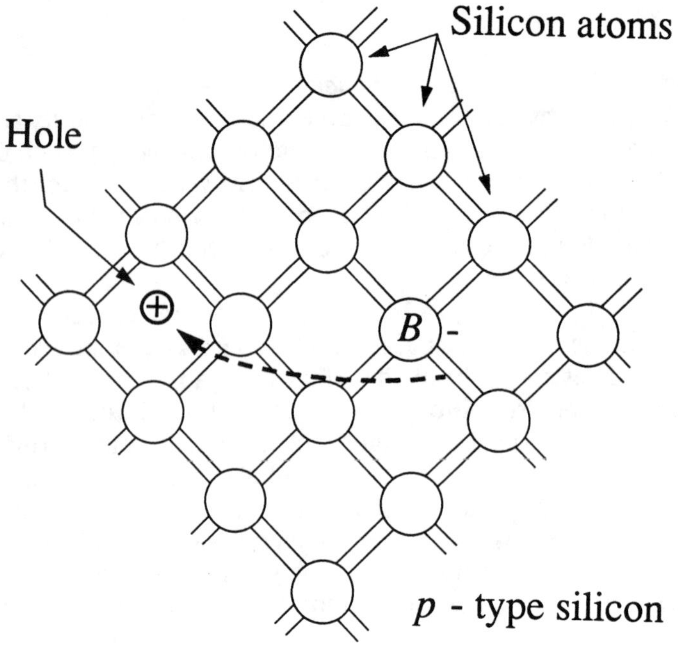

Figure 6.4.4 A substitutional boron impurity atom captures an electron, becoming a negative ion. A free hole appears in the valence band.

electron: $E_A > E_v$, see Fig.6.4.5. The ionization energy of an acceptor is estimated by the hydrogenlike formula (6.4.1), where the mass of the hole m_p replaces the mass of electron m_n. Since $E_A << E_g$, acceptors are also ***shallow impurities***. The energy level of such an electron is just above the top of the valence band by the energy E_A, that is equal $E_v + E_A$, see Fig.6.4.5.

Using the concept of holes, see Section 4.7, one can consider holes to be trapped by the negatively charged acceptor atoms. At low temperatures, acceptors are neutral atoms. At high temperatures, when thermal excitation lifts holes from the acceptor to the hole band, the ionized negatively charged acceptor atom remains in the crystal. The hole is in the band state with the energy given by Eq.(4.7.1). Materials in which holes in the hole band are created by ionization of acceptors are called ***p-type semiconductors.***

It follows from Eq.(6.1.14a) that increasing the hole concentration decreases the chemical potential of the holes μ. As a result, the chemical potential (Fermi level) in p-type semiconductors lies below the middle of the bandgap, see Fig.6.4.3.

To obtain the equilibrium condition in doped semiconductors, we have to find the probability of occupation of the impurity level by an electron or by a hole. The statistics of impurity atom electrons is complicated by the possibility of spin degeneracy of the impurity energy levels, and by the possibility of excited states of impurities. In principle, the calculation of the occupation probability does not allow using the Fermi-Dirac distribution function, and requires applying the Gibbs distribution with a variable number of particles, which contains factors representing the degree of degeneracy of the levels.

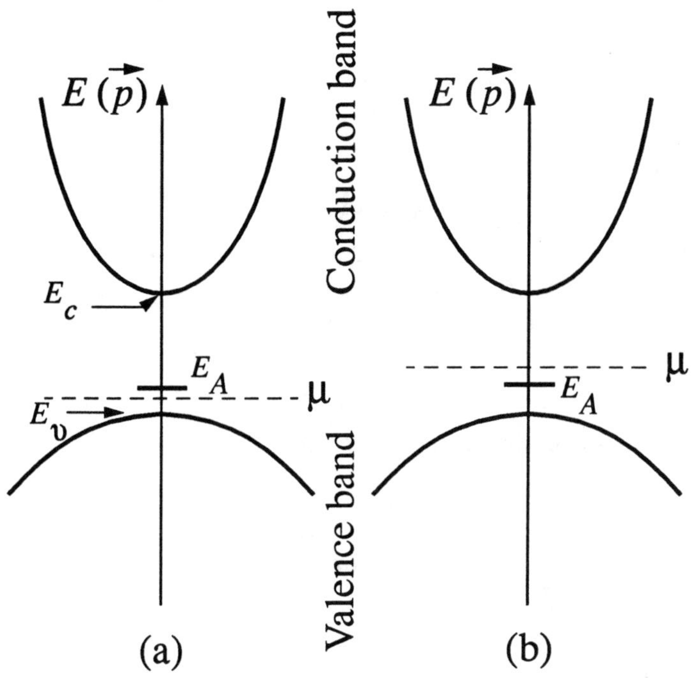

Figure 6.4.5 Fermi-level μ in a *p* –semiconductor. (a) $\mu < E_A$, freezing out of acceptors. (b) $\mu > E_A$, ionized acceptors.

In fact, we approach the problem with some simple intuitive considerations which lead us to correct understanding of the problem. We start with the probability of occupation of a monovalent donor impurity level given by Fermi-Dirac distribution applied to the impurity energy level E_D

$$f_e^{(0)}(E_D) = \frac{1}{e^{\frac{E_D - \mu}{k_B T}} + 1}. \tag{6.4.4}$$

The concentration of nonionized neutral donors n_D is a fraction $f_e^{(0)}$ of the total donor concentration N_D. The donor level E_D is doubly degenerate, because of the two possible values of the electron spin $s_z = \uparrow, \downarrow$. Therefore, there are two ways to fill the bound state of the donor with electrons of two different spin orientations. Therefore, the concentration of neutral donors contains an extra factor of "2",

$$n_D = N_D 2 f_e^{(0)}. \tag{6.4.5}$$

The concentration of ionized donors N_D^+ is defined by the fraction $(1 - f_e^{(0)})$ of N_D,

$$N_D^+ = N_D(1 - f_e^{(0)}). \tag{6.4.6}$$

There is no factor of "2" in Eq.(6.4.6), because there is only one way in which the impurity atom is ionized by losing the electron. Equations (6.4.5) and (6.4.6) enable us to find the ratio n_D/N_D^+ in the case of a doubly degenerate electron level,

$$\frac{n_D}{N_D^+}=\frac{2f_e^{(0)}}{1-f_e^{(0)}}=2e^{\frac{\mu-E_D}{k_BT}}. \tag{6.4.7}$$

Combining Eq.(6.4.7) with the relation

$$n_D+N_D^+=N_D\,, \tag{6.4.8}$$

we find that the electron distribution function n_D/N_D for a monovalent impurity atom with a double spin degeneracy of the electron level is

$$\frac{n_D}{N_D}=\frac{1}{1+\frac{1}{2}e^{\frac{E_D-\mu}{k_BT}}}\,, \tag{6.4.9}$$

and the concentration of ionized donors equals

$$\frac{N_D^+}{N_D}=\frac{1}{1+2e^{-\frac{\mu-E_D}{k_BT}}}. \tag{6.4.10}$$

If the temperature is high enough, $k_BT >> (E_D - \mu)$ Eq.(6.4.9) reduces to the Boltzman approximation

$$n_D=2N_De^{-\frac{E_D-\mu}{k_BT}}. \tag{6.4.11}$$

If Eq.(6.1.13a) is valid for electrons in the conduction band, the relative concentration of electrons in a donor state with respect to the total electron concentration $n_0 + n_D$ equals

$$\frac{n_D}{n_0+n_D}=\frac{1}{1+\frac{N_c}{2N_D}e^{-\frac{E_c-E_D}{k_BT}}}. \tag{6.4.12}$$

The energy $(E_c - E_D)$ here is the ionization energy of a donor. A numerical estimate shows that at room temperature, $T = 300\,K$, and a donor concentration $N_D = 10^{16} cm^{-3}$, there are only a few electrons in the donor states $n_D/(n_0 + N_D) \approx 0.4\%$. At room temperature the donor states are almost completely ionized, and all electrons are in the conduction band, $n_D = 0$ and $N_D{}^+ = N_D$, see Fig.6.4.2a.

At low temperatures, e.g. at about $T = 0\,K$, all electrons of an n-type semiconductor are trapped by donors and we find that $n_D = N_D$ and $N_D{}^+ = 0$. This follows from Eq.(6.4.9), since in this case $\mu > E_D$, and $e^{(E_D-\mu)/k_BT} \approx 0$, see Fig.6.4.2b. The Fermi level μ is above the donor level E_D. Electrons are not excited from the donor level to the conduction band and ***donor states are frozen out***.

The concentration of neutral acceptors in p-type material equals $p_A = N_A - N_A^-$, where N_A is the total concentration of acceptors and N_A^- is the concentration of ionized acceptors. A similar analysis of hole localization by acceptor impurities yields the following concentration of neutral acceptors p_A,

$$\frac{p_A}{N_A} = \frac{1}{1+\frac{1}{g}e^{\frac{\mu-E_A}{k_BT}}}, \tag{6.4.13}$$

and the concentration of ionized acceptors N_A^-,

$$\frac{N_A^-}{N_A} = \frac{1}{1+ge^{\frac{E_A-\mu}{k_BT}}} \quad . \tag{6.4.14}$$

The parameter g in Eq.(6.4.14) is connected with the degeneracy of the valence band. It was shown in Section 4.4 that in the approximation of the Luttinger Hamiltonian, Eq.(4.4.13), the valence band is four-fold degenerate. The parameter g is usually assumed to be four.

At room temperature, almost all acceptors are ionized. The results given by Eqs.(6.4.13) and (6.4.14) lead to $p_A = 0$ and $N_A^- = N_A$ respectively, and $\mu > E_A$, see Fig.6.4.5b. At low temperatures near $T = 0\,K$, electrons are in their lowest states. Therefore, in a p-type semiconductor no electrons are lifted into acceptor states, and $p_A = N_A$ and $N_A^- = 0$. It follows from Eq.(6.3.13) that $e^{(\mu-E_D)/k_BT} \approx 0$ and $\mu < E_A$. This is the ***freezing out of acceptors.***

Between total ionization at $T = 300K$ and freezing out at $T = 0\,K$, there is the region of ***partial ionization*** of donors and acceptors, see Fig.6.5.1.

6.5 CHARGE NEUTRALITY IN DOPED MATERIALS

It is known from classical electrodynamics that in a homogeneous semiconducting medium the bulk charge at any given point should vanish. Semiconductor crystals with charged carriers, donors, and acceptors should be electrically neutral in equilibrium. The charge neutrality condition imposes strict limitations on the possible distribution of electrons and holes over the energy levels of the crystal.

Negative charges in the crystal are electrons and ionized acceptors. The positive charge consists of holes and ionized donors. Therefore, the charge neutrality condition takes the form

$$n_0 + N_A^- = p_0 + N_D^+ \, . \tag{6.5.1}$$

The four quantities n_0, p_0, N_A^-, and N_D^+ depend on the chemical potential (Fermi level) $\mu(T)$ that should be constant in the state of equilibrium according to the rules of statistical physics.

The substitution of n_0, p_0, N_A^-, and N_D^+ from Eqs.(6.1.8), (6.1.11), (6.4.9), and (6.4.10) leads to a general equation which defines the Fermi level of the material μ in terms of the temperature and the band structure parameters. For any temperature and arbitrary level of doping, equation (6.5.1) must be solved numerically.

In doped semiconductors, the concentrations of electrons and holes are no longer equal. Nevertheless, if the electrons in the conduction band and holes in the hole band obey the classical Boltzman statistics, Eqs.(6.1.13a) and (6.1.14a), the product n_0p_0 still depends on the parameters of the semiconductor material and does not depend either on chemical potential μ or on the level of doping. This means that the above derivation of Eq.(6.1.19) is valid for both intrinsic and doped semiconductors in which the carriers obey classical statistics. Equation (6.1.19) is very useful for determining the concentrations of electrons and holes in doped semiconductors.

In order to simplify the calculations, we consider some special cases where the solutions are obtained in a simple analytic form.

When temperatures are high enough, the ionization of donors and acceptors is complete: $n_D = 0$, $p_A = 0$. Equation (6.5.1) becomes

$$n_0 + N_A = p_0 + N_D \,. \qquad (6.5.2)$$

Taking into account Eq.(6.1.19) and solving the resulting quadratic equation for n_0, one obtains

$$n_0 = \frac{N_D - N_A}{2} + \left[\left(\frac{N_D - N_A}{2}\right)^2 + n_i^2\right]^{1/2} \,. \qquad (6.5.3)$$

In the quadratic formula, the positive sign is taken, since n_0 should go to n_i in the limit of an intrinsic material, when $N_A = N_D = 0$.

In an n-type semiconductor $N_D >> N_A$, and Eq.(6.5.3) reduces to

$$n_0 = N_D \,. \qquad (6.5.4)$$

This is the ***majority*** carrier concentration, which greatly exceeds the concentration n_i obtained in intrinsic material at the same temperature. The band-to-band thermal excitation also creates equal numbers of electrons and holes. They contribute to both the majority n_0 and ***minority*** p_0 carrier concentrations. The concentration of minority carriers in the case of an n-type semiconductor is found from Eqs.(6.1.19) and (6.5.4),

$$p_0 = \frac{n_i^2}{n_0} = \frac{n_i^2}{N_D} \,. \qquad (6.5.5)$$

Numerical estimates show that for $n_0 \approx N_D = 10^{16} cm^{-3}$ the hole concentration is very low, $p_0 \approx = 10^4 cm^{-3}$. Increasing the donor concentration N_D decreases the concentration of minor carriers according to Eq.(6.5.5). We note that the concentration of electrons is not simply the sum of the donor concentration N_D and the intrinsic concentration n_i.

Neglecting the holes and considering the interval of temperatures where electrons are not degenerate, Eq.(6.1.13a), and the donors are partly ionized, Eq.(6.4.10), we find instead of Eq.(6.5.4)

$$n_0 = N_D^+ \,. \qquad (6.5.6)$$

Substituting n_0 from Eq.(6.1.13a) and N_D^+ from Eq.(6.4.9), one finds

$$N_c e^{\frac{\mu - E_c}{k_B T}} = \frac{N_D}{1 + 2e^{\frac{\mu - E_D}{k_B T}}} \,. \qquad (6.5.7)$$

It is useful to find from Eq.(6.5.7) the direct dependence of the electron concentration n_0 on the donor ionization energy. To find it, we rearrange Eq.(6.5.7) in the following way,

$$n_0 = \frac{N_D}{1 + 2\frac{n_0}{N_c} e^{\frac{E_c - E_D}{k_B T}}}. \tag{6.5.8}$$

Equation (6.5.8) is a quadratic equation for n_0. Its solution is

$$n_0 = \frac{N_c}{4} e^{-\frac{E_c - E_D}{k_B T}} \left[\left(1 + 8\frac{N_D}{N_c} e^{\frac{E_c - E_D}{k_B T}} \right)^{1/2} - 1 \right]. \tag{6.5.9}$$

Here $E_c - E_D$ is the donor ionization energy.

At temperatures $k_B T << E_c - E_D$, the expansion of the square root in Eq.(6.5.9) results in

$$n_0 = \left(\frac{1}{2} N_D N_c \right)^{1/2} e^{-\frac{E_c - E_D}{k_B T}}. \tag{6.5.10}$$

The electron gas is assumed to be nondegenerate at these temperatures. The electron concentration in the conduction band decreases exponentially with decreasing temperature.

At high temperatures $k_B T >> E_c - E_D$, the expansion of the square root in Eq.(6.5.9) leads to the result Eq.(6.5.4), which corresponds to the complete ionization of donors.

The temperature dependence of the electron concentration in a wide range of temperatures is shown in Fig.6.5.1, where three different regions are seen: partial ionization of impurities, and extrinsic and high temperature intrinsic behavior.

If the level of doping reaches the point where the electron concentration is $n_0 > N_c$ with N_c given by Eq.(6.1.13b), the Fermi level μ shifts to the conduction band and the electron gas becomes a degenerate gas with Fermi-Dirac statistics. The same happens in the hole band when $p_0 > N_v$. The corresponding values of the Fermi energy in the conduction and hole bands are given by Eqs.(6.1.24) and (6.1.25).

6.6 COMPENSATED SEMICONDUCTORS

There are no materials, in fact, where impurities of a single type are present. In Ge, the material with a well developed technology, the concentration of residual impurities reaches $10^9 - 10^{10} cm^{-3}$. Therefore in n –semiconductors, we always have a compensating concentration of acceptors. The small amount of compensating impurities affects the temperature dependence of the carrier concentration and localization at the Fermi-level.

Semiconductors that are doped by both donors and acceptors up to a high level are called ***compensated***. The ratio

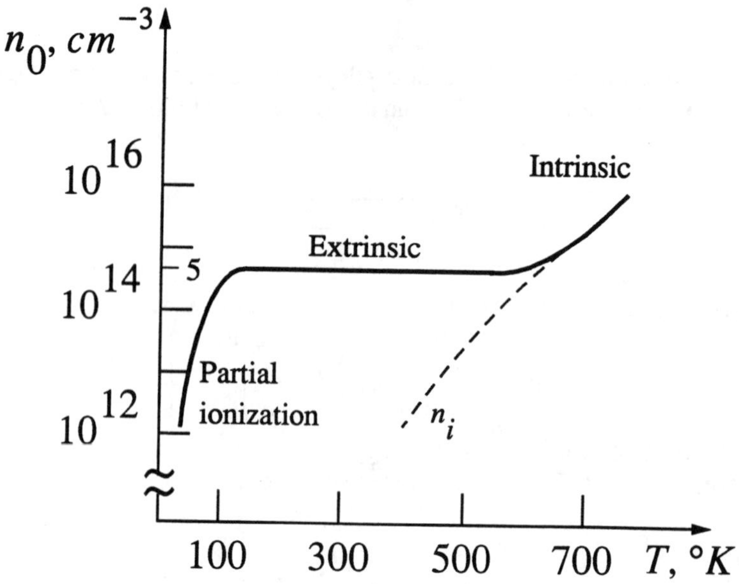

Figure 6.5.1 Electron concentration versus temperature in a wide range of temperatures in silicon doped with 5×10^{14} donors per cm^3. (After D.A. Neamen "Semiconductor Physics and Devices". Irwin, Boston, 1992).

$$K = \begin{cases} \dfrac{N_A}{N_D} & \text{for} \quad N_A < N_D \\ \dfrac{N_D}{N_A} & \text{for} \quad N_A > N_D \end{cases}$$

is referred to as the ***compensation ratio***. The term compensated is used when K is not too small ($K \approx 0.05$ or larger).

In the case of a low level of compensation, we have an extrinsic semiconductor which has been discussed in Section 6.4. When the compensation ratio reaches a value close to unity ($K \sim 1$), a new type of material appears. It does not behave as an extrinsic material. The distribution of the large number of donors and acceptors in the crystal is random. The local density of charged donors and acceptors fluctuates from point to point. This results in fluctuations of the lattice potential from point to point. The bottom of the conduction band and the top of the valence band fluctuate also. Electrons and holes are localized in fluctuation potential wells. A special type of conductivity appears in which the carriers hop from one potential well to another. It is called ***hopping conductivity***.

BIBLIOGRAPHY

D.K. Ferry, *Semiconductors* (Macmillan, New York, 1991).

CHAPTER 7

Electron-Electron Interaction

7.1 SCREENING IN AN ELECTRON GAS

The electron band structure considered so far concerned the behavior of a single electron in a crystal lattice. The Coulomb and exchange interactions were neglected.

In fact, these interactions affect considerably the behavior of electrons in a crystal. Due to the long-range nature of the electron-electron interaction the charge of any electron exerts forces on all the other charges in the crystal and rearranges them. It creates some difficulties in considering many-electron systems. Any individual electron being charged induces a potential which causes the other electrons to move. The charges redistribute in order to counteract this potential. The resulting potential does not behave like the Coulomb potential of the initial charge. The total picture is a self-consistent description of the total charge distribution and of the total potential.

Negatively charged electrons move in the crystal in a uniformly distributed positively charged background created by the positively charged ion cores of the crystal lattice. The total system of electrons and the crystal lattice is neutral and stable. But although it is homogeneous on average, the sea of electrons has fluctuations of electron density and corresponding fluctuations of electronic charge. A charge fluctuation at some point of the crystal creates an electric field at all the other points, and affects the motion of the other electrons.

On the other hand, the Pauli principle does not allow electrons to come very close to each other, and put a limit on the possible size of electron charge fluctuations.

Typical concentrations of electrons in semiconductors vary from $10^{15} cm^{-3}$ up to $10^{19} cm^{-3}$. The corresponding average separation of electrons $\vec{r}$ can be estimated from Eq.(2.3.17). It is of the order of (10 - 100) $\AA$, which is larger than the lattice parameter a. Therefore, the electron-electron interaction in semiconductors can be considered as a small perturbation of the electron gas. In the linear response approximation the time-dependent bare potential $U(\vec{r},t)$ created by an electron charge fluctuation can be taken in the form of a single Fourier harmonic, e.g. in the form of traveling wave with frequency ω and wave vector $\vec{q}$:

$$U(\vec{r},t) = U_{\vec{q}} e^{-i\omega t + i\vec{q}\cdot\vec{r}} + U_{\vec{q}} e^{i\omega t - i\vec{q}\cdot\vec{r}}, \tag{7.1.1}$$

where $U_{\vec{q}}$ is a real amplitude. To find the charge fluctuation that corresponds to the potential energy $U(\vec{r},t)$, we solve the time-dependent Schrödinger equation

$$i\hbar \frac{\partial \phi(\vec{r},t)}{\partial t} = \left[\hat{H}^{(0)} + U(\vec{r},t)\right]\phi(\vec{r},t) \tag{7.1.2}$$

with the help of the time-dependent perturbation approximation, considered in Section 1.16. Here $\hat{H}^{(0)}$ is the unperturbed single-electron Hamiltonian. The zero approximation wave functions have been obtained in Section 2.3, Eq.(2.3.11),

$$\phi_{\vec{p}}^{(0)} = \frac{1}{\sqrt{V}} e^{-\frac{i}{\hbar}E(\vec{p})t} e^{\frac{i}{\hbar}\vec{p}\cdot\vec{r}}. \tag{7.1.3}$$

In the initial state, $|\phi_{\vec{p}}^{(0)}|^2$ is a constant and the electrons are in an equilibrium state. When the small perturbation given by Eq.(7.1.1) is taken into account in Eq.(7.1.2), the small first order time-dependent perturbation $\phi_{\vec{p}}^{(1)}$ can be found by using the time-dependent perturbation calculations from Section 1.16

$$\phi_{\vec{p}}^{(1)}(\vec{r},t) = -\Bigg[\frac{U_{\vec{q}} e^{-i\left(\omega + \frac{E(\vec{p})}{\hbar}\right)t} e^{\frac{i}{\hbar}(\vec{p}+\hbar\vec{q})\vec{r}}}{E^{(0)}(\vec{p}+\hbar\vec{q}) - E^{(0)}(\vec{p}) - \hbar\omega} + \frac{U_{\vec{q}} e^{i\left(\omega - \frac{E(\vec{p})}{\hbar}\right)t} e^{\frac{i}{\hbar}(\vec{p}-\hbar\vec{q})\vec{r}}}{E^{(0)}(\vec{p}-\hbar\vec{q}) - E^{(0)}(\vec{p}) + \hbar\omega}\Bigg]. \tag{7.1.4}$$

The total wave function $\phi_{\vec{p}}(\vec{r},t)$ equals

$$\phi_{\vec{p}}(\vec{r},t) = \phi_{\vec{p}}^{(0)} + \phi_{\vec{p}}^{(1)}. \tag{7.1.5}$$

The local charge density fluctuation $\delta\rho(\vec{r},t)$ connected with the potential $U(\vec{r},t)$, is

$$\delta\rho(\vec{r},t) = (-e)\sum_{\vec{p}} f_e^{(0)}(\vec{p})\,[|\,\phi_{\vec{p}}\,|^2 - 1]. \tag{7.1.6}$$

The second term in the brackets represents the uniformly distributed positive charge of the ion cores which is needed here in order to have a stable, electrically neutral system, $f_e^{(0)}(\vec{p})$ is the Fermi-Dirac occupation probability that selects in the sum over $\vec{p}$ the states occupied by electrons. Substituting the wave function Eq.(7.1.5) in Eq.(7.1.6), one finds in the linear approximation

$$\delta\rho(\vec{r},t) = (-e)\sum_{\vec{p}} f_e^{(0)}(\vec{p})\,\{\phi_{\vec{p}}^{*(0)}\phi_{\vec{p}}^{(1)} + \phi_{\vec{p}}^{(0)}\phi_{\vec{p}}^{*(1)}\}. \tag{7.1.7}$$

The substitution of $\phi_{\vec{p}}^{(1)}$ from Eq.(7.1.4) in Eq.(7.1.7) results in

$$\delta\rho(\vec{r},t) = -e\sum_{\vec{p}} f_e^{(0)}(\vec{p})\,[\frac{1}{E(\vec{p}) - E(\vec{p}+\hbar\vec{q}) + \hbar\omega} +$$

$$\frac{1}{E(\vec{p}) - E(\vec{p}-\hbar\vec{q}) - \hbar\omega}]\,U_{\vec{q}}\,e^{-i\omega t + i\vec{q}\cdot\vec{r}} + c.c. \tag{7.1.8}$$

Here and everywhere below, we have omitted the superscript (0) to the energy $E(\vec{p})$ in order to simplify the notation. Replacing the summation variable $\vec{p}$ by $\vec{p}+\hbar\vec{q}$ in the second term of Eq.(7.1.8) results in

$$\delta\rho(\vec{r},t) = (-e)\sum_{\vec{p}} \frac{f_e^{(0)}(\vec{p}) - f_e^{(0)}(\vec{p}+\hbar\vec{q})}{E(\vec{p}) - E(\vec{p}+\hbar\vec{q}) + \hbar\omega}\,U_{\vec{q}}\,e^{-i\omega t + i\vec{q}\vec{r}} + \text{ c.c.} \tag{7.1.9}$$

The charge density fluctuation $\delta\rho(\vec{r},t)$ gives rise to a fluctuation of the potential $\delta U(\vec{r},t)$ which can be found from the Poisson equation

$$\nabla^2\delta U(\vec{r},t) = -e\frac{\delta\rho}{\varepsilon_0}, \tag{7.1.10}$$

where ε_0 is the absolute permittivity of the crystal. Considering Eq.(7.1.10) in the Fourier space representation, we obtain

$$\delta U_{\vec{q}} = \frac{e^2}{\varepsilon_0 q^2}\sum_{\vec{p}} \frac{f_e^{(0)}(\vec{p}) - f_e^{(0)}(\vec{p}+\hbar\vec{q})}{E(\vec{p}) - E(\vec{p}+\hbar\vec{q}) + \hbar\omega}\,U_{\vec{q}}\,. \tag{7.1.11}$$

Here $\delta U_{\vec{q}}$ is the Fourier transform of the potential fluctuation $\delta U(\vec{r},t)$. Equation (7.1.11) represents the potential energy connected with charge redistribution in the system.

To make our consideration ***self-consistent***, we should assume that the calculated potential energy $\delta U_{\vec{q}}$ is equal to the initial potential energy $U_{\vec{q}}$ given by Eq.(7.1.11), which creates the redistribution of the electron charge. The condition for the ***self-consistent*** behavior of the electron gas is

$$U(\vec{r},t) = \delta U(\vec{r},t). \tag{7.1.12}$$

Equation (7.1.12) defines the intrinsic behavior of the electron gas in the presence of electron-electron interactions, see Section 7.3.

If an external field is applied to the electron gas, the initial potential $U(\vec{r},t)$ Eq.(7.1.1), is equal to the sum of δU and the potential energy due to the external field $W(\vec{r},t)$.

$$U(\vec{r},t) = W(\vec{r},t) + \delta U(\vec{r},t). \tag{7.1.13}$$

Fourier transforming Eq.(7.1.13) and substituting Eq.(7.1.11), one obtains

$$U_{\vec{q}} = W_{\vec{q}} + \frac{e^2}{\varepsilon_0 q^2}\sum_{\vec{p}} \frac{f_e^{(0)}(\vec{p}) - f_e^{(0)}(\vec{p}+\hbar\vec{q})}{E(\vec{p}) - E(\vec{p}+\hbar\vec{q}) + \hbar\omega} U_{\vec{q}}, \tag{7.1.14}$$

where $W_{\vec{q}}$ is the Fourier transform of $W(\vec{r},t)$. Solving Eq.(7.1.14) for $U_{\vec{q}}$, we find

$$U_{\vec{q}} = \frac{\varepsilon_0 W_{\vec{q}}}{\varepsilon(\vec{q},\omega)}, \tag{7.1.15}$$

where

$$\varepsilon(\vec{q},\omega) = \varepsilon_0 - \frac{e^2}{q^2}\sum_{\vec{p}} \frac{f_e^{(0)}(\vec{p}) - f_e^{(0)}(\vec{p}+\hbar\vec{q})}{E(\vec{p}) - E(\vec{p}+\hbar\vec{q}) + \hbar\omega} \tag{7.1.16}$$

is the Lindhardt ***dielectric function*** of the electron gas. According to Eq.(7.1.15), the effective potential $U_{\vec{q}}$ acting on electrons equal the external potential $W_{\vec{q}}$ divided by the dielectric function $\varepsilon(\vec{q},\omega)$, which depends on the frequency ω and the wave vector $\vec{q}$.

As an example, we consider the response of the electron gas with electron-electron interactions to an external static electric field. The response is defined in this case by the dielectric function Eq.(7.1.16) evaluated at $\omega = 0$ and in the long-wave length limit $\varepsilon(\vec{q},0)\,|_{\vec{q}\to 0}$. To find the correct formula for $\varepsilon(\vec{q},\omega)$ in this limit, one should make an expansion of numerator and denominator in Eq.(7.1.15) in powers of $\vec{q}$

$$f_e^{(0)}(\vec{p}) - f_e^{(0)}(\vec{p}+\hbar\vec{q}) \cong -\hbar\vec{q}\cdot\frac{\partial f_e^{(0)}(\vec{p})}{\partial E(\vec{p})}\frac{\partial E(\vec{p})}{\partial \vec{p}}, \tag{7.1.17.a}$$

$$E(\vec{p}) - E(\vec{p}+\hbar\vec{q}) \cong -\hbar\vec{q}\cdot\frac{\partial E(\vec{p})}{\partial \vec{p}}. \tag{7.1.17.b}$$

The substitution of Eq.(7.1.17) in Eq.(7.1.16) results in

$$\varepsilon(q,0)\,|_{\vec{q}\to 0} = \varepsilon_0 + \frac{e^2}{q^2}\sum_{\vec{p}}\left(-\frac{\partial f_e^{(0)}(\vec{p})}{\partial E(\vec{p})}\right). \tag{7.1.18}$$

It follows from Eq.(7.1.18) that $\varepsilon(\vec{q},0) \to \infty$ when $\vec{q} \to 0$. The effective potential energy from (7.1.15) vanishes in this limit

$$U_{\vec{q}} = \frac{\varepsilon_0 W_{\vec{q}}}{\varepsilon(\vec{q},0)}\,|_{\vec{q}\to 0} \to 0. \tag{7.1.19}$$

The result of Eq.(7.1.19) means that an external d.c. field is compensated totally by polarization effects in the electron gas. The effect is called the ***electrostatic screening*** of the d.c. field. Electrostatic screening of this type is required by the general condition for statistical equilibrium.

In an inhomogeneous medium, the concentration of carriers depends on the spatial coordinate $\vec{r}$,

$$n_0 = n(\vec{r}).$$

The corresponding value of the chemical potential for an inhomogeneous material according to Eq.(6.1.16) is also dependent on $\vec{r}$

$$\mu(\vec{r}) = k_B T \ln n(\vec{r}) + k_B T \ln \frac{1}{N_c}. \tag{7.1.20}$$

The condition of equilibrium of an inhomogeneous medium is defined by the constancy of the electrochemical potential

$$\mu(\vec{r}) - e\phi_e(\vec{r}) = Constant. \tag{7.1.21}$$

The variation of the electron concentration is accompanied by the creation of the electric potential energy $(-e)\phi(\vec{r})$. Substituting the chemical potential from Eq.(7.1.20) in Eq.(7.1.21), one obtains the equilibrium condition in the form

$$k_B T \ln n(\vec{r}) - e\phi_n(\vec{r}) = Constant. \tag{7.1.22}$$

It is seen from this result that any ***increase*** in the electron concentration $n(\vec{r})$ is accompanied a corresponding ***increase*** of the electric potential $\phi_e(\vec{r})$, see Fig.7.1.1. The electrochemical potential is maintained constant.

In the case of holes, the equilibrium condition is similar to Eq.(7.1.22)

$$k_B T \ln p(\vec{r}) + e\phi_p(\vec{r}) = Constant. \tag{7.1.23}$$

Here $p(\vec{r})$ is the inhomogeneous concentration of holes and $\phi_p(\vec{r})$ is the electrostatic potential of holes. We see that when $p(\vec{r})$ ***increases***, the electric potential $\phi_p(\vec{r})$ should ***decrease***. In the case of holes the hole concentration decreases to maintain the electrochemical potential of holes constant.

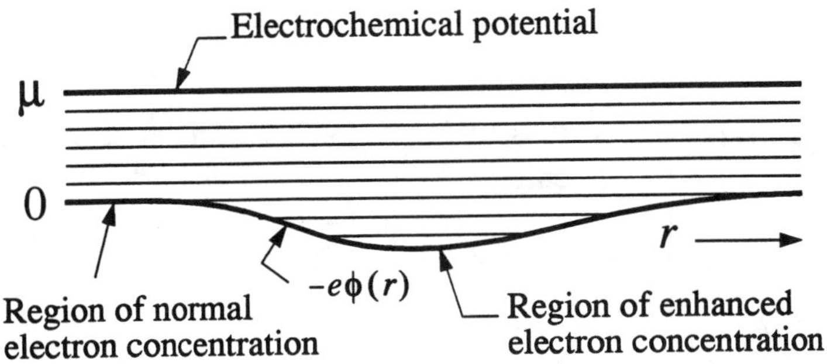

Figure 7.1.1 In thermal and diffusive equilibrium the electrochemical potential $\mu - e\phi$ is constant. To maintain it constant we increase the electron concentration in regions of space where the potential energy is low, and we decrease the concentration in regions where the potential is high (After C. Kittel "Introduction to Solid State Physics" John Wiley & Sons, N.Y., 1986).

7.2 SCREENING OF A CHARGED IMPURITY ATOM

Donors and acceptors in semiconductor crystals are charged impurities which possess a Coulomb-like potential energy, see Fig.7.2.1,

$$W(r)=\frac{e^2}{4\pi\varepsilon_0 |\vec{r}\,|}. \qquad (7.2.1)$$

This potential is a kind of external potential which is strongly influenced by the electron gas when the Coulomb interactions are taken into account. The Fourier transform of this potential energy has the form

$$W_{\vec{q}}=\frac{e^2}{\varepsilon_0 q^2}. \qquad (7.2.2)$$

The effective screened potential from Eq.(7.1.15) equals

$$U_{\vec{q}}=\frac{\varepsilon_0 W_q}{\varepsilon(\vec{q},\omega)}=\frac{e^2}{q^2\varepsilon(\vec{q},\omega)}. \qquad (7.2.3)$$

Since the potential energy Eq.(7.2.1) depends on $\vec{r}$ and does not depend on t, the Fourier transform $U_{\vec{q}}$ from Eq.(7.2.3) has a strong dependence on $\vec{q}$. The electrostatic screening is described by the dielectric function $\varepsilon(\vec{q},0)\,|_{\vec{q}\to 0}$ given by Eq.(7.1.18).

In the case of a nondegenerate electron gas in the crystal, the distribution function $f(\vec{p})$ has a Boltzman form (2.3.14). Its derivative is

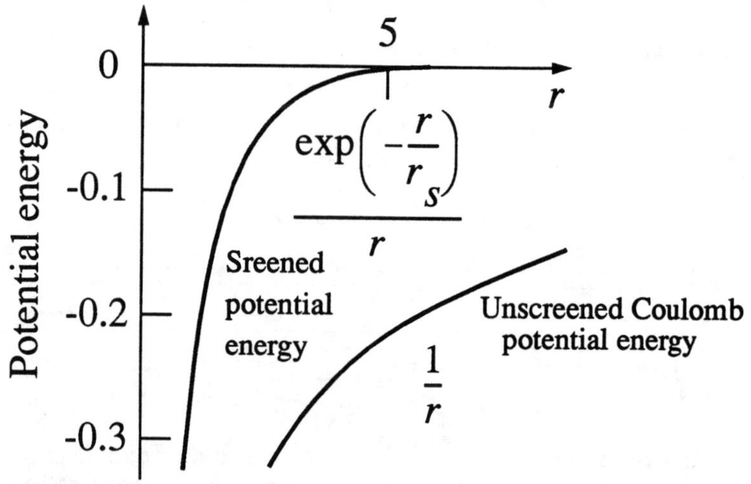

Figure 7.2.1 Comparisons of screened and unscreened Coulomb potentials of a unit positive charge acting on an electron. The screening length r_s has been taken equal to unity.

$$\frac{\partial f_e^{(0)}(\vec{p})}{\partial E(\vec{p})} = -\frac{1}{k_B T} f_e^{(0)}(\vec{p}). \tag{7.2.4}$$

Substituting Eq.(7.2.4) in Eq.(7.1.18), and taking into account that $\sum_p f(\vec{p}) = n_0$, one obtains

$$\varepsilon(\vec{q}) = \varepsilon_0 + \frac{e^2}{q^2}\frac{1}{k_B T}\sum_p f_e^{(0)}(\vec{p}) = \varepsilon_0 + \frac{e^2}{q^2}\frac{n_0}{k_B T} = \varepsilon_0\left(1 + \frac{\lambda^2}{q^2}\right), \tag{7.2.5}$$

where the quantity λ is defined by

$$\lambda^2 = \frac{e^2}{\varepsilon_0}\frac{n_0}{k_B T}. \tag{7.2.6}$$

Combining Eq.(7.2.5) and (7.2.3), we obtain

$$U_{\vec{q}} = \frac{e}{\varepsilon_0(q^2 + \lambda^2)}. \tag{7.2.7}$$

When $q \to 0$, the effective potential $U_{\vec{q}}$ remains finite. Recovering $U(\vec{r})$ by an inverse Fourier transformation, we find

$$U(r) = \frac{e}{4\pi\varepsilon_0 r} e^{-\lambda r}. \tag{7.2.8}$$

It is seen from Eq.(7.2.8) that an impurity Coulomb potential induces a space-dependent charge distribution which surrounds the impurity atom. The modification of the impurity potential created by the redistributed electron cloud is called the electrostatic ***screening*** of the impurity potential.

Equation (7.2.8) shows that the long-range part of the impurity Coulomb potential is compensated by screening. The quantity $\lambda^{-1} \equiv r_s$ is called the ***screening radius***, since the factor $e^{-\lambda r}$ in the potential (7.2.8) decreases to $1/e$ of its initial value at the distance

$$r_s = \lambda^{-1} = \left(\frac{\varepsilon_0 k_B T}{e^2 n_0}\right)^{1/2}. \tag{7.2.9}$$

A comparison of the screened potential Eq.(7.2.8) with the unscreened Coulomb potential Eq.(7.2.1) is presented in Fig.7.2.1. The long-range Coulomb potential of the impurity atom becomes a short range screened potential. It is seen from Eq.(7.2.9) that the screening radius decreases with increasing electron concentration. The effect of screening become stronger. The screening radius r_s, Eq.(7.2.9), was first calculated by Debye and Hückel in their theory of electrolytes. It is called the ***Debye screening radius.*** Order of magnitude estimates show that in typical semiconductors it is much larger than the lattice parameter a.

When the electron gas in a semiconductor is degenerate, the Fermi-Dirac distribution function should be used in Eq.(7.1.18). Similar calculations of corresponding screening radius lead to

$$\lambda^2 = \frac{3e^2}{2\varepsilon_0}\frac{n_0}{E_F}, \tag{7.2.10}$$

and the screening radius is defined by

$$r_s^2 = \lambda^{-2} = \left(\frac{4\varepsilon_0 E_F}{3e^2 n_0}\right)^{1/2}. \qquad (7.2.11)$$

We see that in a degenerate electron gas the screening radius decreases with increasing electron concentration n_0, and screening effects become more prominent.

Impurity centers in semiconductors scatter the conduction electrons. When the long-range impurity potential is reduced due to the screening to the screened impurity potential, the scattering of carriers by impurities is reduced as well. A comparison of the screened and unscreened potentials has already been made in Fig.7.2.1.

Effects of screening in semiconductor nanostructures with low dimensionality (quantum wells, quantum wires, and superlattices) differ from what we have in three dimensions. The differences follow from the different energy dependence of the densities of states in these systems. As a result, the electron-impurity interaction in low-dimensional systems is not as strongly modified by the presence of the other electrons as in three dimensions. This happens because of the strong anisotropy between the responses of the electron gas to an external perturbation parallel or transverse to the planar interfaces in the system. The in-plane response is similar to the bulk response. But in the transverse directions the carriers are strongly localized, and they are not able to distort their wave functions and create a change in the distribution of electrons. It was shown in this Section that in three-dimensional bulk material the screening radius decreases with increasing electron concentration n_0. In a two-dimensional quantum well the screening radius remains constant with the increase of n_0, and screening effects are considerably suppressed. The more the dimensionality is reduced, the stronger is the reduction of the screening.

7.3 PLASMA OSCILLATIONS

An important consequence of the electron-electron interaction is the existence of ***collective plasma oscillations*** in an electron system. To find the frequency of these oscillations, we consider the intrinsic behavior of the electron system, which is defined by the self-consistency condition, Eq.(7.1.12). Combining Eq.(7.1.12) with $\delta U_{\vec{q}}$ from Eq.(7.1.11), one finds

$$U_{\vec{q}} = \frac{e^2}{\varepsilon q^2} \sum_{\vec{p}} \frac{f_e^{(0)}(\vec{p}) - f_e^{(0)}(\vec{p} + \hbar\vec{q})}{E(\vec{p}) - E(\vec{p} + \hbar\vec{q}) + \hbar\omega} U_{\vec{q}}. \qquad (7.3.1)$$

If we introduce the dielectric permittivity given by Eq.(7.1.16), Eq.(7.3.1) takes the form

$$U_{\vec{q}}\,\varepsilon(\vec{q}, \omega) = 0. \qquad (7.3.2)$$

Because $U_{\vec{q}}$ does not vanish, Eq.(7.3.2) means that the intrinsic behavior of the electrons is defined by the equation $\varepsilon(\vec{q}, \omega) = 0$, e.g.

$$\varepsilon(\vec{q},\omega)=\varepsilon_0-\frac{e^2}{q^2}\sum_{\vec{p}}\frac{f_e^{(0)}(\vec{p})-f_e^{(0)}(\vec{p}+\hbar\vec{q})}{E(\vec{p})-E(\vec{p}+\hbar\vec{q})+\hbar\omega}=0. \tag{7.3.3}$$

The frequency ω which is the solution of Eq.(7.3.3) is called ***plasma*** frequency. It is convenient to replace the summation variable $\vec{p}$ by $\vec{p}-\hbar\vec{q}$ in the third term of Eq.(7.3.3), which yields for $\varepsilon(\vec{q},\omega)$ the representation

$$\varepsilon(\vec{q},\omega)=\varepsilon_0+\frac{e^2}{q^2}\sum_{\vec{p}}f_e^{(0)}(\vec{p})\left[\frac{1}{E(\vec{p}+\hbar\vec{q})-E(\vec{p})+\hbar\omega}+\frac{1}{E(\vec{p}-\hbar\vec{q})-E(\vec{p})-\hbar\omega}\right]. \tag{7.3.4a}$$

Replacing now $\vec{p}$ by $-\vec{p}$ and using the relations $E(-\vec{p})=E(\vec{p})$ and $f_e^{(0)}(E(-\vec{p}))=f_e^{(0)}(E(\vec{p}))$ in the third term of Eq.(7.3.4a), we obtain

$$\varepsilon(\vec{q},\omega)=\varepsilon_0+\frac{e^2}{q^2}\sum_{\vec{p}}2f_e^{(0)}(\vec{p})\frac{E(\vec{p}+\hbar\vec{q})-E(\vec{p})}{[E(\vec{p}+\hbar\vec{q})-E(\vec{p})]^2-(\hbar\omega)^2}. \tag{7.3.4b}$$

Experiments and theoretical estimates show that plasma collective oscillations occur at high frequencies $\omega > E(\vec{p})/\hbar$. This allows neglecting the electron energies in the denominator and making an expansion in powers of $\hbar\vec{q}$ in the numerator,

$$E(\vec{p}+\hbar\vec{q})-E(\vec{p})\cong\hbar\vec{q}\frac{\partial E(\vec{p})}{\partial\vec{p}}+\frac{1}{2}\hbar^2q^2\frac{\partial^2E(\vec{p})}{\partial\vec{p}^2}. \tag{7.3.5}$$

We must keep the term of the second order in q in this expansion, because the first term gives a vanishing contribution to the dielectric permittivity Eq.(7.3.4b) due to the condition $\sum_{\vec{p}}f_e^{(0)}(\vec{p})\frac{\partial E(\vec{p})}{\partial\vec{p}}=0$. Substituting Eq.(7.3.5) in Eq.(7.3.4b) results in the following expression for the dielectric function

$$\varepsilon(\vec{q},\omega)=\varepsilon_0-\frac{n_0e^2}{m^*\omega^2}, \tag{7.3.6}$$

where $\frac{1}{m^*}=\frac{\partial^2E(\vec{p})}{\partial\vec{p}^2}$ is the reciprocal effective mass, and $\sum_{\vec{p}}f_e^{(0)}(\vec{p})=n_0$ is the electron concentration. Finally, Eq.(7.3.3) reduces to

$$\omega^2=\omega_p^2, \tag{7.3.7a}$$

where

$$\omega_p=\left(\frac{n_0e^2}{\varepsilon_0m^*}\right)^{\frac{1}{2}} \tag{7.3.7b}$$

is the ***plasma oscillation*** frequency.

Plasma oscillations in an electron gas with Coulomb interactions are collective oscillations of the charge density. An order of magnitude estimate shows that $\hbar\omega_p$ in semiconductors is of the order of millielectronvolts (meV), i.e. of the order of an optical vibration frequency.

Including the next order terms in the expansion (7.3.5), one obtains the dependence of the plasma frequency on the wave vector $\vec{q}$

$$\omega_p(\vec{q}) = \omega_p\left(1 + \alpha\frac{(v_F q)^2}{\omega_p^2}\right).$$

Here α is a coefficient of the order of unity, and v_F is the electron velocity at the Fermi surface.

Quantization of the plasma oscillations according to the rules of quantum mechanics leads to the concept of quanta of these oscillations called ***plasmons.***

It follows from Eq.(7.3.6) that for $\omega < \omega_p$, the dielectric susceptibility of electron gas is negative, $\varepsilon < 0$. The corresponding refractive index $\kappa = \varepsilon^{1/2}$ is imaginary. This results in the total reflection of an electromagnetic wave from the surface of the semiconductor. When $\omega > \omega_p$, ε has a positive value and the material is transparent. The effect is well seen in optical spectra of semiconductors.

7.4 HEAVILY DOPED SEMICONDUCTORS

A very important group of semiconductor materials is called ***heavily doped*** semiconductors. The concentration of donors (or acceptors) can reach the level where the Bohr radii of impurity atoms a_0 overlap, that is when the average separation of impurities $\bar{r}$ is smaller than the Bohr radius of an impurity atom, see Fig.7.4.1. Then

$$\bar{r} \ll a_0 . \tag{7.4.1}$$

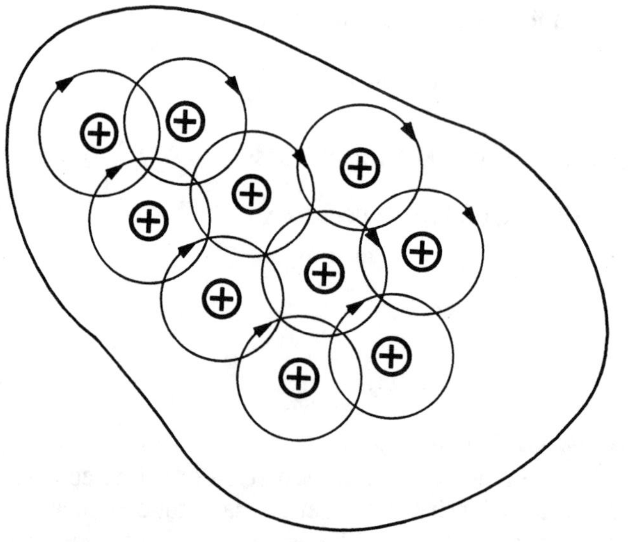

Figure 7.4.1 Overlapping impurity states in a heavily doped semiconductor.

Since the Pauli exclusion principle allows for any electron state to be occupied by only two electrons of opposite spins, some electrons from the regions of overlap go into the conduction band. These electrons form in the conduction band a highly degenerate electron gas with Fermi-energy E_F.

The high concentration of electrons results in the screening of impurity atoms. The average separation of impurity atoms in the crystal is connected with their concentration by the relation Eq.(2.3.17): $\bar{r} \sim n^{-1/3}$. The Bohr radius of an impurity atom has been defined by Eq.(6.4.2): $a_0 = \frac{4\pi\varepsilon_r\varepsilon_0\hbar^2}{m_n e^2}$. Substituting $\bar{r}$ and a_0 into Eq.(7.4.1), one obtains

$$\bar{r} \ll \frac{4\pi\varepsilon_r\varepsilon_0\hbar^2}{m_n e^2},$$

or

$$\frac{e^2}{4\pi\varepsilon_r\varepsilon_0\bar{r}} \ll \frac{\hbar^2}{m_n\bar{r}^2} \quad \text{or} \quad \frac{e'^2}{\varepsilon_r\bar{r}} \ll \frac{\hbar^2}{m_n\bar{r}^2}. \tag{7.4.2}$$

The quantity on the right hand side of inequality Eq.(7.4.2) is the kinetic energy of an electron in the degenerate gas in the conduction band. The quantity on the left hand side is the Coulomb potential energy of the interacting electrons.

The inequality Eq.(7.4.2) means that the potential energy of an electron is less than the kinetic energy, that is, the electron gas in the conduction band is a nearly ideal gas. This behavior of a degenerate electron gas has been discussed before, at the end of Section 2.3. It has been shown that in a degenerate gas the potential energy $e'^2/(\varepsilon_r\bar{r})$ increases with concentration n slower than the kinetic energy $\hbar^2/(m_n\bar{r}^2)$. As a result, the degenerate electron gas is more ideal when it is more dense.

A numerical estimate shows that the condition Eq.(7.4.2) holds for typical electron concentrations in semiconductors given by

$$n > \frac{1}{a_0^3}\left(\frac{m_n}{\varepsilon_0 m_0}\right)^3 \approx 10^{17} - 10^{19} cm^{-3}.$$

The inequality (7.4.2) is referred to in the literature as the ***heavy doping condition***.

When doping impurities are present in the crystal, the crystal lattice potential energy is not translationally invariant. It fluctuates from one lattice site to another. The bottom of the conduction band and the top of the valence band fluctuate in a similar way, see Fig.7.4.2. An estimate of the impurity potential in heavily doped materials allows finding the fluctuations of the crystal lattice potential energy.

It is known from statistical physics that the root mean square deviation of the total number of impurities $\sqrt{<(\Delta N)^2>}$ is proportional to the square root of the total number of impurities N

$$\sqrt{<(\Delta N)^2>} = \sqrt{<(N - <N>)^2>} \sim \sqrt{N}, \tag{7.4.3}$$

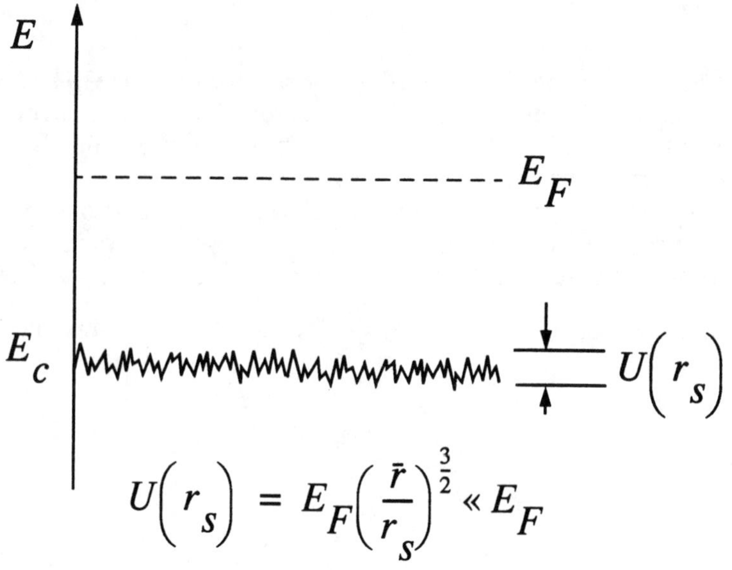

Figure 7.4.2 Conduction band of heavily doped semiconductor. Small fluctuations of electron potential energy do not affect degenerate electrons in the conduction band.

where $<N>$ is the average number of impurities. We consider the fluctuation of the total number of impurities in some volume of the crystal V_0. For simplicity, we take a spherical volume $V_0 = (4/3)\pi R^3$, where R is the radius of the sphere.

We take an n-type semiconductor with totally ionized impurities, where $n_0 = N_D$. The average number of charged impurities in the volume V_0 equals in order of magnitude to

$$n_0 R^3, \tag{7.4.4}$$

where n_0 is equal to the concentration of impurities N_D. The average fluctuation of this number according to Eq.(7.4.3) is

$$\sqrt{n_0 R^3}.$$

The charged impurities in the volume V_0 create a fluctuation of the electrostatic Coulomb potential energy

$$U(R) = \frac{e^2}{4\pi\varepsilon_r\varepsilon_0 R}\sqrt{nR^3}. \tag{7.4.5}$$

The linear size of the fluctuation R and the value of the fluctuation potential $U(R)$ are limited by the screening radius r_s, that is

$$R \le r_s\,. \tag{7.4.6}$$

Fluctuations $U(R)$ with $R > r_s$ are screened out. This means that in the crystal there are no potential wells with $R > r_s$. Taking $R_{\max} = r_s$ in Eq.(7.4.5), one finds for the maximum fluctuation of the potential energy $U(r_s)$

$$U(r_s) = \frac{e^2}{4\pi\varepsilon_r\varepsilon_0 r_s}\sqrt{n_0 r_s^3}\,. \tag{7.4.7}$$

Taking into account the value of r_s for a degenerate electron gas given by Eq.(7.2.10) results in

$$U(r_s) = E_F\left(\frac{1}{n_0 r_s^3}\right)^{1/2} = E_F\left(\frac{\bar{r}}{r_s}\right)^{3/2}. \tag{7.4.8}$$

For the effects of screening to be observable, we must have many electrons, so that $\bar{r} < r_s$. This means that the fluctuation potential energy Eq.(7.4.8) is much smaller than the Fermi-energy E_F as shown in Fig.7.4.2. The potential energy fluctuation $U(r_s)$ creates small fluctuations of the bottom of the conduction band which result in a tail of the density of states in the energy gap E_g.

The transition from the light doping region, where

$$n_0 a_B^3 \ll 1, \tag{7.4.9}$$

to the heavy doping region

$$n_0 a_B^3 \gg 1, \tag{7.4.10}$$

is an example of a transition from dielectric (semiconductor) to a metal with nearly free electrons. The possibility of having an ideal gas of carriers in a heavily doped semiconductor is very important for constructing semiconductor devices.

BIBLIOGRAPHY

J.M. Ziman *Principles of the Theory of Solids* (Cambridge University Press, 1972).

CHAPTER **8**

Transport in Semiconductors

8.1 KINETIC EQUATION

We have so far considered current carriers in semiconductors in thermodynamic equilibrium. In equilibrium, the electron distribution function $f_e^{(0)}(\vec{p})$ is, in general, the Fermi-Dirac function, which reduces to the Boltzmann distribution for a nondegenerate electron gas. It is a function of the electron energy $E(\vec{p})$ only. Since $E(\vec{p})$ is an even function of $\vec{p}$, the equilibrium distribution function is also an even function of $\vec{p}$. The probabilities of having the momentum $\vec{p}$ or the momentum $(-\vec{p})$ are equal in equilibrium.

In equilibrium, the electrons in a semiconductor perform random thermal motion. When an external electric field $\vec{\mathcal{E}}$ is applied to the crystal, the system becomes a nonequilibrium one, since there appears a force $\vec{F} = -e\vec{\mathcal{E}}$ which acts on each electron. The distribution function becomes a nonequilibrium function $f(\vec{p}, t)$. The corresponding acceleration of an electron equals

$$\vec{w} = (-e)\frac{\vec{\mathcal{E}}}{m_n}. \tag{8.1.1}$$

Since the accelerated motion of electrons is superimposed on their random thermal motion, this acceleration occurs during the time Δt between electronic collisions with the crystal lattice, impurities, or themselves. The electron is able to increase its velocity by the amount

$$\vec{u} = \vec{w}\Delta t = (-e)\frac{\vec{\mathcal{E}}}{m_n}\Delta t \ . \tag{8.1.2}$$

The velocity $\vec{u}$ is called the ***drift velocity*** of an electron. Velocities of the random thermal motion of an electron are of the order of $10^6 cm/s = 10^4 m/s$ for low electric field. They are several orders of magnitude larger than drift velocities $u \approx 10^2 cm/s = 1\ m/s$.

When an electron gains the directed drift velocity $\vec{u}$, it always collides either with the crystal lattice or with an impurity atom. The electron loses its direction of motion and the drift velocity. Then, the velocity again increases during the time Δt and reaches the value $\vec{u}$. The next collision of the electron again destroys both the direction and the magnitude of the velocity $\vec{u}$.

A regime is possible in which the accelerated directed motion of electrons in an external field $\vec{\mathcal{E}}$ and random collisions of electrons with the lattice or with impurity atoms, lead to a certain dynamic equilibrium balance. This balance can be described in terms of the nonequilibrium electron distribution function $f(\vec{p})$. The system reaches equilibrium when the rate of change of the distribution function due to the external field $\vec{\mathcal{E}}$ is equal to the rate of change of $f(\vec{p})$ due to electronic collisions. The equilibrium condition is

$$\left(\frac{df}{dt}\right)_{field} = \left(\frac{df}{dt}\right)_{collisions} \tag{8.1.3}$$

This equation was written for the first time by Boltzmann, and it is called the ***Boltzmann equation*** or the ***kinetic equation.*** The kinetic equation defines a stationary flux of electrons. If we put a semiconductor wire into a circuit and look at a certain cross section of this wire, we find that the number of electrons going from left to right is less than the number of electrons going from right to left, see Fig.8.1.1. The electron gas as a whole drifts with the constant velocity $\vec{u}$ in the direction opposite to $\vec{\mathcal{E}}$.

We start with a study of the left hand term in the kinetic equation (8.1.3) which describes the variation of the distribution function in the presence of the electric field $\vec{\mathcal{E}}$.

In order to find the dependence of $f(\vec{p})$ on $\vec{\mathcal{E}}$, we consider the effect of an external field $\vec{\mathcal{E}} = (\mathcal{E}_x,0,0)$ on the electrons in the elemental volume $dp_x dp_y dp_z$ at the point $\vec{p} = (p_x,p_y,p_z)$ of $\vec{p}$-space, see Fig.8.1.2. The number of electrons with the energy $E(\vec{p})$ in this volume is defined by the nonequilibrium distribution function $f(\vec{p})$. Being accelerated by the external field $\vec{\mathcal{E}}$, these electrons leave the volume $dp_x dp_y dp_z$ at $\vec{p}$ in the mean free path time Δt. Some other electrons which increase their quasimomenta from the value $(p_x - m_n w_x \Delta t, p_y, p_z)$ to (p_x, p_y, p_z), enter this volume.

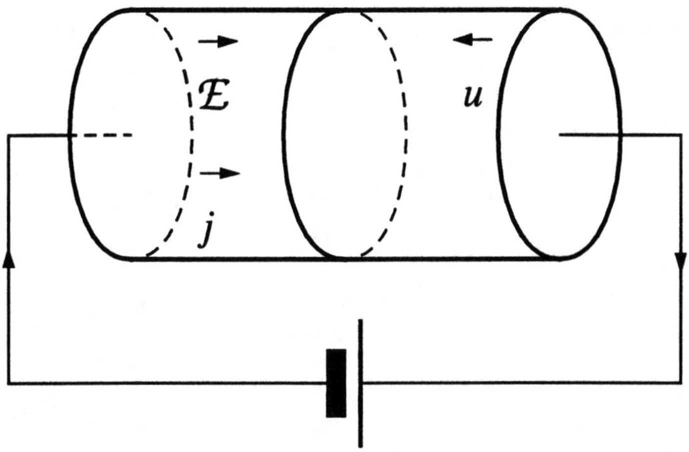

Figure 8.1.1 The directions of drift velocity and electron current density in the circuit.

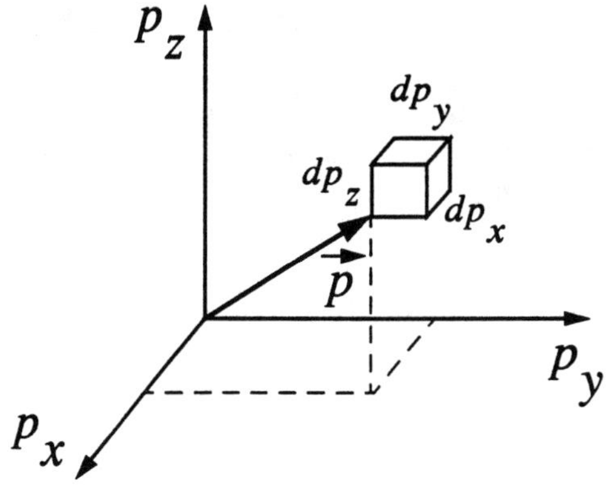

Figure 8.1.2 The elemental volume in $\vec{p}$-space.

Their number is proportional to the distribution function $f(p_x - m_n w_x \Delta t, p_y, p_z)$. The change in the number of electrons in the volume $dp_x dp_y dp_z$ is equal to the difference of the number of electrons leaving it and the number of electrons entering it,

$$f(p_x, p_y, p_z) - f(p_x - m_n w_x \Delta t, p_y, p_z).$$

Making here an expansion in the small quantity $\Delta p_x = m_n w_x \Delta t$, we find

$$f(p_x, p_y, p_z) - f(p_x - m_n w_x \Delta t, p_y, p_z) \approx \frac{df}{dp_x} m_n w_x \Delta t. \tag{8.1.4}$$

Dividing Eq.(8.1.4) by Δt, passing to the limit $\Delta t \to 0$ on the left hand side of this equation and substituting the acceleration Eq.(8.1.1) in the right hand side results in the rate of change of the distribution function

$$\left(\frac{df(\vec{p})}{dt}\right)_{field} = (-e)\frac{df(\vec{p})}{dp_x}\mathcal{E}_x. \tag{8.1.5}$$

The right hand side of the Boltzmann equation (8.1.5) is determined by the collisions of electrons from the elemental volume $dp_x dp_y dp_z$ with lattice vibrations or some other defects of the crystal, for example, impurities. Each collision changes either the direction or the magnitude of the electron quasimomentum $\vec{p}$, and removes the electron from the volume $dp_x dp_y dp_z$. At the same time, some scattering processes induce the electrons to leave the state $\vec{p}'$ and enter the volume $dp_x dp_y dp_z$. We denote the probability of an electron collision that changes its quasimomentum from $\vec{p}$ to $\vec{p}'$ by $W(\vec{p},\vec{p}')$.

The transition $\vec{p} \rightarrow \vec{p}'$ occurs in the case where the initial state $\vec{p}$ is filled by an electron. The probability of the state $\vec{p}$ being filled is given by the distribution function $f(\vec{p})$. The final state into which the electron goes should be empty. The probability of the state $\vec{p}'$ being empty equals $[1-f(\vec{p}')]$.

The scattering rate (the number of transitions per second) ***out of the state*** $\vec{p}$ into some other state $\vec{p}'$ is given by the product of all three factors

$$W(\vec{p},\vec{p}')f(\vec{p})[1-f(\vec{p}')]. \tag{8.1.6}$$

There are also collisions that bring some electrons ***into the state*** $\vec{p}$ from all the other states $\vec{p}'$. The rate of scattering into the state $\vec{p}$ equals

$$W(\vec{p}',\vec{p})f(\vec{p}')[1-f(\vec{p})]. \tag{8.1.7}$$

The total rate of change of the distribution function due to the scattering processes is equal to the difference of the rates at which electrons are scattered into and out of the state $\vec{p}$,

$$\left(\frac{df}{dt}\right)_{collisions} = \sum_{\vec{p}'}\{W(\vec{p}',\vec{p})f(\vec{p}')[1-f(\vec{p})] - W(\vec{p},\vec{p}')f(\vec{p})[1-f(\vec{p}')]\}. \tag{8.1.8}$$

Here the sum is taken over the all possible states $\vec{p}'$.

Substituting Eqs.(8.1.5) and Eq.(8.1.8) into the Boltzmann kinetic equation (8.1.3), we find

$$(-e)\frac{df(\vec{p})}{dp_x}\mathcal{E}_x = \sum_{\vec{p}'}\{W(\vec{p}',\vec{p})f(\vec{p}')[1-f(\vec{p})] - W(\vec{p},\vec{p}')f(\vec{p})[1-f(\vec{p}')]\}. \tag{8.1.9}$$

In equilibrium ($\vec{\mathcal{E}}=0$), the number of electrons leaving the state $\vec{p}$ and going into the state $\vec{p}'$, and the number leaving the state $\vec{p}'$ and going into the state $\vec{p}$ should be equal. Therefore,

$$W(\vec{p}',\vec{p})f_e^{(0)}(\vec{p}')[1-f_e^{(0)}(\vec{p})] = W(\vec{p},\vec{p}')f_e^{(0)}(\vec{p})[1-f_e^{(0)}(\vec{p}')]. \tag{8.1.10}$$

Substituting here Fermi-Dirac distribution function, one obtains the useful relation

$$W(\vec{p}',\vec{p})e^{\frac{E(\vec{p})}{k_BT}} = W(\vec{p},\vec{p}')e^{\frac{E(\vec{p}')}{k_BT}}. \tag{8.1.11}$$

The quantum mechanical perturbation approximation from Section 1.16 provides the procedure for calculating $W(\vec{p},\vec{p}')$ in every particular case of electron scattering

$$W(\vec{p}',\vec{p}) = \frac{2\pi}{\hbar}|\hat{H}_{fi}^{(1)}|^2\,\delta(E_f-E_i). \tag{8.1.12}$$

Here $|\hat{H}_{fi}^{(1)}|^2$ is the square of a matrix element which depends on the particular scattering mechanism, and $\delta(E_f - E_i)$ shows that transitions occur only between initial and final states of equal total energy, which means that the energy conservation law holds for all transitions. Possible scattering mechanisms will be discussed in Section 8.4.

8.2 RELAXATION TIME APPROXIMATION

In many cases electron scattering occurs due to different imperfections in the crystal lattice, (e.g. impurities, dislocations, or the surface). The scattering is elastic in these cases, that is the electron energy does not change in the process of scattering

$$E(\vec{p}) = E(\vec{p}'). \tag{8.2.1}$$

This means that the magnitude of the electron quasimomentum remains constant in the scattering process. Only the direction of the quasimomentum changes. Combining Eq.(8.2.1) with Eq.(8.1.11), one obtains the relation

$$W(\vec{p}', \vec{p}) = W(\vec{p}, \vec{p}'), \tag{8.2.2}$$

which allows transforming the kinetic equation (8.1.9) into

$$(-e)\frac{df(\vec{p})}{dp_x}\mathcal{E}_x = \sum_{\vec{p}'} W(\vec{p}, \vec{p}')[f(\vec{p}') - f(\vec{p})]. \tag{8.2.3}$$

Assuming the external electric field $\vec{\mathcal{E}} = (\mathcal{E}_x, 0, 0)$ to be a small perturbation, we seek the solution of the kinetic equation (8.2.4) in the form of the expansion

$$f(\vec{p}) = f_e^{(0)}(\vec{p}) + f^{(1)}(\vec{p}), \tag{8.2.4}$$

where $f_e^{(0)}(\vec{p})$ is the equilibrium Fermi-Dirac distribution function and $f^{(1)}(\vec{p})$ is a small nonequilibrium perturbation of it which depends linearly on the external field $\vec{\mathcal{E}}$. Because in elastic scattering the quasimomentum of an electron $\vec{p}$ changes only its direction, it is convenient to write $f^{(1)}(\vec{p})$ in the following form

$$f^{(1)}(\vec{p}) = \frac{df_e^{(0)}(\vec{p})}{dE}\vec{\chi}(E) \cdot \vec{p}, \tag{8.2.5}$$

where the vector $\vec{\chi}(E)$ depends only on the energy of the electron E. Substitution of Eq.(8.2.5) in the difference $[f(\vec{p}') - f(\vec{p})]$ appearing in Eq.(8.2.3) results in

$$[f(\vec{p}') - f(\vec{p})] = \frac{df_e^{(0)}}{dE}\vec{\chi}(E) \cdot [\vec{p}' - \vec{p}] = -\frac{df_e^{(0)}}{dE}\vec{\chi}(E) \cdot \vec{p}\left[1 - \frac{p_\chi'}{p_\chi}\right] =$$

$$-f^{(1)}(\vec{p})\left[1 - \frac{p_\chi'}{p_\chi}\right], \tag{8.2.6}$$

where p_χ is the component of $\vec{p}$ in the direction of the vector $\vec{\chi}$. Combining Eq.(8.2.6) with Eq.(8.2.3), results in

$$(-e)\frac{df(\vec{p})}{dp_x}\mathcal{E}_x = -\frac{f^{(1)}(\vec{p})}{\tau(\vec{p})}, \tag{8.2.7}$$

where the quantity

$$\frac{1}{\tau(\vec{p})} = \sum_{\vec{p}'} W(\vec{p},\vec{p}')\left[1-\frac{p_x'}{p_x}\right] \tag{8.2.8}$$

is called the relaxation rate and τ is the ***relaxation time.*** To clarify the meaning of this term we consider Eq.(8.2.7) in the case where the electric field $\vec{\mathcal{E}}$ is switched off. The kinetic equation takes the form

$$\frac{\partial f(\vec{p})}{\partial t} = -\frac{f^{(1)}(\vec{p})}{\tau(\vec{p})}. \tag{8.2.9}$$

Substituting Eq.(8.2.4) in the left hand side of Eq.(8.2.9), separating variables and integrating, one can find

$$f^{(1)}(\vec{p}) = f(\vec{p}) - f_e^{(0)}(\vec{p}) = Const\, e^{-\frac{t}{\tau(\vec{p})}} \tag{8.2.10}$$

This result means that when the electric field is switched off, the equilibrium distribution of electrons $f_e^{(0)}(\vec{p})$ is restored after a time t of the order of τ according to the exponential law. This is the reason why τ is called the relaxation time.

It should be noted that the relaxation time approach, Eq.(8.2.8), is correct if the electron scattering is elastic. In case of inelastic anisotropic scattering of electrons, the relaxation time approximation does not work.

8.3 ELECTRICAL CONDUCTIVITY

Knowledge of the nonequilibrium distribution function $f(\vec{p})$ allows finding several important kinetic properties of electrons (or holes) in the crystal. One of them is the electrical current which flows in a semiconductor in the presence of the external electric field $\vec{\mathcal{E}} = (\mathcal{E}_x, 0, 0)$.

In order to find the density of electric current, we consider an imaginary elemental cylinder with cross sectional area dS that are normal to the x-direction of the electric field and whose height is $v_x dt$, see Fig.8.3.1. Here v_x is the x-component of the electron velocity defined in Eq.(2.5.12). The volume of the cylinder is $dS\, v_x dt$. The number of electrons with momenta between $\vec{p}$ and $\vec{p}+d\vec{p}$ inside the cylinder is

$$f(\vec{p})dp_x dp_y dp_z dS\, v_x dt.$$

Being accelerated by the external electric field, all these electrons will leave the cylinder within the time interval dt, crossing the bottom of the cylinder dS. If we take into account both the electrons crossing the area dS from the left to the right and from the right to the left, the net number of electrons crossing dS in the direction of the field $\mathcal{E}_x$ and in the time interval dt is

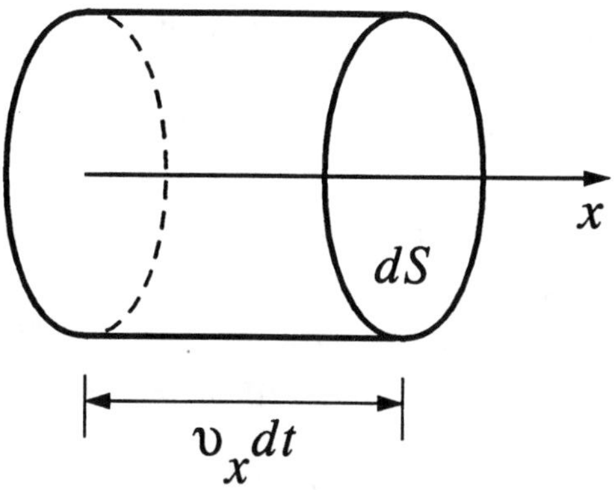

Figure 8.3.1 Elemental cylinder with cross-section dS and height $v_x dt$.

$$dt dS \int f(\vec{p}) v_x dp_x dp_y dp_z \,. \tag{8.3.1}$$

The charge of each electron is $(-e)$. Multiplying Eq.(8.3.1) by $(-e)$, dividing by $dt dS$, and using Eq.(2.3.39), we find the net flow of charged particles or the electric current density

$$j_x = (-e)\frac{2}{(2\pi\hbar)^3}\int f(\vec{p}) v_x dp_x dp_y dp_z \,. \tag{8.3.2}$$

Substituting $f(\vec{p})$ in the form of Eq.(8.2.4) and taking into account that $f_e^{(0)}(\vec{p})$ is an even function of $\vec{p}$ and does not contribute to the current density Eq.(8.3.2), one obtains

$$j_x = (-e)\frac{2}{(2\pi\hbar)^3}\int f^{(1)}(\vec{p}) v_x dp_x dp_y dp_z \,. \tag{8.3.3}$$

In order to find $f^{(1)}(\vec{p})$, we should substitute Eq.(8.2.4) into the kinetic equation (8.2.7):

$$(-e)\mathcal{E}_x \frac{d[f_e^{(0)} + f^{(1)}(\vec{p})]}{dp_x} = -\frac{f^{(1)}(\vec{p})}{\tau(\vec{p})}. \tag{8.3.4}$$

Since the left hand side of Eq.(8.3.4) contains the small electric field $\mathcal{E}_x$, we should neglect $f^{(1)}(\vec{p})$ with respect to $f_e^{(0)}(\vec{p})$ there. Only the first order terms in $\mathcal{E}_x$ should be kept in the equation for $f^{(1)}(\vec{p})$. The kinetic equation (8.3.4) in this approximation takes the form

$$(-e)\mathcal{E}_x \frac{df_e^{(0)}(\vec{p})}{dp_x} = -\frac{f^{(1)}(\vec{p})}{\tau(\vec{p})}. \tag{8.3.5}$$

It follows from Eq.(8.3.5) that

$$f^{(1)}(\vec{p}) = e\mathcal{E}_x \frac{df_e^{(0)}}{dp_x}\tau(\vec{p}). \tag{8.3.6}$$

Taking into account that $f_e^{(0)}(\vec{p})$ depends, in fact, only on the energy of the electron and calculating the chain derivative, we can represent Eq.(8.3.6) in the form

$$f^{(1)}(\vec{p}) = e\tau(\vec{p})\mathcal{E}_x v_x \frac{df_e^{(0)}(E)}{dE}, \tag{8.3.7}$$

where $v_x = dE(\vec{p})/dp_x$. Combining Eq.(8.3.7) with Eq.(8.3.3) results in

$$j_x = -e^2\mathcal{E}_x \frac{2}{(2\pi\hbar)^3}\int v_x^2 \frac{df_e^{(0)}(E)}{dE}\tau(\vec{p})dp_x dp_y dp_z \,. \tag{8.3.8}$$

The total velocity of electron is the sum of drift velocity $\vec{u}$ from Eq.(8.1.2) and the thermal velocity. Since the drift velocity is usually very small in comparison with the thermal velocity, the drift velocity can be neglected in the integrand of Eq.(8.3.8). Then, it may be assumed that for the random thermal motion of electrons

$$v_x^2 = \frac{v^2}{3} = \frac{2E}{3m_n}. \tag{8.3.9}$$

Substituting Eq.(8.3.9) into Eq.(8.3.8) and introducing the electron density of states $D(E)$, we find the density of current in the form

$$j_x = -e^2\mathcal{E}_x \frac{2}{3m_n}\int D(E)E\tau(E)\frac{df_e^{(0)}(E)}{dE}dE, \tag{8.3.10}$$

where the integral over $dp_x dp_y dp_z$ is replaced by the density of states and an integral over the electron energy E.

The concentration of electrons in the band is determined through the density of states as

$$n_0 = \sum_p f_e^{(0)} = \int D(E)f_e^{(0)}(E)dE. \tag{8.3.11}$$

Combining Eq.(8.3.10) and Eq.(8.3.11) one obtains the final expression for the current density

$$j_x = \frac{n_0 e^2}{m_n} < \tau > \mathcal{E}_x \,, \tag{8.3.12}$$

where the average relaxation time is

$$< \tau > = -\frac{2}{3}\frac{\int D(E)E\tau(E)\frac{df_e^{(0)}(E)}{dE}dE}{\int D(E)f_e^{(0)}(E)dE}. \tag{8.3.13}$$

In a three-dimensional electron gas the electron density of states $D(E)$ varies as a square root of the energy, see Eq.(6.1.3). Substituting this form for $D(E)$ in Eq.(8.3.13), and integrating by parts in denominator one finds

$$< \tau >_{3D} = \frac{\int E^{3/2}\tau(E)\frac{df_e^{(0)}}{dE}dE}{\int E^{3/2}\frac{df_e^{(0)}}{dE}dE}. \tag{8.3.14}$$

Semiconductor nanostructures with low dimensionality of the electron gas have some other energy dependences of density of states which have been discussed in Section 6.3. In the two-dimensional electron gas of a quantum well, the density of states has the constant value given by Eq.(6.3.5). Substituting this density of states in Eq.(8.3.13), results in

$$<\tau>_{2D} = \frac{\int E\tau(E)\frac{df_e^{(0)}}{dE}dE}{\int E\frac{df_e^{(0)}}{dE}dE}. \tag{8.3.15}$$

In the one-dimensional electron gas of a quantum wire, the electron density of states is proportional to the inverse square root of the energy, see Eq.(6.3.10). The average relaxation time $\tau(E)$ takes the form

$$<\tau>_{1D} = \frac{\int E^{1/2}\tau(E)\frac{df_e^{(0)}}{dE}dE}{\int E^{1/2}\frac{df_e^{(0)}}{dE}dE}. \tag{8.3.16}$$

The integration in the integrals of Eqs.(8.3.14), (8.3.15), and (8.3.16) can be taken from zero up to infinity, because it is accepted that the conduction band is wide enough and the electron distribution function is zero at the upper edge of the conduction band. Note that for $\tau(E) = \tau_0 = const(E)$ we get from Eqs.(8.3.14)-(8.3.16) that $<\tau> = \tau_0$.

If the external electrical field has an arbitrary direction, $\vec{\mathcal{E}} = (\mathcal{E}_x, \mathcal{E}_y, \mathcal{E}_z)$, a similar calculation leads to the vector form of Eq.(8.3.12)

$$\vec{j} = \frac{n_0 e^2 <\tau>}{m_n}\vec{\mathcal{E}} = \sigma\vec{\mathcal{E}}. \tag{8.3.17}$$

Equation (8.3.17) shows that the density of electric current $\vec{j}$ is proportional to the external electric field $\vec{\mathcal{E}}$. This is ***Ohm's law in differential form***. The proportionality coefficient σ is called ***electrical conductivity***

$$\sigma = \frac{n_0 e^2 <\tau>}{m_n}. \tag{8.3.18}$$

Sometimes the reciprocal quantity $\rho = \sigma^{-1}$ is used, which is called the ***resistivity***.

The differential form of Eq.(8.3.17) means that this relation holds at each point inside the semiconductor. The conductivity σ is a characteristic of the material, and depends only on the physical state of the material, for example, on the type of chemical bond in the material, or on the temperature.

The electric current density Eq.(8.3.17) can be represented in the form

$$\vec{j} = (-e)n_0(-e)\left[\frac{<\tau>}{m_n}\right]\vec{\mathcal{E}} = -(-e)n_0 b\vec{\mathcal{E}}, \tag{8.3.19}$$

where the quantity b equals

$$b = e\frac{<\tau>}{m_n}. \tag{8.3.20}$$

b is called the ***electron mobility***. Comparing the right hand side of Eq.(8.3.19) with electric current density $\vec{j} = (-e)n_0 < \vec{v} >$ from physics of electromagnetism, we see that the mobility is the average velocity of an electron in a unit electric field. In some books one minus from Eq.(8.3.19) is introduced into Eq.(8.3.20) to indicate that the electronic mobility is negative, i.e. electrons move opposite to the electric field.

Similar consideration leads to the mobility of holes, which is a positive quantity. Because the change of the charge sign of the carrier is accompanied by the change of the sign of mobility, the conductivity σ does not depend on the type of the carrier. Like the conductivity, the mobility is an intrinsic property of a semiconductor material: it does not depend on $\vec{\mathcal{E}}$ and depends on the temperature and the doping concentration. The mobility describes how well a carrier moves in an external electric field.

Our expansion Eq.(8.2.4) has some limits on its validity. It is assumed in Eq.(8.3.4) that $f^{(1)} \ll f_e^{(0)}$. An order of magnitude estimate shows

$$f^{(1)} = e\mathcal{E}_x \frac{df_e^{(0)}}{dp_x}\tau \approx \frac{e\mathcal{E}_x}{m_n v_{th}} f_e^{(0)}\tau = \frac{u}{v_{th}} f_e^{(0)}, \tag{8.3.21}$$

where v_{th} is the thermal velocity of an electron. If the drift velocity $u \ll v_{th}$, our approach is valid. A limitation on the allowed value of the external field $\vec{\mathcal{E}}$ follows from the inequality

$$u = \frac{e\mathcal{E}\tau}{m_n} \ll v_{th} \quad \text{or} \quad \mathcal{E} \ll \frac{m_n v_{th}}{e\tau}. \tag{8.3.22}$$

The inequality (8.3.22) places a limit on the magnitude of the electric field $\mathcal{E}$.

In metals at room temperature, the concentration of electrons reaches the value $n \approx 8.5 \times 10^{28} m^{-3}$. The thermal velocity of an electron is $v_{th} \approx 10^6 m/s$, and the relaxation time equals $\tau \approx 2.5 \times 10^{-14} s$. Substituting these values in Eq.(8.3.16) results in

$$\mathcal{E} < 2 \times 10^8\ V/m.$$

Electric fields obtained in laboratories are orders of magnitude lower than this value. We see that Ohm's law works very well in metals.

The situation is quite different in semiconductors where the concentrations of electrons are considerably lower. Deviations from Ohm's law occur very often. Most semiconductor devices, for example, *pn* junctions, work due to violations of Ohm's law.

8.4. MECHANISMS OF SCATTERING IN SEMICONDUCTORS

In a perfect crystal lattice, where there are no lattice vibrations, electrons or holes would move with constant quasimomentum and energy, see Chapter 2. There would be no collisions, and the relaxation time τ and the carrier mobility b would be infinite. In reality, the unavoidable carrier scattering from lattice vibrations and from various defects and imperfections of the crystal lattice randomize the direction of the electron (hole) quasimomentum, and change its magnitude on a time scale $10^{-12}s$. This leads to a finite conductivity and finite mobility. Many mechanisms of electron scattering in semiconductors are known.

a. In an ***ideal crystal*** with no impurities, current carriers are scattered by ***lattice vibrations***. This is one of the main scattering mechanisms, and depends strongly on the type of crystal and on its band structure.

In many-valley semiconductors, the conduction band has several equivalent ellipsoidal minima. In the case of silicon, the number of such minima is six, see Fig.4.1.2c. Within each minimum electrons are scattered from ***acoustic*** phonons. Long-wavelength phonons are involved in this ***intravalley*** scattering. Scattering by acoustic phonons results from the modulation of the energy bands by the strains induced in the crystal by the long wavelength acoustic phonons. This is why this scattering mechanism is called the ***deformation potential mechanism***. Short-wavelength phonons with quasiwave vectors near the boundary of the Brillouin zone contribute to the ***intervalley*** scattering.

Optical phonons are connected with relative displacement of the atoms of different sublattices. The number of optical phonons decreases exponentially with decreasing temperature, see Eq.(5.3.13). As a result, the scattering of an electron with the annihilation of an optical phonon is significant only at high temperatures. However, in the process of scattering a new optical phonon can be created by an energetic electron at any temperature. This process has a weak temperature dependence.

The scattering in a direct gap materials of the GaAs-type is different. For the electric field below 3 kV/cm all the scattering processes are in the Γ-valley, i.e. intravalley scattering. For the higher electric fields electrons undergo intervalley scattering into the higher energetic X- and L-valleys. This scattering leads to the negative differential conductivity which is used in higher frequency generators.

Since the effective mass of electrons in GaAs is rather small, $m_n = 0.067\ m_0$, their scattering by acoustic phonons is weak. This is the reason why the electron mobilities in GaAs are five times larger than in Si. Higher mobilities mean big advantage from the viewpoint of technical applications (higher speed and lower power dissipation).

The maxima of the valence bands of Si and GaAs occur at the Γ-point of the Brillouin zone. The interaction with both acoustic and optical phonons are equally important for holes in both materials.

b. In ***doped materials*** donors and acceptors are ***substitutional impurities*** which replace atoms of the host lattice. Whether neutral or charged they are also sources of carrier scattering. Ionized impurities are important at room temperatures when impurity concentrations are of the order of $10^{17}cm^{-3} = 10^{23}m^{-3}$. In bulk semiconductors, mobilities reach values of about $2\ 10^3 cm^2/Vs = 0.2\ m^2/Vs$. Mobilities of several typical semiconductors are given in the Table 8.4.1.

c. The number of scattering mechanisms is enlarged by the presence of ***imperfections of the crystal lattice***. Shallow dopant impurity atoms do not affect significantly the cores of the atoms they replace. But there are some other defects

Crystal Material	μ_n (Electrons)	μ_p (Holes)
Ge	3900	1900
Si	1500	480
InAs	33,000	460
InSb	80,000	1250
GaAs	8500	400
GaP	110	75
CdS	340	50

Table 8.4.1 Room temperature mobilities of several typical semiconductors in cm^2/Vs.

which create large lattice distortions around their lattice sites. The atomic potential of these impurities is of the short range type. Electrons can be trapped by these impurities with localization energies in the middle of the energy gap E_g. These are ***deep level impurities*** which work as ***traps*** for electrons. Au and Ge impurities are defects of this type in III-V compounds.

In addition to all shallow and deep substitution impurities, there are also ***interstitial impurities*** and ***vacancies*** which also destroy the translational symmetry of the crystal, see Fig.8.4.1. These defects are sources of considerable electronic scattering, which reduces the quality of semiconductors.

Serious restrains for applications are connected with one-dimensional linear imperfections called ***dislocations***. They are created as a result of the gliding of one atomic plane with respect to another (shear deformation). The distribution of atoms in a plane normal to a dislocation is shown in Fig.8.4.2. Due to the shear deformation, an extra half plane of atoms appears, where edge is a linear dislocation. Dislocations have only negative aspect for devices. The efforts of technologists are applied to avoiding these defect by well developed growth technologies. The art of crystal growing has reached the level when dislocation-free crystals of Si and GaAs can be grown.

d. The ***surface*** of a crystal, or the ***interface*** between two different materials, are defects in the sense that they break the symmetry of the crystal, and they can also scatter or capture current carriers. At the surface or interface, unsaturated chemical ***dangling bonds*** appear, which weaken the interatomic forces acting on the atoms of the surface or interface. This disbalance results in the phenomenon of ***surface*** or ***interface reconstruction***. The symmetry of the surface or interface layer becomes different from that of any layer in the bulk. Dangling bonds create special surface or interface electron states, whose energy lie in the middle of the band gap. These states capture electrons and perturb the operation of semiconductor devices.

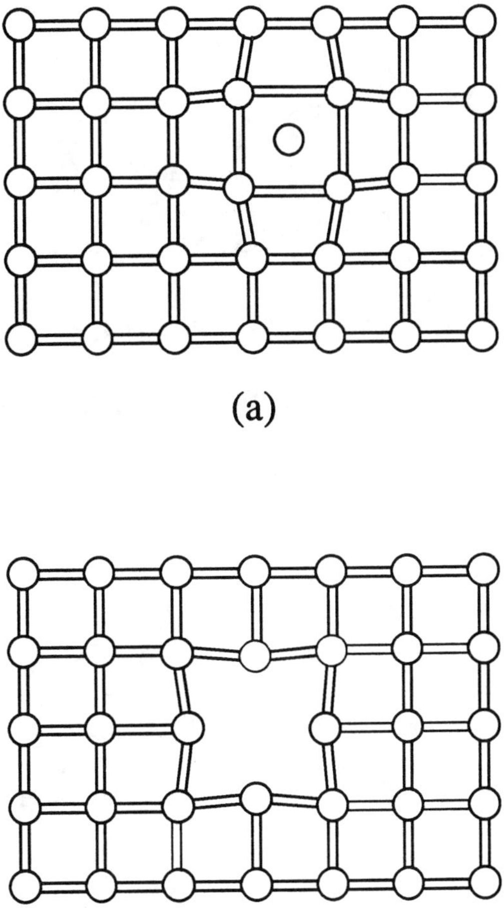

(a)

(b)

Figure 8.4.1 Defects in a crystal: (a) Interstitial impurity. (b) Vacancy.

e. A special type of disorder exists in semiconductor alloys, e.g. $A_xB_{1-x}C$. In these alloys one kind of atom, say C, occupies the sites of one of the face centered cubic lattices of the zinc blend (T_d-symmetry) structure. The other kinds of atoms, A and B, are randomly distributed on the sites of the second face-centered cubic lattice. Although the underlying lattice is periodic, the crystal potential is now non-periodic. In principle, there is no quasimomentum, no Bloch functions, no energy gap E_g, since their existence requires the periodicity of a crystal.

A way to overcome this difficulty is to use the ***virtual crystal approximation***. It consists of replacing the non-periodic potential of the A and B atoms by the periodic potential of an "average" atom. For an $A_xB_{1-x}C$ alloy, the true potential is replaced by the virtual crystal potential $xU_A + (1-x)U_B + U_C$, which is periodic. One can find quasimomentum, Bloch wave functions, energy gap. The difference between the true potential and the virtual crystal potential is treated as a perturbation which scatters the Bloch waves. This scattering leads to band tailing in the density

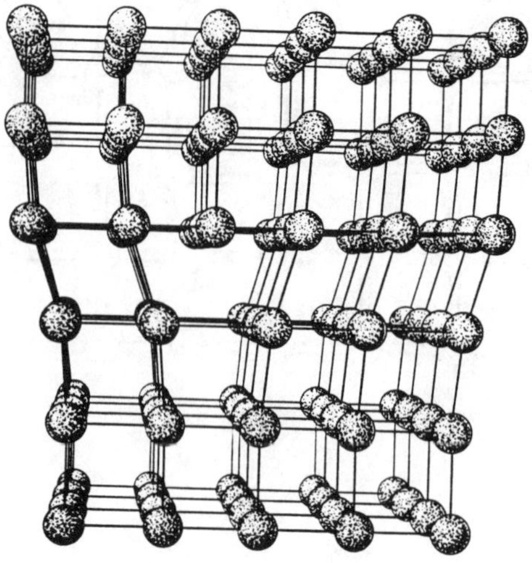

Figure 8.4.2 Dislocation in the crystal (After C. Kittel "Introduction to Solid State Physics", John Wiley & Sons, N.Y., 1986).

of states, and also makes the electrical conductivity finite.

Since the A and B atoms are isovalent, the scattering potential is effective almost within the unit cell. This means that it is of short range type.

f. Transport properties of ***semiconductor nanostructures*** are different from those of bulk semiconductors, because the degrees of freedom of the carriers are reduced by the confining potential. The mobilities of electrons in these structures are considerably higher than bulk mobilities at low temperatures, due to the special technique of doping, called ***modulation doping***, that is employed in their fabrication.

Modulation doped structures are obtained by introducing donor impurities in one part of the system, whereupon all the electrons leave their parent donors and go to reside in another part of the system, where they are confined by a potential barrier at the interface between these two parts of the system. An example is the superlattice made by alternating layers of the higher energy gap material of $Al_xGa_{1-x}As$ and layers of the lower energy gap material GaAs and it was discussed earlier in Section 2.8 (see Fig.2.8.1). Dopants are introduced in the high energy gap material, see Fig.4.6.1b. Electrons leave these donors and reside some distance away, in the low gap material, where there are no impurities, and to which they are confined by the potential at the interface. The scattering of electrons by impurities is therefore significantly reduced. Very large mobilities, above $10^6 cm^2/Vs = 10^2 m^2/Vs$, can be achieved in this way.

The number of scattering mechanisms is further enlarged by the presence of interfaces in semiconductor nanostructures. In the construction of semiconductor devices, special attention is given to finding materials with almost perfect interfaces. One of the advantages of Si for the fabrication of devices is that it forms an almost perfect interface with its dioxide SiO_2. Lattice matched structures of GaAs/AlGaAs are also considerably better from the view point of the interface quality.

8.5 CONDUCTIVITY OF ANISOTROPIC MEDIA

Ohm's law in the form of Eq.(8.1.17) is valid for isotropic media. This means that if the electric field is applied in the x-direction in an isotropic medium, $\vec{\mathcal{E}}=(\mathcal{E}_x,0,0)$, the electric current flows in the same x-direction. But most crystals are anisotropic. The dependence of the current density $\vec{j}$ on the electric field $\vec{\mathcal{E}}$ has a more general form in anisotropic media than in isotropic media. An electric field in the x-direction produces an electric current density in all three directions x,y,z:

$$j_x=\sigma_{xx}\mathcal{E}_x;\ j_y=\sigma_{yx}\mathcal{E}_x;\ J_z=\sigma_{zx}\mathcal{E}_x\ . \tag{8.5.1}$$

When the electric field has an arbitrary direction, $\vec{\mathcal{E}}=(\mathcal{E}_x,\mathcal{E}_y,\mathcal{E}_z)$, the relations between the components of the current density and of the field become

$$\begin{aligned} j_x&=\sigma_{xx}\mathcal{E}_x+\sigma_{xy}\mathcal{E}_y+\sigma_{xz}\mathcal{E}_z\ ,\\ j_y&=\sigma_{yx}\mathcal{E}_x+\sigma_{yy}\mathcal{E}_y+\sigma_{yz}\mathcal{E}_z\ ,\\ j_z&=\sigma_{zx}\mathcal{E}_x+\sigma_{zy}\mathcal{E}_y+\sigma_{zz}\mathcal{E}_z\ . \end{aligned} \tag{8.5.2}$$

One can see that in a crystal of arbitrary symmetry, instead of the single coefficient of proportionality σ in Eq.(8.3.17), there appears the quantity represented by the matrix

$$\sigma=\begin{pmatrix}\sigma_{xx} & \sigma_{xy} & \sigma_{xz}\\ \sigma_{yx} & \sigma_{yy} & \sigma_{yz}\\ \sigma_{zx} & \sigma_{zy} & \sigma_{zz}\end{pmatrix}. \tag{8.5.3}$$

The particular form of σ depends on the symmetry of the crystal. Physical quantities with two indices of coordinate components x,y,z are tensors of the second rank. We have met the quantity of this type when considering in Chapter 4 the second rank tensor of reciprocal effective mass, $m_{\alpha\beta}^{-1}$. Equation (8.5.2) can be rewritten in the compact form

$$j_\alpha=\sum_\beta\sigma_{\alpha\beta}\mathcal{E}_\beta\ , \tag{8.5.4}$$

where $\alpha,\beta=x,y,z$. Since the conductivity is a second rank tensor, it should transform in going from one coordinate system to another according to the rule introduced in Appendix 1.

Since it is a macroscopic property of the crystal, the conductivity σ should remain the same under any symmetry transformation of the crystal. Applying different symmetry operations of the crystal to Eq.(8.5.4), one can find the zero components of the tensor σ and simplify the relation (8.5.4).

Applying some symmetry operation S to the current density vector $\vec{j}'$ given in some coordinate system, one can use the transformation law given by Eq.(3.3.4) to obtain

$$S\vec{j}'=\vec{j},$$

where $\vec{j}$ is the current density in the new coordinate system. Expressing $\vec{j}'$ in terms of the electric field $\vec{\mathcal{E}}'$ with the use of Ohm's law $\vec{j}' = \sigma' \vec{\mathcal{E}}'$, one obtains

$$\vec{j} = S\sigma' \vec{\mathcal{E}}'. \tag{8.5.5}$$

The next step is to apply the transformation S to the electric field $\vec{\mathcal{E}}'$ in order to find $\vec{\mathcal{E}}$ in the new coordinate system. Since $\vec{\mathcal{E}}'$ is a vector, it is related to $\vec{\mathcal{E}}$ by the transformation (3.3.5),

$$\vec{\mathcal{E}}' = S^{-1}\vec{\mathcal{E}}, \tag{8.5.6}$$

Substitution of Eq.(8.5.6) in Eq.(8.5.5) results in

$$\vec{j} = S\sigma' S^{-1}\vec{\mathcal{E}}. \tag{8.5.7}$$

Here the quantity $S\sigma' S^{-1}$ is the usual product of matrices. Equation (8.5.7) is Ohm's law in the new coordinate system. Since σ is a macroscopic property of the crystal, it should remain the same under a symmetry transformation of the crystal

$$\sigma = S\sigma' S^{-1}. \tag{8.5.8}$$

Condition (8.5.8) imposes some restrictions on the elements of the tensor σ.

As an example*) of the application of the preceding results we consider the conductivity of a two-dimensional electron gas at the interface of Si and SiO_2 in a metal-oxide-semiconductor transistor (MOS-transistor). Depending on the orientation of the Si substrate, the two-dimensional symmetry of the interface is different. We consider below the (100), (110), and (111) surfaces of the Si substrate. The different symmetries of these surfaces are shown in Fig.8.5.1.

In a two-dimensional electron gas, Ohm's law has the two-dimensional form

$$j_x = \sigma_{xx}\mathcal{E}_x + \sigma_{xy}\mathcal{E}_y, \tag{8.5.9a}$$

$$j_y = \sigma_{yx}\mathcal{E}_x + \sigma_{yy}\mathcal{E}_y, \tag{8.5.9b}$$

or in the matrix form

$$\begin{pmatrix} j_x \\ j_y \end{pmatrix} = \begin{pmatrix} \sigma_{xx} & \sigma_{xy} \\ \sigma_{yx} & \sigma_{yy} \end{pmatrix} \begin{pmatrix} \mathcal{E}_x \\ \mathcal{E}_y \end{pmatrix}, \tag{8.5.9c}$$

where

$$\sigma = \begin{pmatrix} \sigma_{xx} & \sigma_{xy} \\ \sigma_{yx} & \sigma_{yy} \end{pmatrix} \tag{8.5.10}$$

is the two-dimensional conductivity tensor.

We begin the analysis with the (100) surface of Si. The symbol (100) means that the surface is perpendicular to the x-direction, see Fig.8.5.1a. It has the symmetry of a square. All symmetry elements of the square have already been given in Table 3.2.1, Section 3.2. We choose for analysis the operation C_4, which is a rotation through 90° about the axis perpendicular to the square. The rotation C_4 sends the square into itself. The matrix of this rotation is

$$S(C_4) = \begin{pmatrix} 0 & -1 \\ 1 & 0 \end{pmatrix}. \tag{8.5.11}$$

Applying Eq.(8.5.8) with the matrix S given by Eq.(8.5.11), one obtains the relation

*) The authors express their sincere gratitude to Professor Karl Hess for this example, K. Hess "Advanced Theory of Semiconductor Devices", Prentice-Hall, N.J., 1988.

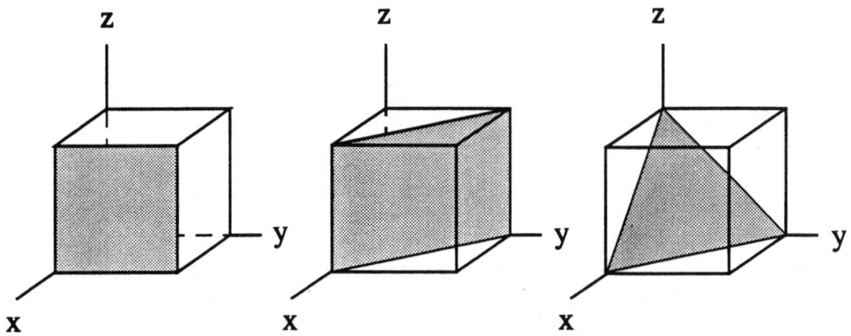

(100) (110) (111)

(a) (b) (c)

Figure 8.5.1 (100), (110), and (111) surfaces shown in a cube.

$$\begin{pmatrix} \sigma_{xx} & \sigma_{xy} \\ \sigma_{yx} & \sigma_{yy} \end{pmatrix} = \begin{pmatrix} 0 & -1 \\ 1 & 0 \end{pmatrix} \begin{pmatrix} \sigma_{xx} & \sigma_{xy} \\ \sigma_{yx} & \sigma_{yy} \end{pmatrix} \begin{pmatrix} 0 & 1 \\ -1 & 0 \end{pmatrix}.$$

Direct multiplication of the matrices on the right hand side of this equation results in the relation

$$\begin{pmatrix} \sigma_{xx} & \sigma_{xy} \\ \sigma_{yx} & \sigma_{yy} \end{pmatrix} = \begin{pmatrix} \sigma_{yy} & -\sigma_{yx} \\ -\sigma_{xy} & \sigma_{xx} \end{pmatrix}. \tag{8.5.12}$$

Equating the corresponding elements of the matrices in Eq.(8.5.12) results in

$$\sigma_{yy} = \sigma_{xx} \quad \text{and} \quad \sigma_{xy} = -\sigma_{yx}. \tag{8.5.13}$$

We see that the diagonal elements of the matrix σ are equal, and the nondiagonal elements are connected by the relation (8.5.13).

Another symmetry transformation from Table 3.2.1 is the rotation U'_2 which transforms $xy \to yx$. The matrix of this transformation is

$$S(U'_2) = \begin{pmatrix} 0 & 1 \\ 1 & 0 \end{pmatrix}. \tag{8.5.14}$$

Substituting rotation Eq.(8.5.14) in Eq.(8.5.8), one obtains

$$\sigma = S(U'_2)\,\sigma\, S(U'_2), \tag{8.5.15}$$

or

$$\begin{pmatrix} \sigma_{xx} & \sigma_{xy} \\ \sigma_{yx} & \sigma_{yy} \end{pmatrix} = \begin{pmatrix} \sigma_{yy} & \sigma_{yx} \\ \sigma_{xy} & \sigma_{xx} \end{pmatrix}. \tag{8.5.16}$$

Equating the corresponding elements of the matrices results in

$$\sigma_{yy} = \sigma_{xx} \quad \text{and} \quad \sigma_{yx} = \sigma_{xy}. \tag{8.5.17}$$

Equations (8.5.13) and (8.5.17) are consistent, if

$$\sigma_{xx} = \sigma_{yy}$$

and

$$\sigma_{xy} = -\sigma_{xy} \,. \tag{8.5.18}$$

Equation (8.5.18) means that

$$\sigma_{xy} = \sigma_{yx} \equiv 0. \tag{8.5.19}$$

A similar symmetry analysis is possible for the (110) surface of the Si substrate. This surface is shown in Fig.8.5.1b. It has the symmetry of a rectangle. A rectangle has C_2-axis (180° rotation) about an axis perpendicular to the surface, and the second operation of the symmetry is, say, the reflection in this plane through x-axis. Reflection in this plane is given by the matrix

$$S(\sigma) = \begin{pmatrix} 1 & 0 \\ 0 & -1 \end{pmatrix}. \tag{8.5.20}$$

Substituting Eq.(8.5.20) into Eq.(8.5.8), one finds

$$\begin{pmatrix} \sigma_{xx} & \sigma_{xy} \\ \sigma_{yx} & \sigma_{yy} \end{pmatrix} = \begin{pmatrix} \sigma_{xx} & -\sigma_{xy} \\ -\sigma_{yx} & \sigma_{yy} \end{pmatrix}. \tag{8.5.21}$$

Therefore, $\sigma_{xy} = \sigma_{yx} = 0$. No symmetry operation interchanges the x- and y-axis so that, in general, $\sigma_{xx} \neq \sigma_{yy}$. The conductivity is anisotropic.

The (111) surface of the Si substrate is transverse to the body diagonal of the cube, see Fig.8.5.1c. It has the symmetry of the honeycomb, that is, it has six elements of symmetry of the bilateral triangle.

There is a general rule of group theory, which states that if the number of symmetry rotations or reflections is larger than the rank of the tensor, the corresponding matrix is diagonal. Six elements of symmetry of bilateral triangle are enough to make the conductivity isotropic.

In the case of the (100)-surface, we have eight elements of symmetry. It is also enough to make conductivity isotropic. But the (110) surface has only two elements of symmetry, which is not enough to diagonalize the tensor σ.

In a semiconductor material with cubic symmetry, like Ge or Si, and tetrahedral symmetry, like GaAs or InP, there are 48 or 24 operations of symmetry, respectively. This number of symmetry operations is enough to reduce the conductivity tensor σ given by Eq.(8.5.3) to a scalar. This means that $\sigma_{xx} = \sigma_{yy} = \sigma_{zz} = \sigma$ and $\sigma_{xy} = \sigma_{xz} = \sigma_{yz} = 0$. It is convenient to write the conductivity tensor for crystals with O_h and T_d symmetry in the general form

$$\sigma_{\alpha\beta} = \sigma\delta_{\alpha\beta} \,, \tag{8.5.22}$$

where $\alpha, \beta = x, y, z$ and $\delta_{\alpha\beta}$ is the Kronecker delta.

BIBLIOGRAPHY

D.K. Ferry *Semiconductors* (MacMillan, New York, 1991).

CHAPTER 9

Nonequilibrium Electrons and Holes

9.1 DIFFUSION

We have considered in Chapter 8 the nonequilibrium state of a semiconductor which is produced by an external electric field. The concentrations of electrons and holes remained in equilibrium throughout the entire ***uniform*** material.

In addition to the drift of electrons, there is a second reason which can induce an electric current in a semiconductor. Another type of nonequilibrium state is created when the crystal is subjected to some external influence which creates a nonuniform distribution of carriers. For example, it can be produced by the local illumination of the crystal, or when the electric current flows in a nonuniform layered semiconductor structure. The carrier concentrations become functions of the spatial coordinate $\vec{r}$: $n(\vec{r})$ and $p(\vec{r})$. The ***nonuniform*** distribution of carriers also takes the crystal out of equilibrium. In order to reach equilibrium, the electrons or holes start to move from regions of high concentration to regions of low concentration. This process is called ***diffusion***. Since the carriers are charged, a net flow of charged particles results in a ***diffusion current***.

Consider, as an example, a linear dependence of the electron concentration

$n(x)$ on the spatial coordinate x, shown in Fig.9.1.1. The diffusion of electrons and the flux of particles is in the negative x-direction. The conventional electron current flows in the positive x-direction. The spatial variation of the carrier concentration in terms of $n(\vec{r})$ is given by the gradient

$$\frac{dn(\vec{r})}{d\vec{r}} = \nabla_{\vec{r}}\, n(\vec{r}). \tag{9.1.1}$$

The electron distribution function $f(\vec{p},\vec{r})$ in a nonuniform material depends not only on the momentum of electrons but also on the coordinate $\vec{r}$. The time derivative of f which has been calculated in Section 8.1.1, should now be supplemented with a second contribution which results from the time dependence of $\vec{r}$. Using the chain rule of differentiation gives the total time derivative of $f(\vec{p},\vec{r})$

$$\frac{df(\vec{p},\vec{r})}{dt} = \frac{\partial f}{\partial \vec{p}}\frac{\partial \vec{p}}{\partial t} + \frac{\partial f}{\partial \vec{r}}\frac{\partial \vec{r}}{\partial t} + \frac{\partial f}{\partial t} = -e\vec{\mathcal{E}}\cdot\frac{\partial f}{\partial \vec{p}} + \vec{v}\cdot\frac{\partial f}{\partial \vec{r}} + \frac{\partial f}{\partial t}. \tag{9.1.2}$$

Here $\partial\vec{p}/\partial t = -e\vec{\mathcal{E}}$, and $\vec{v} = d\vec{r}/dt$ is the electron velocity. Further we consider steady state distribution when $\partial f/\partial t = 0$.

The kinetic equation (8.1.5) in this case takes the form

$$-e\vec{\mathcal{E}}\cdot\frac{\partial f}{\partial \vec{p}} + \vec{v}\cdot\frac{\partial f}{\partial \vec{r}} = \left(\frac{\partial f}{\partial t}\right)_{collision}. \tag{9.1.3}$$

Concentrating now on effects of the nonuniform electron distribution, we neglect for a moment the electric field $\vec{\mathcal{E}}$ on the left hand side of Eq.(9.1.3). Assuming for simplicity the relaxation time approximation Eq.(8.3.5), and using the expansion (8.2.4), one obtains the kinetic equation in the form

$$\vec{v}\cdot\frac{\partial f}{\partial \vec{r}} = -\frac{f^{(1)}(\vec{p},\vec{r})}{\tau(\vec{p})}. \tag{9.1.4}$$

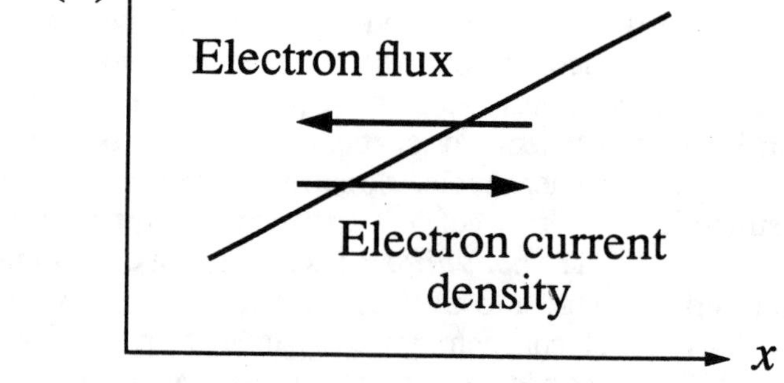

Figure 9.1.1 Diffusion of electrons due to an electron density gradient.

If the gradient of electron concentration exists in the x-direction only, Eq.(9.1.4) becomes one-dimensional

$$v_x \cdot \frac{\partial f_e^{(0)}}{\partial x} = -\frac{f^{(1)}(\vec{p},\vec{r})}{\tau(\vec{p})}, \tag{9.1.5}$$

where the distribution function $f_e^{(0)}(x)$ depends on x through $\mu(x)$. The solution of Eq.(9.1.56) is

$$f^{(1)}(\vec{p},\vec{r}) = -\tau(\vec{p}) v_x \frac{\partial f_e^{(0)}}{\partial x}. \tag{9.1.6}$$

Substituting Eq.(9.1.6) in to the expression for electric current, Eq.(8.3.3), we find the contribution of diffusion to the current density j_x

$$j_x = e\frac{2}{(2\pi\hbar)^3}\int v_x^2 \tau(\vec{p}) \frac{\partial f_e^{(0)}}{\partial x} dp_x dp_y dp_z. \tag{9.1.7}$$

Similarly to what we have done in Section 8.1, we introduce the electron density of states $D(E)$ in the integral in Eq.(9.1.7)

$$j_x = e\frac{2}{3m_n}\int E\tau(E)\frac{\partial f_e^{(0)}}{\partial x} D(E) dE = e\frac{\partial}{\partial x}(D_n n) = eD_n \frac{\partial n(x)}{\partial x}, \tag{9.1.8}$$

where

$$D_n = \frac{\frac{2}{3m_n}\int E\tau(E)\frac{\partial f_e^{(0)}}{\partial n} D(E) dE}{\int D(E) f_e^{(0)} dE} \tag{9.1.9}$$

is called the ***diffusion coefficient*** of the electrons. D_n is a macroscopic characteristic of the electron gas. It has the dimensions of cm^2/s, and is a positive quantity. The electron diffusion coefficient indicates how well an electron moves in a semiconductor as a result of a concentration gradient.

When there is an external electric field $\mathcal{E} = (\mathcal{E}_x, 0, 0)$, the total current of electrons is the sum of Eq.(8.1.21) and Eq.(9.1.8)

$$j_x = neb_n\mathcal{E}_x + eD_n\frac{\partial n}{\partial x},$$

or in vector form

$$\vec{j}_n = neb_n\vec{\mathcal{E}} + eD_n\nabla_{\vec{r}} n. \tag{9.1.10}$$

A kinetic equation similar to Eq.(9.1.5) exists for holes. When the concentration of holes depends on $\vec{r}$, this kinetic equation allows the introduction of the diffusion coefficient of holes D_p. The total current in the case of holes equals

$$\vec{j}_p = peb_p\vec{\mathcal{E}} - eD_p\nabla_{\vec{r}} p, \tag{9.1.11}$$

where $b_p = e\tau(\vec{p})/m_p$ is the mobility of the holes, and $\tau(\vec{p})$ is the relaxation time of the holes. The first term in Eq.(9.1.11) arises from drift in the field $\vec{\mathcal{E}}$, and the second term from the diffusion of holes.

In equilibrium, the total currents vanish $j_p = 0$, $j_n = 0$, and

$$neb_n\vec{\mathcal{E}} = -eD_n \nabla_{\vec{r}} n,$$
$$peb_p\vec{\mathcal{E}} = eD_p \nabla_{\vec{r}} p. \qquad (9.1.12)$$

The inhomogeneous distribution of electrons in the x –direction results in an electric field $\mathcal{E}_x$ with potential $\phi(x)$

$$\mathcal{E}_x = -\frac{\partial \phi}{\partial x}. \qquad (9.1.13)$$

According to statistical thermodynamics, the Fermi-level (the chemical potential) of the electrons is replaced in an external electric field by the electrochemical potential

$$\mu \to \mu + e\phi(x). \qquad (9.1.14)$$

The inhomogeneous electron concentration in the case of Boltzman statistics equals

$$n(x) = N_c e^{\frac{\mu - E_c + e\phi(x)}{k_B T}}. \qquad (9.1.15)$$

The gradient of the concentration is

$$\frac{dn(x)}{dx} = \frac{ne}{k_B T}\frac{d\phi}{dx} = -\frac{ne}{k_B T}\mathcal{E}_x. \qquad (9.1.16)$$

Substituting Eq.(9.1.16) into the equilibrium conditions (9.1.12), one obtains

$$neb_n\mathcal{E}_x = eD_n \frac{ne}{k_B T}\mathcal{E}_x,$$

and a similar procedure for holes results in

$$peb_p\mathcal{E}_x = eD_p \frac{pe}{k_B T}\mathcal{E}_x.$$

Finally, we have for electrons

$$D_n = \frac{b_n k_B T}{e}, \qquad (9.1.17)$$

and holes

$$D_p = \frac{b_p k_B T}{e}. \qquad (9.1.18)$$

Equations (9.1.17) and (9.1.18) are called the ***Einstein relations***, which connect the mobility and the diffusion coefficient. We have obtained the Einstein relation for the classical statistics of carriers. The connection of mobility and diffusion coefficient can be calculated also for the degenerate statistics of carriers.

9.2 EXCESS CARRIERS

It has been shown in Section 6.1 that in equilibrium the thermal excitation of electrons leads to the generation of electrons in the conduction band and holes in the hole band. This process has a characteristic called the ***rate of generation*** G_{th}, which gives the number of electron-hole pairs created per unit volume and per unit time.

There is also the reverse process in which an electron makes a transition from the conduction band into an empty state in the valence band. This process is called electron-hole ***recombination***. When the energy is transferred to the crystal lattice vibrations, or to other electrons, the process is called ***nonradiative recombination***. If the energy released in recombination is emitted in the form of a photon, we have ***radiative recombination***.

The recombination of electrons and holes occur through different channels:

a. The process of recombination in direct gap materials goes through the energy gap E_g, see Fig.9.2.1a. Because a rather large energy E_g should be released, the energy can be emitted in the form of a photon with frequency

$$\omega = \frac{E_g}{\hbar}. \tag{9.2.1}$$

This process of radiative recombination is used in light emitting diodes (LED) and in semiconductor lasers.

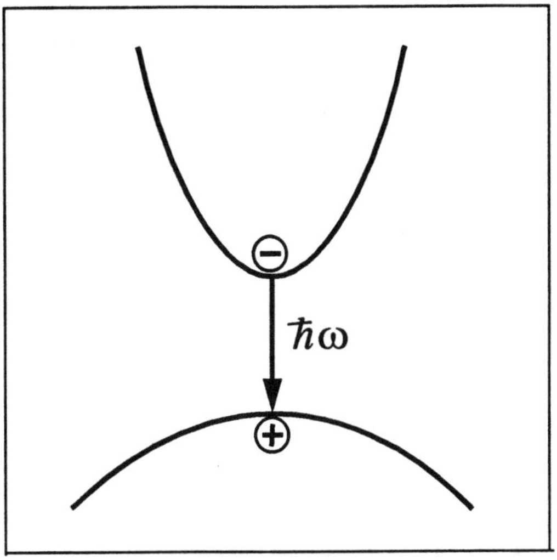

Figure 9.2.1a Direct recombination (band-to band recombination).

b. Another possibility for recombination consists of the creation of many phonons with energies

$$\hbar\omega_i(\vec{q}) \sim k_B T \ll E_g, \quad i = 1,2,\ldots,n \tag{9.2.2}$$

see Fig.9.2.1b. This process has very low probability due to the large number of phonons involved.

c. Impurities or other imperfection of the crystal lattice can help both electrons and holes facilitate the process of recombination by means of the localization of electrons and holes. This process may be either radiative or nonradiative, see Fig.9.2.1c.

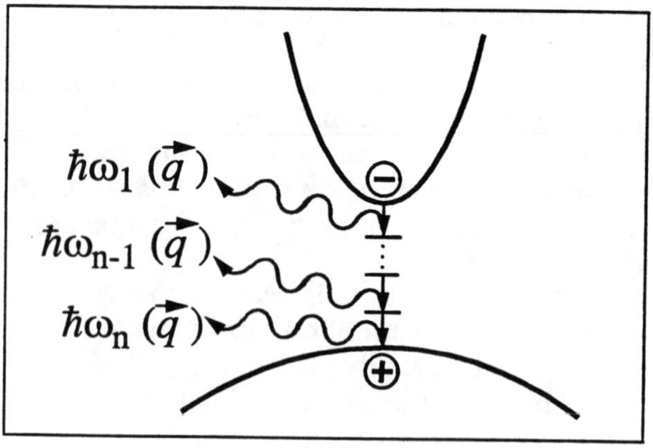

Figure 9.2.1b Nonradiative recombination with phonons.

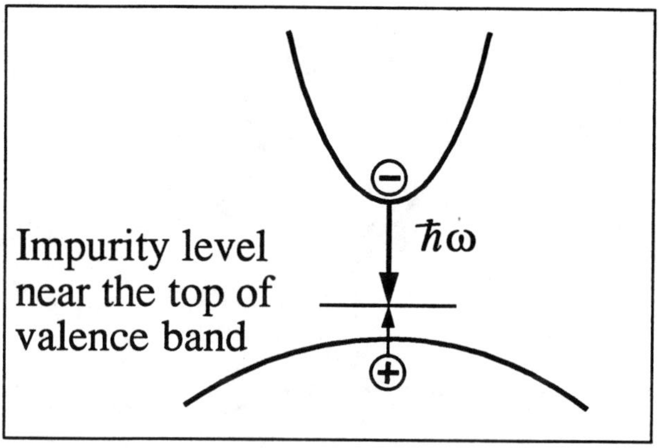

Figure 9.2.1c Radiative band-to impurity recombination.

d. Some recombination processes involve a third particle (electron or hole), see Fig.9.2.1d. In this triple collision, energy is transferred to the kinetic energy of the third particle. Recombination of this type is called ***Auger recombination***.

Since two particles, an electron and a hole, participate in the recombination process, the rate of thermal recombination R_{th} is proportional to the product of the equilibrium concentrations of the electrons and holes

$$R_{th} = \kappa n_0 p_0\,, \tag{9.2.3}$$

where the proportionality coefficient κ depends only on properties of the material and the temperature. The dimensions of κ are cm^3/s.

The condition for the thermal equilibrium of a semiconductor is

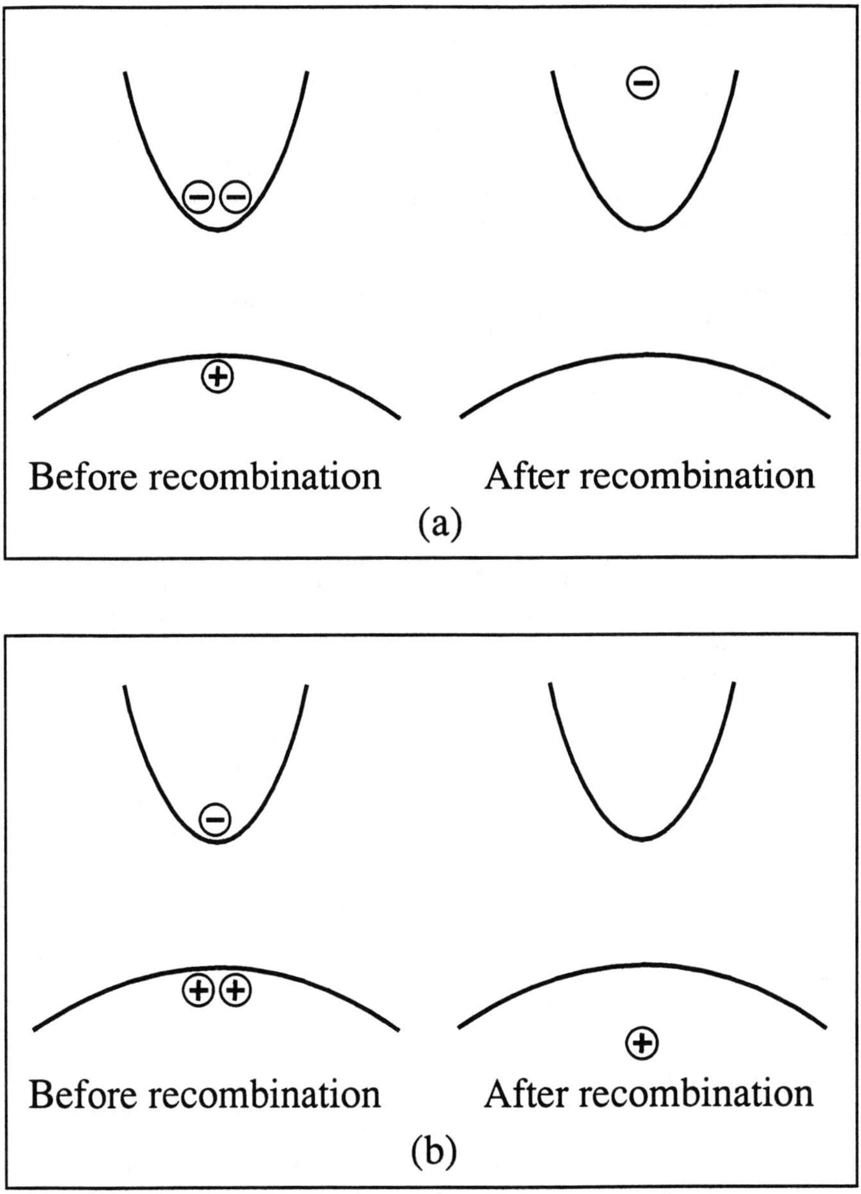

Figure 9.2.1d Auger recombination (a) n –type sample, and (b) p –type sample.

$$G_{th} = R_{th} = \kappa n_0 p_0 \,. \tag{9.2.4}$$

Many semiconductor devices work due to deviations of the carrier concentrations from their equilibrium values caused by the creation of nonequilibrium ***excess carriers***. This optical generation of carriers results from the absorption of light from an external source. The rate of optical generation is proportional to the absorption coefficient of the crystal. Since electrons and holes appear in pairs, we speak about ***bipolar injection*** of carriers. The experimentally observed effect of this injection is ***photoconductivity***.

Excess carriers can also be created by an electric current in nonuniform semiconductor structures. The electric current takes electrons (or holes) from one part of the structure to another part where they are excess carriers. The process is called ***injection*** of excess carriers. The injection is ***monopolar*** in this case.

One method for the production of excess carriers is the impact ionization of impurity atoms by electrons accelerated in a strong electric field.

The nonequilibrium concentrations n and p of electrons and holes differ from the equilibrium concentrations n_0 and p_0 considered in Chapter 6. The excess carrier concentrations Δn and Δp are defined by the relations

$$n = n_0 + \Delta n \quad \text{and} \quad p = p_0 + \Delta p, \tag{9.2.5}$$

where, in the case of bipolar injection,

$$\Delta n = \Delta p. \tag{9.2.6}$$

The injection of excess electrons and holes in equal concentrations produces different effects on majority and minority carriers. The equilibrium concentrations n_0 and p_0 satisfy the law of mass action, Eq.(6.1.21). The quantity n_i reaches the values $10^{10} cm^{-3}$ in Si. If the concentration of impurities is $N_D = 10^{16} cm^{-3}$ and all impurities are ionized, the concentration of major carriers reaches the value $n_0 = N_D$. The minority carrier concentration p_0 is defined by Eq.(6.5.5) and equals $10^4 cm^{-3}$. The bipolar injection of excess carriers, say $10^{12} cm^{-3}$, practically does not change n_0, but the minority carrier concentration p_0 increases by several orders of magnitude.

In a crystal with excess carriers, a certain steady state is possible in which the generation rate $G_{th} + G_{steady}$ is equal to the recombination rate R. In the case of bipolar injection, R is proportional to the product of electron and hole concentrations

$$R = \kappa n p. \tag{9.2.7}$$

The condition for a steady state is

$$G_{th} + G_{steady} = R. \tag{9.2.8}$$

Substituting Eqs.(9.2.4), and (9.2.7) in Eq.(9.2.8) results in

$$G_{steady} = \kappa(np - n_0 p_0) = \kappa(np - n_i^2). \tag{9.2.9}$$

We see from Eq.(9.2.9) that the product of nonequilibrium concentrations does not satisfy the law of mass action, Eq.(6.1.21). If $np > n_i^2$, we have the injection of excess carriers. When $np < n_i^2$, the ***extraction*** of carriers occurs.

Combining the concentrations Eqs.(9.2.5) with Eq.(9.2.9), and taking into account Eq.(9.2.6), results in

$$G_{steady} = \kappa \Delta n [(n_0 + p_0) + \Delta n]. \tag{9.2.10}$$

Equation (9.2.10) allows introducing the average excess carrier recombination time τ, which is defined by

$$G_{steady} = \frac{\Delta n}{\tau}, \tag{9.2.11}$$

where

$$\tau = \frac{1}{\kappa[(n_0 + p_0) + \Delta n]}. \quad (9.2.12)$$

If the level of injection is low, $\Delta n \ll n_0 + p_0$ the excess carrier recombination time becomes a constant quantity

$$\tau = \frac{1}{\kappa(n_0 + p_0)}. \quad (9.2.13)$$

The quantity τ^{-1} is the probability of annihilation of a single electron hole pair per unit time.

In n-type material where $n_0 >> p_0$, the recombination lifetime $\tau_n = 1/(\kappa n_0)$ is inversely proportional to the electron concentration. Similarly, for p-type material with $p_0 >> n_0$, we obtain $\tau_p = 1/(\kappa p_0)$.

In the nonequilibrium state, the time dependence of the excess carrier concentrations is defined by the difference of the corresponding generation rate $G_{n,p}$ due to the external source and the steady-state value G_{steady} given by Eq.(9.2.11). One obtains for the rate of the excess electron concentration variation

$$\frac{\partial \Delta n}{\partial t} = G_n - \frac{\Delta n}{\tau_n}, \quad (9.2.14a)$$

and for excess hole concentration

$$\frac{\partial \Delta p}{\partial t} = G_p - \frac{\Delta p}{\tau_p}. \quad (9.2.14b)$$

Here thermal processes are taken into account in G_{steady}. In the steady state $\partial \Delta n / \partial t = \partial \Delta p / \partial t = 0$ and $G_{n,p} = G_{n,p} = \Delta n / \tau_{n,p}$.

If at the moment $t = t_0$ the generation mechanism is switched off ($G_{n,p} = 0$), the integration of Eqs.(9.2.14) results in

$$\Delta n(t) = \Delta n(t_0) e^{-\frac{t - t_0}{\tau_n}} \quad (9.2.15a)$$

and

$$\Delta p(t) = \Delta p(t_0) e^{-\frac{t - t_0}{\tau_p}}. \quad 9(.2.15b)$$

We see from Eq.(9.2.15) that ΔnandΔp decay exponentially with time constant $\tau_{n,p}$. It is clear why $\tau_{n,p}$ is sometimes called the ***electron-hole recombination time.***

If the level of injection is high, $\Delta n \sim n_0, p_0$, the recombination time depends on Δn according to Eq.(9.2.12). The rate Eq.(9.2.11) becomes a nonlinear function of ΔnandΔp, and the dependence of Δn and Δp on t is no longer exponential.

An important parameter of a nonequilibrium semiconductor with excess carriers is the ratio of the radiative electron-hole recombination rate to the total, radiative and nonradiative, recombination rate. It is called the ***internal quantum efficiency*** η_i. This parameter defines the efficiency of light generation in a semiconductor material. The parameter κ in Eq.(9.2.13) can be split into a sum of radiative and nonradiative parts

$$\kappa = \kappa_r + \kappa_{nr}\,. \tag{9.2.16}$$

The internal quantum efficiency equals

$$\eta_i = \frac{\kappa_r}{\kappa_r + \kappa_{nr}}\,. \tag{9.2.17}$$

This quantity can be written in terms of a recombination lifetime τ that is inversely proportional to κ, see Eq.(9.2.13). Defining the radiative τ_r and the nonradiative τ_{nr} lifetimes by the equation

$$\frac{1}{\tau} = \frac{1}{\tau_r} + \frac{1}{\tau_{nr}}\,,$$

one can find the internal quantum efficiency in the form

$$\eta_i = \frac{\tau}{\tau_r} = \frac{\tau_{nr}}{\tau_r + \tau_{nr}}\,. \tag{9.2.18}$$

The radiative recombination lifetime τ_r is governed by the processes of photon absorption and emission. It depends on the carrier concentrations and material parameters.

The nonradiative recombination lifetime depends significantly on the concentration of defect centers with energy levels in the bandgap. Order of magnitude estimate of τ and η_i for Si and GaAs at $n_0 = 10^{17} cm^{-3}$ and $T = 300K$ yield

$$\tau(\mathrm{Si}) = 100\, ns; \quad \eta_i = 10^{-5},$$

$$\tau(\mathrm{GaAs}) = 50\, ns; \quad \eta_i = 0.5.$$

The radiative lifetime of Si is five orders of magnitude larger than that of GaAs, because it is an indirect semiconductor, where nonradiative processes are important. This results in a small value of η_i. In direct gap GaAs the decay of the excess carriers is through radiative recombination. The internal quantum efficiency is larger. That is why GaAs is more useful in fabricating light emitting structures than Si.

9.3 CONTINUITY EQUATIONS

The description of the process of deviation from equilibrium and the approach to equilibrium is provided, in general, by the kinetic equation (8.1.3). But in the special case of semiconductors, several energy bands (conduction and valence bands) are involved in these processes, which makes calculations based on Eq.(8.1.3) difficult. This is the reason, why a more general approach is used for the description of processes with excess carriers. The background of this approach is the conservation law of particles in the form of a ***continuity equation***. If there is an electric current in a semiconductor, the change of the carrier concentrations is defined by the balance between the particles entering some volume and leaving this volume and it obeys the law of particle conservation which has the form of a continuity equation

$$\frac{\partial n}{\partial t} = \frac{1}{e}\nabla \cdot \vec{j}_n \quad \text{and} \quad \frac{\partial p}{\partial t} = -\frac{1}{e}\nabla \cdot \vec{j}_p, \tag{9.3.1}$$

where $\vec{j}_n$ and $\vec{j}_p$ are the current densities given by Eqs.(9.1.10) and (9.1.11).

The number of excess electrons and holes must balance each other to maintain space charge neutrality. If the total charge density of a doped semiconductor with excess carriers is $\rho = e(p - n + N_D^+ - N_A^-)$, the time derivative of ρ is defined by the continuity equation

$$\frac{\partial \rho}{\partial t} + \nabla \cdot \vec{j} = 0, \tag{9.3.2}$$

where $\vec{j} = \vec{j}_n + \vec{j}_p$ is the total current density. Combining $\vec{j}$ from Eq.(8.3.17) with the Poisson equation $\nabla \cdot \vec{\mathcal{E}} = e\rho/\varepsilon$, where ε is the absolute permittivity, one obtains

$$\frac{\partial \rho}{\partial t} = -\frac{\sigma}{\varepsilon}\rho.$$

Separating variables, we find

$$\frac{d\rho}{\rho} = -\frac{\sigma}{\varepsilon}dt.$$

Integration of this equation results in

$$\rho = \rho_0 e^{-\frac{t}{\tau_M}}, \tag{9.3.3}$$

where

$$\tau_M = \frac{\varepsilon}{\sigma} \tag{9.3.4}$$

is the relaxation time of the charge density ρ, and the initial value of the charge density is $\rho(t = 0) = \rho_0$.

A numerical estimate shows that a deviation of the charge density from neutrality decays in a time of the order of a few picoseconds,

$$\tau_M \approx 10^{-12} s.$$

In a uniform semiconductor, the diffusion and drift currents tend to make a volume charge fluctuation vanish within the relaxation time τ_M. In nonuniform semiconductor structures, a certain equilibrium charge distribution appears.

On the other hand, there should also be a recombination equilibrium between electrons and holes which is defined by the recombination times $\tau_{n,p}$ from Section 9.2.

The total rate of concentration change is obtained by combining Eqs.(9.2.14) and (9.3.1),

$$\frac{\partial n}{\partial t} = G_n - \frac{\Delta n}{\tau_n} + \frac{1}{e}\nabla \cdot \vec{j}_n, \tag{9.3.5}$$

$$\frac{\partial p}{\partial t} = G_p - \frac{\Delta p}{\tau_p} - \frac{1}{e}\nabla \cdot \vec{j}_p. \tag{9.3.6}$$

These are ***continuity equations*** for electrons and holes, which take into account both the charge relaxation and the recombination relaxation.

In many important semiconductors the conditions

$$\tau_M \ll \tau_n, \tau_p \tag{9.3.7}$$

hold which mean that the charge relaxation is faster than the recombination relaxation. Charge relaxation realizes the quasineutrality condition

$$\rho \cong 0 \tag{9.3.8a}$$

in the bulk of the material. Substituting $\rho \cong 0$ in Eq.(9.3.2), one can find another form of the quasineutrality condition,

$$\nabla \cdot \vec{j} = 0, \tag{9.3.8b}$$

which means that in the neutral semiconductor there is no source of electric current.

When thermal equilibrium is broken by the injection of excess carriers, bulk charge with a density

$$\rho = e(N_D^+ - N_A^- - n + p) \tag{9.3.9}$$

appears. The subtraction of Eq.(9.3.5) from (9.3.6) leads to

$$\frac{\partial}{\partial t}(p-n) + \frac{1}{e}\nabla \cdot (\vec{j}_p + \vec{j}_n) - \frac{\Delta p}{\tau_p} + \frac{\Delta n}{\tau_n} = 0. \tag{9.3.10}$$

If the rate of change of the bound charges is slower than the rate of change of the free charges,

$$\frac{\partial}{\partial t}(N_D^+ - N_A^-) \ll \frac{\partial}{\partial t}(p-n), \tag{9.3.11}$$

one finds with the help of Eq.(9.3.9) that

$$e\frac{\partial(p-n)}{\partial t} = \frac{\partial \rho}{\partial t}. \tag{9.3.12}$$

If the quasineutrality condition, Eq.(9.3.8a), holds, it follows from Eqs.(9.3.12) and (9.2.5) that the concentrations of the excess carriers are equal,

$$\Delta n = \Delta p. \tag{9.3.13}$$

In near equilibrium, the numbers of excess electrons and holes must balance one another to maintain charge neutrality.

Taking into account the other form of the neutrality condition, Eq.(9.3.8b), and Eq.(9.3.13), we obtain from Eq.(9.3.10) the conclusion that

$$\tau_p = \tau_n .$$

This means that under quasiequilibrium conditions, there is a single, common lifetime of electrons and holes,

$$\tau_p = \tau_n = \tau. \tag{9.3.14}$$

9.4 AMBIPOLAR DIFFUSION AND DRIFT

When an external electric field $\vec{\mathcal{E}}$ is applied to a semiconductor with excess carriers, the rate of change of the carrier concentrations is governed by both the generation-recombination rate and the drift of carriers. The excess electrons and holes move in the same electric field $\vec{\mathcal{E}}$, which creates a specific correlation in their motion. ***Ambipolar diffusion*** and ***ambipolar drift*** occur, which are different from the monopolar diffusion considered in Section 9.1.

The simple example of excess carriers injected by the illumination of a semiconductor surface allows one to understand the main features of these phenomena. If the diffusion coefficient of the electrons is larger than that of the holes, $D_n > D_p$, electrons diffuse faster than holes. A nonequilibrium distribution of charge appears which creates the electric field of ambipolar diffusion $\vec{\mathcal{E}}$. This field accelerates the holes and retards the electrons. In the steady state the field $\vec{\mathcal{E}}$ makes the electron and hole fluxes compensate each other, and creates a neutral system.

We are interested in considering injected carriers for which the condition of charge neutrality in the form $\Delta n = \Delta p$ is assumed to hold in the steady state. We also assume that carriers are created in the process of band-to-band optical transitions, in which case $G_n = G_p$.

In order to consider the diffusion and drift explicitly, the continuity equations from Section 9.3 can be used. One-dimensional motion of carriers is assumed for simplicity. If the currents given by Eqs.(9.1.10), and (9.1.11) are introduced in the continuity equations (9.3.5) and (9.3.6), one obtains in the one-dimensional case

$$\frac{\partial p}{\partial t} = G - \frac{\Delta p}{\tau} - \frac{1}{e}\mathcal{E}_x\frac{\partial \sigma_p}{\partial x} - \frac{1}{e}\sigma_p\frac{\partial \mathcal{E}_x}{\partial x} + D_p\frac{\partial^2 p}{\partial x^2}, \tag{9.4.1}$$

and

$$\frac{\partial n}{\partial t} = G - \frac{\Delta n}{\tau} + \frac{1}{e}\mathcal{E}_x\frac{\partial \sigma_n}{\partial x} + \frac{1}{e}\sigma_n\frac{\partial \mathcal{E}_x}{\partial x} + D_n\frac{\partial^2 n}{\partial x^2}. \tag{9.4.2}$$

Multiplying the first equation by σ_n and the second equation by σ_p, and taking the sum of the resulting equations, yields

$$\frac{\partial p}{\partial t} = G - \frac{\Delta p}{\tau} - \frac{1}{e\sigma}\left(\sigma_n\frac{\partial \sigma_p}{\partial x} - \sigma_p\frac{\partial \sigma_n}{\partial x}\right)\mathcal{E}_x + \frac{\sigma_n D_p + \sigma_p D_n}{\sigma}\frac{\partial^2 p}{\partial x^2}, \tag{9.4.3}$$

where $\sigma = \sigma_n + \sigma_p$. The quantity

$$D_a = \frac{\sigma_n D_p + \sigma_p D_n}{\sigma} \tag{9.4.4}$$

is called the ***coefficient of ambipolar diffusion***. It is convenient to express D_a in terms of $\sigma_p = epb_p$ and $\sigma_n = enb_n$,

$$D_a = \frac{p+n}{[p/D_n + n/D_p]}. \tag{9.4.5}$$

The coefficient in the third term of Eq.(9.4.3) has the form

$$\frac{1}{e\sigma}\left[\sigma_n \frac{\partial \sigma_p}{\partial x} - \sigma_p \frac{\partial \sigma_n}{\partial x}\right] = \frac{n-p}{[p/b_p + n/b_n]}\frac{\partial p}{\partial x}. \tag{9.4.6}$$

Combining Eqs.(9.4.4) and (9.4.6) with Eq.(9.4.3), one obtains the ***continuity equation of ambipolar diffusion***

$$\frac{\partial p}{\partial t} = G - \frac{\Delta p}{\tau} - v_a \frac{\partial p}{\partial x} + D_a \frac{\partial^2 p}{\partial x^2}, \tag{9.4.7}$$

where v_a equals

$$v_a = \frac{n-p}{[p/b_p - n/b_n]}\mathcal{E}_x. \tag{9.4.8}$$

Because $dp/dt = (dp/dx)(dx/dt) = v(dp/dx)$ the third term of Eq.(9.4.7) represents the time rate of change of the concentration p due to the common motion of the excess electrons and holes with the ***ambipolar drift velocity*** v_a. The fourth term in Eq.(9.4.7) is the rate of change of p due to the ambipolar diffusion of the excess carriers with the diffusion coefficient D_a.

It follows from Eq.(9.4.5) that in an n-type semiconductor where $n \gg p$, $D_a \cong D_p$. For p-type material, $p \gg n$, $D_a \cong D_n$. This means that in extrinsic semiconductors D_a reduces to the diffusion coefficient of the ***minority carriers***. In an intrinsic semiconductor, $n_0 = p_0 = n_i$, and

$$D_a = \frac{2D_p D_n}{D_p + D_n}.$$

The qualitative dependence of D_a on the electron concentration is shown in Fig.9.4.1.

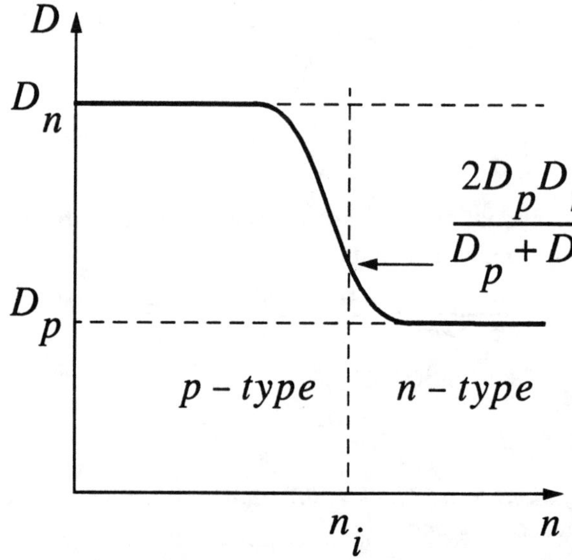

Figure 9.4.1 Dependence of ambipolar diffusion coefficient on the electrons density n.

Because v_a is proportional to $\mathcal{E}_x$, we can obtain the ***ambipolar mobility***

$$b_a = \frac{v_a}{\mathcal{E}_x} = \frac{n-p}{[p/b_n - n/b_p]}. \qquad (9.4.9)$$

It is seen from Eq.(9.4.8) that the ambipolar mobility b_a does not vanish only when $n \neq p$. This means that the group of the excess carriers in an extrinsic semiconductor can move with some velocity in spite of the neutrality condition. In n-type material, $n \gg p$, $b \cong b_p > 0$. This means that the excess carriers move in the direction of the field $\mathcal{E}_x$, that is in the direction of the minority carriers (positive holes). In p-type material $p \gg n$, $b = -b_n < 0$. This means that the excess carriers move against $\mathcal{E}_x$, that is again as minority carriers (negative electrons). In an intrinsic semiconductor ($n_0 = p_0 = n_i$), $b_a = 0$ and the motion of the excess carriers is governed by $\mathcal{E}_x$.

The preceding results are very useful in analyzing semiconductor devices: pn-junctions, bipolar transistors, and solar cells are devices with minority carrier conductivity. The majority carriers in them contribute only to the built-in-field in the interface regions.

9.5 DIFFUSION IN EXTRINSIC MATERIALS

The concentration of excess majority carriers is relatively small in comparison with the background majority carrier concentration, but the concentration of minority carriers can be much larger than the background minority carrier concentration. Therefore, it is important to treat only the continuity equation for minority carriers when considering diffusion in extrinsic materials.

In n-type material ($n \gg p$), we consider the behavior of the minority holes. Let us consider the case when the excess hole concentration Δp is created on the surface of semiconductor, and the distribution of the minor carriers is defined by the boundary conditions on the surface,

$$\Delta p = (\Delta p)_0 \quad \text{at} \quad x = 0, \qquad (9.5.1a)$$

and in the bulk of the crystal

$$\Delta p = 0 \quad \text{for } x \to \infty. \qquad (9.5.1.b)$$

We apply the continuity equation (9.4.7) in order to find the stationary distribution of minority carriers in the bulk region where there is no generation. In one dimension, the continuity equation for excess holes takes the form

$$\frac{\partial p}{\partial t} = -\frac{\Delta p}{\tau} - b_p \mathcal{E}_x \frac{\partial \Delta p}{\partial x} + D_p \frac{\partial^2 \Delta p}{\partial x^2} = 0, \qquad (9.5.2)$$

where $\vec{v}_a = b_p \vec{\mathcal{E}}$. The excess concentration is taken to be small, and D_p, $b_p \mathcal{E}_x$ do not depend on x. It is convenient to introduce in Eq.(9.5.2) coefficients with dimensions of length

$$\lambda = \frac{2D_p}{b_p\mathcal{E}_x} \quad \text{and} \quad L = (D_p\tau_p)^{1/2}. \tag{9.5.3}$$

Equation (9.5.2) in terms of these coefficients takes the form

$$\frac{\partial^2}{\partial x^2}\Delta p - \frac{2}{\lambda}\frac{\partial}{\partial x}\Delta p - \frac{\Delta p}{L^2} = 0. \tag{9.5.4}$$

The general solution of this simple second order differential equation has the form

$$\Delta p = Ae^{s_1 x} + Be^{s_2 x}, \tag{9.5.5}$$

where

$$s_{1,2} = \frac{1}{\lambda} \pm \left(\frac{1}{\lambda^2} + \frac{1}{L^2}\right)^{1/2} = \frac{1}{\lambda}\left[1 \pm \left(1 + \frac{\lambda^2}{L^2}\right)^{1/2}\right]. \tag{9.5.6}$$

If the positive direction of the x-axis points into the crystal, a positive electric field $\mathcal{E}_x > 0$ and $\lambda > 0$ pulls the minority carriers into the sample. Then, $s_1 > 0$ and $s_2 < 0$. Substituting Eq.(9.5.5) in the boundary conditions (9.5.1a) leads to

$$A = 0 \quad \text{and} \quad B = (\Delta p)_0. \tag{9.5.7}$$

The excess minority carrier concentration Δp decreases in the positive x direction exponentially,

$$\Delta p = (\Delta p)_0 e^{-\frac{x}{l_2}}, \tag{9.5.8}$$

where $l_2 = -1/s_2$ is the distance over which Δp decreases to $1/e$ of its initial value,

$$\frac{1}{l_2} = -s_2 = \frac{1}{\lambda}\left[\left(1 + \frac{\lambda^2}{L^2}\right)^{1/2} - 1\right]. \tag{9.5.9}$$

When the electric field is weak, so that $\lambda^2/L^2 >> 1$, or $\mathcal{E}_x << \mathcal{E}_{cr} = L/(b_p\tau_p)$, Eq.(9.5.9) leads to

$$l_2 = L = \sqrt{D_p\tau_p}, \tag{9.5.10}$$

where L is the ***diffusion length*** of holes. We see that the minority carriers move by simple diffusion.

In the opposite case of a strong electric field, when $\lambda^2/L^2 << 1$, or $\mathcal{E}_x >> \mathcal{E}_{cr} = L/(b_p\tau_p)$, one can make the expansion $\left(1 - \frac{\lambda^2}{L^2}\right)^{1/2} \cong 1 + \frac{\lambda^2}{2L^2}$. The quantity expressed by Eq.(9.5.9) yields the ***downstream diffusion length***

$$l_d(\mathcal{E}_x) = \frac{2L^2}{\lambda} = b_p\mathcal{E}_x\tau. \tag{9.5.11}$$

In a strong electric field, the ***drift length*** l_d is equal to the distance which the injected excess carriers travel during the lifetime τ.

A negative electric field $\mathcal{E}_x < 0$ and $\lambda < 0$, pulls the majority carriers into the crystal and opposes the penetration of the minority carriers. In this case $s_1 < 0$ and $s_2 > 0$, and the boundary conditions (9.5.1b) give

$$A = (\Delta p)_0 \quad \text{and} \quad B = 0. \tag{9.5.12}$$

The excess minority carrier concentration Δp decreases exponentially with increasing x,

$$\Delta p = (\Delta p)_0 e^{-\frac{x}{l_1}},$$

with the characteristic length

$$\frac{1}{l_1} = -s_1 = \frac{1}{|\lambda|}\left[1+\left(1+\frac{\lambda^2}{L^2}\right)^{1/2}\right]. \tag{9.5.13}$$

It is seen from Eq.(9.5.13) that $l_1 < L$, that is the electric field opposes the penetration of excess carriers into the crystal. For $E_x << E_{cr}$ Eq.(9.5.13) coincides with Eq.(9.5.9), and $l_1 = l_2 = L$. In a strong electric field for which $\lambda^2/L^2 >> 1$, the ***upstream diffusion length*** is

$$l_u = \frac{2}{|\lambda|} = \frac{D_p}{b_p \mathcal{E}_x}. \tag{9.5.14}$$

When the electric field increases, $\mathcal{E}_x \to \infty$, $l_u \to 0$.

The downstream "diffusion" length l_d can be quite large, because it is the drift length defined by the distance which the excess holes drift in the field before recombining. The upstream excess carrier concentration decays quite rapidly over the distance given by Eq.(9.5.14) defined by the electric field.

The dependence of the minority excess carrier concentration on the distance under "downstream" and "upstream" conditions is shown in Fig.9.5.1.

9.6 ELECTROCHEMICAL POTENTIAL OR QUASI-FERMI LEVEL

A nonequilibrium semiconductor can not be characterized by a Fermi-level (chemical potential) μ. The expressions for the concentrations of electrons and holes given by Eqs.(6.1.8) and (6.1.11), and the law of mass action, Eq.(6.1.21), are also no longer valid.

Nevertheless, we are able to consider electrons and holes to be in ***local equilibrium***. The charge relaxation time for each group of carriers is of the order of $\tau_M \cong 10^{-12}s$, see Eq.(9.3.4). The average time τ for a recombination process is larger, and varies from 10^{-3} to $10^{-10}s$. As a result, electrons and holes inside their bands reach equilibrium before they recombine.

When electrons in the conduction band and holes in the valence band are separated in local thermodynamic equilibrium, we can introduce different Fermi levels (chemical potentials) μ_n and μ_p. The corresponding distribution functions in local electron and hole equilibrium are Fermi-Dirac distributions with the same temperature T:

$$f_n = \frac{1}{1+e^{\frac{E_n-\mu_n}{k_BT}}}, \quad \text{where} \quad E_n = E_c + \frac{p^2}{2m_n} \tag{9.6.1}$$

and

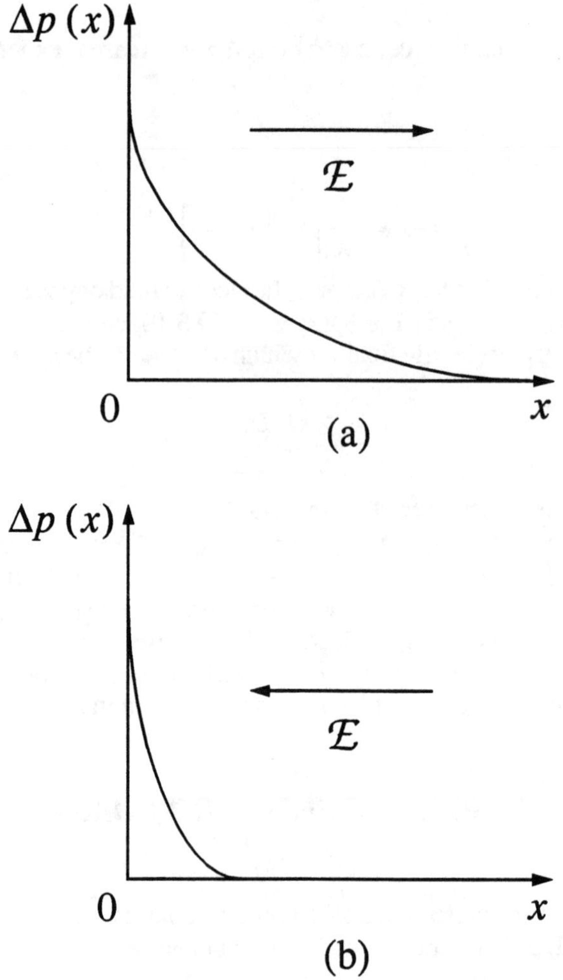

Figure 9.5.1 Diffusion (a) along the field, and (b) against the field. A constant excess density $(\Delta p)_0$ is assumed to exist at $x = 0$.

$$f_p = \frac{1}{1 + e^{\frac{\mu_p - E_p}{k_B T}}}, \quad \text{where} \quad E_p = E_v - \frac{p^2}{2m_p}. \tag{9.6.2}$$

These equations define μ_n and μ_p. In total equilibrium, the chemical potentials of the electrons and holes are equal,

$$\mu_n = \mu_p = \mu. \tag{9.6.3}$$

When the electrons and holes are nondegenerate, the distributions (9.6.1) and (9.6.2) reduce to the Boltzmann distributions

$$f_n \cong e^{\frac{\mu_n - E_n}{k_B T}} \quad \text{and} \quad f_p \cong e^{\frac{E_p - \mu_p}{k_B T}}. \tag{9.6.4}$$

The nonequilibrium concentrations of electrons and holes in the nondegenerate case equal

$$n(\vec{r}) = \frac{2}{(2\pi\hbar)^3}\int f_n(E_n)d\vec{p} = N_c e^{\frac{\mu_n - E_c}{k_B T}}, \tag{9.6.5}$$

$$p(\vec{r}) = \frac{2}{(2\pi\hbar)^3}\int f_p(E_p)d\vec{p} = N_v e^{\frac{E_p - \mu_p}{k_B T}}. \tag{9.6.6}$$

Combining the nonequilibrium concentrations, Eqs (9.6.5) and (9.6.6), with the equilibrium concentrations, Eqs.(6.1.12) and (6.1.13), one obtains

$$\mu_n(\vec{r}) - \mu = k_B T \ln\frac{n(\vec{r})}{N_c} \tag{9.6.7a}$$

and

$$\mu - \mu_p(\vec{r}) = k_B T \ln\frac{p(\vec{r})}{N_v}. \tag{9.6.7b}$$

The product of concentration is equal to

$$n(\vec{r})p(\vec{r}) = n_i^2 e^{\frac{\mu_n - \mu_p}{k_B T}}. \tag{9.6.8}$$

Comparing Eq.(9.6.8) with law of mass action, Eq.(6.1.21), one can see that the difference μ_n - μ_p is a measure of the departure from equilibrium. In equilibrium μ_n = μ_p, and Eq.(9.6.8) reduces to the law of mass action, Eq.(6.1.21).

When an electric field $\vec{\mathcal{E}} = -\nabla\phi$ is applied, the energies of the electrons and holes become $E_n(\vec{p})$ - $e\phi_n(\vec{r})$ and $E_p(\vec{p})$ - $e\phi_p(\vec{r})$, respectively, where ϕ_n and ϕ_p are the electric potentials of the electrons and holes, respectively. The concentrations of electrons and holes are

$$n(\vec{r}) = N_c e^{\frac{\mu_n(\vec{r}) - E_c + e\phi_n(\vec{r})}{k_B T}}, \tag{9.6.9a}$$

$$p(\vec{r}) = N_v e^{\frac{E_v - \mu_p(\vec{r}) - e\phi_p(\vec{r})}{k_B T}}. \tag{9.6.9b}$$

The quantities $\mu_n(\vec{r})$ + $e\phi_n(\vec{r})$ and $\mu_p(\vec{r})$ - $e\phi_p(\vec{r})$ appearing in these expressions are the electrochemical potentials of the electrons and holes defined in Eqs.(7.1.21). The substitution of these concentrations in the combined conductivity and diffusion currents Eqs.(9.1.10) and (9.1.11), and the use of the Einstein relations, Eqs.(9.1.17) and (9.1.18), results in

$$\vec{j}_n = nb_n\nabla[\mu_n(\vec{r}) - e\phi_n(\vec{r})] \tag{9.6.10a}$$

$$\vec{j}_p = pb_p\nabla[\mu_p(\vec{r}) + e\phi_p(\vec{r})]. \tag{9.6.10b}$$

We see that for bipolar nonequilibrium conductivity, the gradients of the electrochemical potentials are responsible for both the current due to the electric field $\vec{E} = -\nabla\phi$, and the diffusion current created by the concentration gradients.

BIBLIOGRAPHY

D.K. Ferry *Semiconductors* (MacMillan, New York, 1991).

CHAPTER **10**

Semiconductor Devices

10.1 *p-n* JUNCTION

Any semiconductor device has junctions of different materials. An important example is the junction between differently doped p- and n- regions of the same semiconductor. A junction of this type is called a ***homojunction***. A p-n junction can be fabricated by doping a single crystal with acceptors on one side of it, say the left side, and with donors on the right side. As a result the p-type material is in metallurgical contact with n-type material.

We assume that in the bulk of the p-material the acceptors are ionized $N_A^- = N_A$, and the concentration of holes (major carriers of p-material) is

$$p_{p0} = N_A^- = N_A \, . \tag{10.1.1}$$

A certain concentration of electrons (minority carriers in p-material) is also present in the p-material with the concentration defined by Eq.(6.1.9),

$$n_{p0} = \frac{n_i^2}{N_A} , \tag{10.1.2}$$

where n_i is the intrinsic electron concentration: $n_i = p_i$. Similarly, in the n-material, donors are ionized, $N_D^+ = N_D$, and the concentration of electrons (majority carriers in n-material) equals

$$n_{n0} = N_D^+ = N_D \,. \qquad (10.1.3)$$

The concentration of holes (minority carriers in n-material) in accordance with Eq.(6.1.9) is

$$p_{n0} = \frac{n_i^2}{N_D} \,. \qquad (10.1.4)$$

The band structure of p- and n-materials that are not in the contact is shown in Fig.10.1.1a. When n- and p-materials are put in contact, concentration gradients appear. The gradient of hole concentration $\nabla p(x)$ occurs because the concentration of holes on the p-side, $p_{p0} = N_A$, is larger than the concentration of holes on the n-side, p_{n0}. In a similar way the gradient $\nabla n(x)$ occurs, because the concentration of electrons on the n-side, n_{n0}, is larger than the concentration of electrons on the p-side, n_{p0}. Gradients of concentration initiate the diffusion current of majority holes from p- to the n-region, which results in recombination of holes with local electrons. The concentration of holes decreases in the n-region down to the low value of the minority carrier concentration p_{n0}. The diffusion of holes leaves behind negatively charged acceptors in the vicinity of the junction as is shown in Fig.10.1.1b.

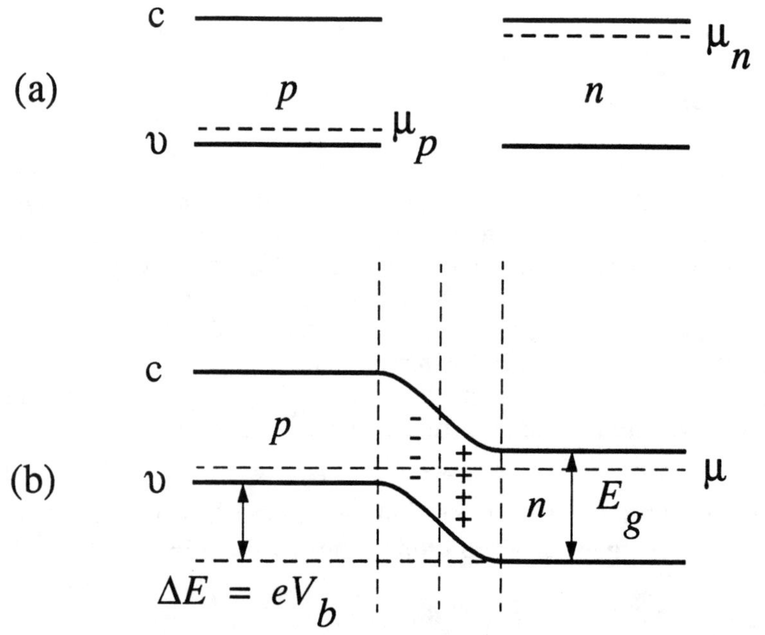

Figure 10.1.1 (a) Separated p- and n- type semiconductors, (b) schematics of the energy-band diagram of $p-n$ junction.

The diffusion of majority electrons from the n-side to the p-side also results in their recombination, and the concentration of electrons decreases to the low value of the minority carrier concentration n_{p0}. The positively charged donor atoms remain on the right side of the junction, Fig. 10.1.1b. Since charged acceptors and donors in the vicinity of the junction are immobile, a narrow charge-space layer with negative ions (-) on the p-side and positive ions (+) on the n-side appears, as is depicted in Fig. 10.1.2. This layer is almost totally depleted of mobile charge carriers. It is called the ***depletion layer***. The thickness of the depletion layer is determined by the diffusion lengths of holes and electrons in n- and p-materials, respectively.

The doubly charged layer results in an electric field $\vec{\mathcal{E}}_b$ directed from (+) to (-), Fig. 10.1.2. It is called a ***built-in field***. This electric field opposes diffusion of mobile carriers across the junction, and stops it. An equilibrium occurs in the presence of the built-in field. When an electron moves from the n-side to the p-side in the built-in potential, it is accelerated. The energy of the electron increases. As a result, the electron energy levels and the bottom of the conduction band, as well as the top of the valence band, move upward, creating the band bending shown in Fig. 10.1.1b. In equilibrium there appears a built-in potential difference V_b between the two sides of the charged layer. The p-side has a lower potential (hence lower potential energy for a hole) than the n-side. An electron has lower potential energy on the n-side than on the p-side.

Band bending lines up the Fermi levels in the p- and n-materials,

$$\mu_p = \mu_n = \mu \,. \tag{10.1.5}$$

Equation (10.1.5) is the condition for the p- and n-regions to be in equilibrium. There is no current across the junction in equilibrium. The diffusion and drift currents cancel each other (for electrons and holes independently).

Let us consider an abrupt junction with constant impurity concentrations on each side of it. The simplest approach to calculating the built-in field and potential consists of neglecting the electron and hole concentrations in the depletion layer. We assume that there are no current carriers in the depletion region,

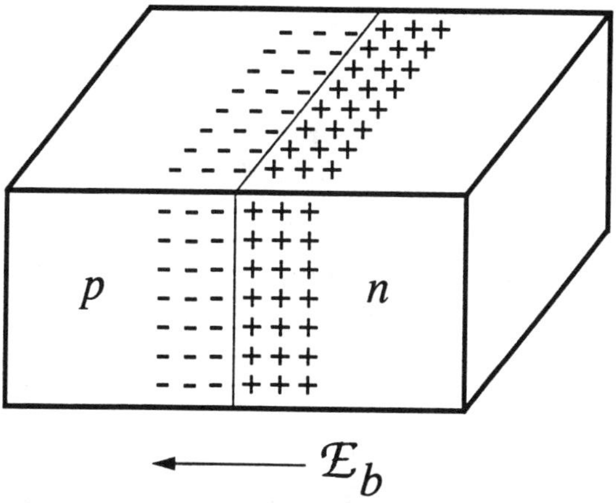

Figure 10.1.2 Built-in electric field $\vec{\mathcal{E}}_b$ in $p-n$ junction.

$$n \ll N_D, \quad p \ll N_A . \tag{10.1.6}$$

This approximation is called the ***depletion approximation***. A p-n junction of this type is shown in Fig. 10.1.3. We shall consider the motion of holes and electrons across the junction in the x-direction. In the one-dimensional approximation the space charge densities are:

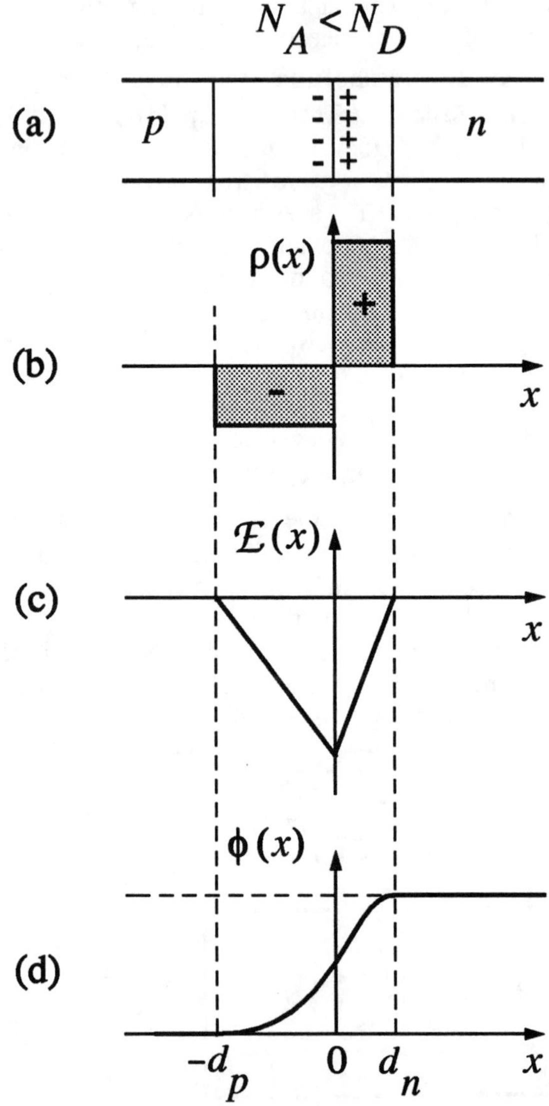

Figure 10.1.3 Schematics of $p-n$ junction in the depletion approximation (a) with its charge (b), electric field (c), and potential (d) distributions at equilibrium.

$$\rho = \begin{Bmatrix} -eN_A & \text{for} & -d_p \le x \le 0 \\ eN_D & \text{for} & 0 \le x \le d_n \end{Bmatrix}, \qquad (10.1.7)$$

where d_p and d_n are the widths of the depletion layers in the p- and n-regions. Since the semiconductor should be electrically neutral, the total acceptor charge is equal to the total donor charge

$$eN_A d_p = eN_D d_n \, . \qquad (10.1.8a)$$

This step-like charge distribution is shown in Fig.10.1.3b. The neutrality condition defines the relative widths of the depletion layers in the p- and n-regions:

$$\frac{d_p}{d_n} = \frac{N_D}{N_A}, \qquad (10.1.8b)$$

i.e. each width is inversely proportional to the doping of the corresponding region. If we have $N_A < N_D$, this leads to $d_p > d_n$, and vice versa.

Using the macroscopic equations of electrostatics, one can calculate the built-in electric field and the potential. The Poisson equation is

$$\nabla \vec{\mathcal{E}} = \frac{\rho}{\varepsilon} \, . \qquad (10.1.9a)$$

In the one-dimensional approximation the only non zero component of the electric field is $\mathcal{E}_x$, so that Eq.(10.1.9a) takes the simpler form

$$\frac{d\mathcal{E}_x(x)}{dx} = \frac{\rho(x)}{\varepsilon}, \qquad (10.1.9b)$$

where ρ in accordance with Eq.(10.1.7) equals $\rho(x) = \rho_p = -eN_A$ in the p-region and $\rho(x) = \rho_n = eN_D$ in the n-region, i.e.

$$\frac{d\mathcal{E}_x(x)}{dx} = -\frac{eN_A}{\varepsilon} \quad \text{for} \quad -d_p \le x \le 0 \, , \qquad (10.1.10a)$$

and

$$\frac{d\mathcal{E}_x(x)}{dx} = \frac{eN_D}{\varepsilon} \quad \text{for} \quad 0 \le x \le d_n \, . \qquad (10.1.10b)$$

Integration of Eqs.(10.1.10a,b) results in

$$\mathcal{E}_x(x) = -\frac{eN_A x}{\varepsilon} + C_1 \qquad \text{for} \quad -d_p \le x \le 0 \, ,$$

$$E_x(x) = \frac{eN_D x}{\varepsilon} + C_2 \qquad \text{for} \quad 0 \le x \le d_n \, ,$$

where C_1 and C_2 are the constants of integration. Because the electric field lines begin on the positive charges and end on the negative charges, and in view of the neutrality condition (10.1.8a,b) the field $\mathcal{E}_x$ must vanish at $x = -d_p$ and $x = d_n$ that is

$$0 = \frac{eN_A d_p}{\varepsilon} + C_1 \, , \quad C_1 = -\frac{eN_A d_p}{\varepsilon} \, ; \qquad (10.1.11a)$$

$$0 = \frac{eN_D d_n}{\varepsilon} + C_2 \, , \quad C_2 = -\frac{eN_D d_n}{\varepsilon} \, . \qquad (10.1.11b)$$

Substituting Eqs.(10.1.11a,b) in the result of integration of Eqs.(10.1.10a,b) one finds a linear dependence of $\mathcal{E}_x$ on x:

$$\mathcal{E}_x(x) = -\frac{eN_A x}{\varepsilon} - \frac{eN_A d_p}{\varepsilon} = -\frac{eN_A}{\varepsilon}(x + d_p) \quad \text{for} \quad -d_p \leq x \leq 0 \qquad (10.1.12a)$$

and

$$\mathcal{E}_x(x) = \frac{eN_D x}{\varepsilon} - \frac{eN_D d_n}{\varepsilon} = \frac{eN_D}{\varepsilon}(x - d_n) \quad \text{for} \quad 0 \leq x \leq d_n \,. \qquad (10.1.12b)$$

The electric field intensity $\mathcal{E}_x(x)$ is plotted as a function of x in Fig.10.1.3c. Being negative, the electric field reaches its maximum absolute value at the metallurgical junction of the p- and n-regions, and vanishes at the ends of the depletion region.

It is convenient to introduce the electrostatic potential ϕ for the built-in field by the equation $\vec{\mathcal{E}}(x) = -\nabla\phi(x)$. Equations (10.1.12a,b) take the following forms in terms of the electrostatic potentials

$$\frac{d\phi(x)}{dx} = \frac{eN_A}{\varepsilon}(x + d_p) \qquad \text{for} \quad -d_p \leq x \leq 0 \qquad (10.1.13a)$$

and

$$\frac{d\phi(x)}{dx} = -\frac{eN_D}{\varepsilon}(x - d_n) \qquad \text{for} \quad 0 \leq x \leq d_n \,. \qquad (10.1.13b)$$

The integration of Eqs.(10.1.13) results in a parabolic dependence of ϕ on x in the p-region,

$$\phi(x) = \frac{eN_A}{e\varepsilon}(x + d_p)^2 + C_3 \,, \qquad (10.1.14a)$$

and in the n-region,

$$\phi(x) = \frac{eN_D}{e\varepsilon}(x - d_n)^2 + C_4 \,, \qquad (10.1.14b)$$

where C_3 and C_4 are constants of integration. If we take the potential ϕ at $x = -d_p$ to be zero, i.e.

$$\phi(-d_p) = 0 \,, \qquad (10.1.15)$$

then $C_3 = 0$. It follows from Eq.(10.1.12a,b) that the electric field $\mathcal{E}_x(x)$ at $x = 0$ is finite. This means that the potential ϕ is continuous at $x = 0$, that is

$$\frac{eN_A d_p^2}{2\varepsilon} = -\frac{eN_D d_n^2}{2\varepsilon} + C_4 \,.$$

Therefore

$$C_4 = \frac{e}{2\varepsilon}(N_A d_p^2 + N_D d_n^2)$$

Substituting C_3 and C_4 into Eqs.(10.1.14a,b) one obtains

$$\phi(x) = \frac{eN_A}{2\varepsilon}(x + d_p)^2 \qquad \text{for} \quad -d_p \leq x \leq 0 \,, \qquad (10.1.16a)$$

$$\phi(x) = -\frac{eN_D}{2\varepsilon}(x^2 - 2xd_n)^2 + \frac{eN_A}{2\varepsilon}d_p^2 \quad \text{for} \quad 0 \le x \le d_n . \qquad (10.1.16b)$$

The dependence of ϕ on x is shown in Fig.10.1.3d. The potential ϕ increases monotonically from $\phi = 0$, defined by Eq.(10.1.15) up to the built-in potential drop across the depletion region, V_b, which follows from Eq.(10.1.16b) for $x = d_n$:

$$V_b = \phi(d_n) = -\frac{eN_D}{2\varepsilon}(d_n^2 - 2d_n d_n) + \frac{eN_A}{2\varepsilon}d_p^2 = \frac{eN_D}{2\varepsilon}d_n^2 + \frac{eN_A}{2\varepsilon}d_p^2 . \qquad (10.1.17)$$

Introducing unity into the last of Eqs.(10.1.17) in the form $1 = (N_A + N_D) / (N_A + N_D)$ transforms it into

$$V_b = \phi(d_n) = \frac{e}{2\varepsilon}\left\{\frac{N_D(N_A + N_D)}{N_A + N_D}d_n^2 + \frac{N_A(N_A + N_D)}{N_A + N_D}d_p^2\right\} =$$

$$= \frac{e}{2\varepsilon}\left\{\frac{N_A N_D}{N_A + N_D}d_n^2 + \frac{N_D^2}{N_A + N_D}d_n^2 + \frac{N_A^2}{N_A + N_D}d_p^2 + \frac{N_A N_D}{N_A + N_D}d_p^2\right\},$$

and applying the neutrality condition Eq.(10.1.8) $(N_A d_p = N_D d_n)$ yields

$$V_b = \phi(d_n) = \frac{e}{2\varepsilon}\left\{\frac{N_A N_D}{N_A + N_D}d_n^2 + 2\frac{N_A N_D}{N_A + N_D}d_p d_n + \frac{N_A N_D}{N_A + N_D}d_p^2\right\} =$$

$$\frac{e}{2\varepsilon}\frac{N_A N_D}{N_A + N_D}(d_n + d_p)^2. \qquad (10.1.18)$$

V_b is built-in potential drop across the depletion region. Comparing Figures 10.1.1 and 10.1.3 we obtain

$$V_b = \frac{(\mu_n - \mu_p)}{e} \approx \frac{E_g}{e}, \qquad (10.1.19)$$

i.e. the built-in potential is of the order of the energy band gap divided by the electron charge. This is due to the fact that μ_n in the n-type material is close to the conduction band edge, whereas μ_p is near the valence band edge.

Equation (10.1.16a) evaluated at $x = 0$ defines the voltage drop, V_p, in the p-region,

$$V_p = \frac{eN_A d_p}{2\varepsilon}d_p . \qquad (10.1.20a)$$

The difference between the potentials V_b, Eq.(10.1.17), and V_p, Eq.(10.1.20a), gives the voltage drop, V_n, in the n-region

$$V_n = \frac{eN_D d_n}{2\varepsilon}d_n . \qquad (10.1.20b)$$

Taking into account the neutrality condition (10.1.8a) one can see that

$$\frac{V_p}{V_n}=\frac{d_p}{d_n}=\frac{N_D}{N_A}. \tag{10.1.21}$$

Thus, the voltage drop V_b is distributed between the p- and the n-regions in proportion to their widths $d_{p,n}$ (or inversely proportionally to their doping $N_{D,A}$). From Eqs.(10.1.18) and (10.1.21) the width of the depletion layer, d, and its parts d_n and d_p are

$$d_n=\frac{1}{N_D}\left(\frac{2\varepsilon V_b}{e}\right)^{1/2}, \quad d_p=\frac{1}{N_A}\left(\frac{2\varepsilon V_b}{e}\right)^{1/2},$$

$$d=d_n+d_p=\left(\frac{1}{N_A}+\frac{1}{N_D}\right)\left(\frac{2\varepsilon V_b}{e}\right)^{1/2}. \tag{10.1.22}$$

An important feature of this result is that the width of the depletion region, d, is proportional to $\sqrt{V_b}$ and inversely proportional to the doping. The heavier the doping, the narrower is the depletion region. In tunnel diodes the doping is of the order of $10^{18} cm^{-3}$, which gives a value of d smaller than $10^{-6} cm$. In ordinary p-n junctions with the doping level around $10^{15} cm^{-3}$, the width d is of the order of $10^{-4} cm$.

Thermally generated electrons in the p-region and thermally generated holes in the n-region flow easily to the opposite sides of the junction. The microscopic picture of the equilibrium of a p-n junction shows that it is a kind of dynamic equilibrium. Thermal excitation transfers electrons from the valence to the conduction band. Thermally excited carriers contribute to the ***thermal generation current***. On the other hand, some electrons from the conduction band can gain energy and move into the p-region (above the potential barrier) where they recombine with holes. These electrons contribute to the ***recombination current***. In equilibrium the generation and recombination currents compensate each other.

When an external voltage V is applied to a p-n junction, the distribution of charge and electric field change, which results in a current across the p-n junction. We can find the width of the depletion layer $d = d_p + d_n$ by assuming that the resistivities of both the n- and p-regions are equal to zero, so that the voltage drops within the depletion region only. In this approximation the electric field distribution is defined by Eqs.(10.1.12a,b), and the potential distribution by Eqs.(10.1.16a,b), with d_n, d_p, and d given by Eq.(10.1.22), the only difference being that V_b has to be replaced by the new value of the voltage drop across the depletion region.

When the positive battery terminal is connected to the p-region and the negative battery terminal is connected to the n-region we create in the depletion region an external electric field $\vec{\mathcal{E}}$ which is opposite to the built-in field $\vec{\mathcal{E}}_b$, see Fig.10.1.4a. The net field in the junction, $(\vec{\mathcal{E}}_b - \vec{\mathcal{E}})$, is smaller than the field $\vec{\mathcal{E}}_b$ in the unbiased junction. The corresponding applied voltage is positive, $V_f > 0$. It is called a ***forward bias***. The presence of an external bias voltage causes the misalignment of the Fermi levels in the p- and n-regions. They become separated by eV_f as is shown in Fig.10.1.4b. A forward bias increases the potential in the p-region with

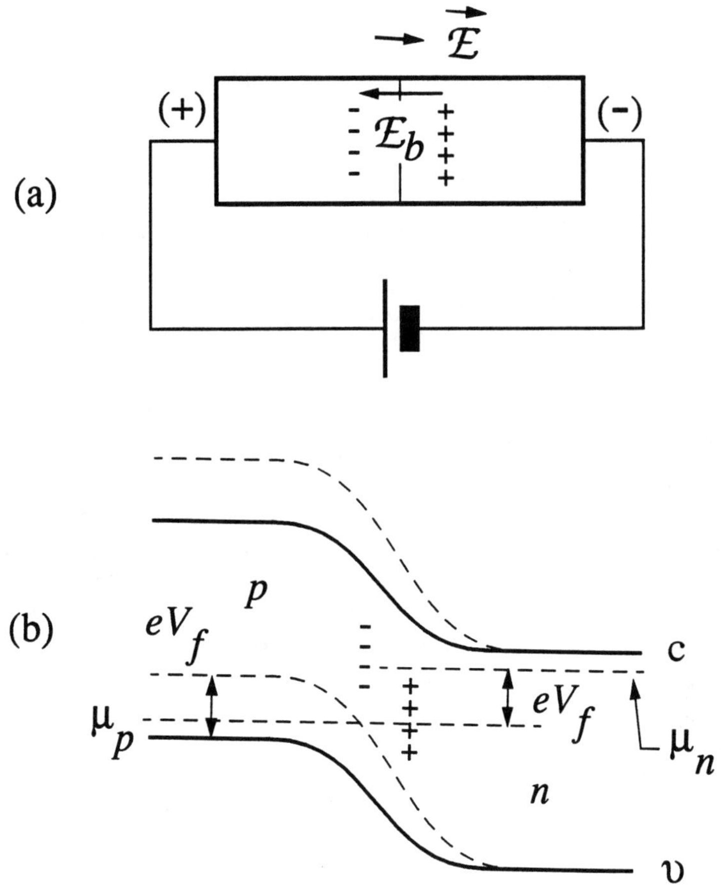

Figure 10.1.4 Electric field (a), and energy-band diagram (b) for a forward biased $p-n$ junction.

respect to the potential in the n-region. A reduction of the negative built-in potential V_b given by Eqs.(10.1.18) and (10.1.22) occurs, $-V_b \to -V_b + V_f$. Replacing V_b by $V_b - V_f$ in Eq.(10.1.22) one obtains

$$d_n = \frac{1}{N_D}\left(\frac{2\varepsilon(V_b - V_f)}{e}\right)^{1/2}, \quad d_p = \frac{1}{N_A}\left(\frac{2\varepsilon(V_b - V_f)}{e}\right)^{1/2},$$

$$d = \left(\frac{1}{N_A} + \frac{1}{N_D}\right)\left(\frac{2\varepsilon(V_b - V_f)}{e}\right)^{1/2}. \tag{10.1.23}$$

The band bending and the height of the potential barrier between the p- and n-regions decrease, and the width of the depletion region decreases as well. This leads to additional majority carrier fluxes, i.e. the flow of holes from the p- to the n-region and the flow of electrons from the n- to the p-region. As a result, the concentration of electrons at the end of the depletion region on the p-side of the junction becomes

$$n(-d_p) = n_{n0} e^{-\frac{eV_b}{k_B T}} e^{\frac{eV_f}{k_B T}}, \tag{10.1.24a}$$

while the concentration of holes at the end of the depletion region on the n-side of the junction is

$$p(d_n) = p_{p0} e^{-\frac{eV_b}{k_BT}} e^{\frac{eV_f}{k_BT}} . \tag{10.1.24b}$$

Since more holes are able to flow from the p- to the n-region, and more electrons go from the n- to the p-region the corresponding total recombination current, $I_r^{(f)}$, increases by the Boltzmann factor in Eqs.(10.1.24a,b)

$$I_r^{(f)} = I_r^{(0)} e^{\frac{eV_f}{k_BT}} , \tag{10.1.25}$$

where $I_r^{(0)}$ is the recombination current at equilibrium, when $V_f = 0$.

The process of excess hole penetration into the n-type material and excess electron penetration into the p-type material is called ***minority carrier injection.*** Since electrons thermally generated in the p-region and holes thermally generated in the n-region flow easily to opposite sides of the junction, the applied bias does not change the thermal generation current, I_g, i.e. $I_g^{(0)}$, where $I_g^{(0)}$ is the generation current at equilibrium. The total forward current, $I^{(f)}$, is equal to $I_r^{(f)}$, Eq.(10.1.25), minus the generation current $I_g^{(0)}$. Since $I_r^{(0)} = I_g^{(0)}$, one finds

$$I^{(f)} = I_g^{(0)} \left(e^{\frac{eV_f}{k_BT}} - 1 \right) . \tag{10.1.26}$$

Here $I_g^{(0)}$ is the sum of hole and electron contributions. When the applied voltage is small, $eV_f << k_BT$, the Taylor expansion of the exponent leads to a linear current-voltage characteristic,

$$I^{(f)} = I_g^{(0)} \frac{e}{k_BT} V_f \quad \text{for} \quad V_f \ll \frac{k_BT}{e} . \tag{10.1.27}$$

At higher voltages, when eV_f exceeds k_BT, the current increases exponentially with the applied voltage. We would like to recall here that at 300 K the value of k_BT/e corresponds to 26 meV, so that the nonlinearity of the current-voltage characteristic starts at a low applied voltage.

When the negative terminal of the battery is connected to the p-region and the positive terminal is connected to the n-region, an external electric field $\vec{\mathcal{E}}$ parallel to $\vec{\mathcal{E}}_b$ is created in the depletion region (see Fig.10.1.5a). The total field in the junction increases up to $\vec{\mathcal{E}}_b + \vec{\mathcal{E}}$. The corresponding applied voltage is negative and we denote it by $-V_r$ with $V_r > 0$. It is called a ***reverse bias.*** The reverse bias decreases the potential of the p-region with respect to that of the n-region as shown in Fig.10.1.5b. An increase of the negative built-in potential occurs, $-V_b \to -V_b - V_f$. Replacing V_b by $V_b + V_r$ in Eq.(10.1.22) one obtains

$$d_n = \frac{1}{N_D} \left(\frac{2\varepsilon(V_b + V_r)}{e} \right)^{1/2} , \quad d_p = \frac{1}{N_A} \left(\frac{2\varepsilon(V_b + V_r)}{e} \right)^{1/2} ,$$

$$d = \left(\frac{1}{N_A} + \frac{1}{N_D} \right) \left(\frac{2\varepsilon(V_b + V_r)}{e} \right)^{1/2} . \tag{10.1.28}$$

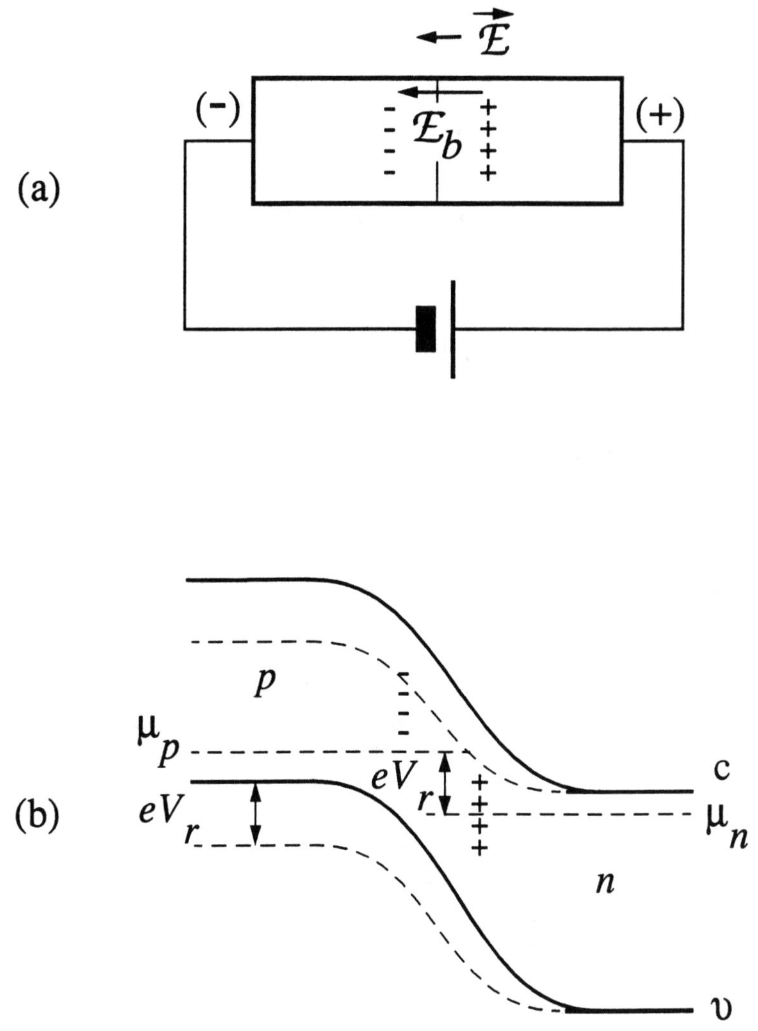

Figure 10.1.5 Electric field (a), and energy-band diagram (b) for a reverse biased $p-n$ junction.

The band bending, the height of the potential barrier between the p- and n-regions, and the depletion region width increase as V_r increases. In contrast to the case of the forward bias, now fewer electrons and holes can flow over the barrier. The concentrations of electrons and holes decrease by the new Boltzmann factor

$$n(d_p) = p_p\, e^{-\frac{eV_b}{k_B T}} e^{-\frac{eV_r}{k_B T}}, \tag{10.1.29a}$$

and

$$p(d_n) = p_n\, e^{-\frac{eV_b}{k_B T}} e^{-\frac{eV_r}{k_B T}}. \tag{10.1.29b}$$

The total recombination current is reduced by the same Boltzmann factor

$$I_r^{(r)} = I_r^{(0)} e^{-\frac{eV_r}{k_BT}}. \quad (10.1.30)$$

The Boltzmann factor in Eq.(10.1.28) controls the small number of holes and electrons which are still able to go over the potential barrier. Taking into account the generation current in the same way this was done in deriving Eq.(10.1.26), we obtain

$$I^{(r)} = I_g^{(0)}\left(e^{-\frac{eV_r}{k_BT}} - 1\right), \quad (10.1.31)$$

where $I^{(r)}$ is the total reverse current. The generation current dominates the recombination current at reverse bias. When the applied voltage is small, $eV_r << k_BT$, the current is proportional to the voltage, while at high voltages, $eV_r >> k_BT$, the first term in Eq.(10.1.31) is exponentially small so that $I^{(r)} = - I_g^{(0)}$. The only current flowing in the junction under a high reverse bias is $I_g^{(0)}$. The dependence of the total current on the voltage is shown in Fig.10.1.6. The p-n junction acts as a rectifier or a diode. The p-n junction is a non-ohmic element with a nonlinear current-voltage characteristic. For $V_f = V_r >> k_BT/e$ the forward current exceeds the reverse current by several orders of magnitude. This is why the reverse current can be assumed to be equal to zero in comparison with the forward current.

One important parameter of a diode is the speed at which the diode can be operated. The speed is defined by a capacitance which can be attributed to the p-n junction. The junction capacitance is connected with the doubly charged layer in the junction. The junction capacitance is responsible for the time necessary to change the positive and negative charges fixed in the depletion layer when the applied voltage changes. The thickness of a depletion layer is proportional to $(V_b - V)^{1/2}$. It increases under reverse bias and decreases under forward bias. Thus, the junction capacitance of a reverse biased diode is smaller (RC response time shorter) then that of a forward biased diode. Semiconductor rectifiers (diodes) have some advantages over vacuum type rectifiers. They are smaller in size and have a longer lifetime.

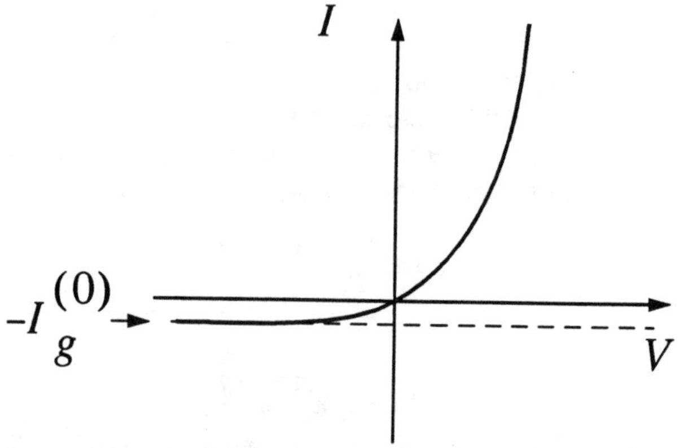

Figure 10.1.6 Current-voltage characteristic of $p-n$ junction.

A second important application of semiconductor diodes is to the photodetection of light. The absorption of photons in a depletion region results in the generation of electron-hole pairs. The generation current exceeds the thermal equilibrium value $I_g^{(0)}$ which is compensated by $I_r^{(0)}$. This is why the total current $I = I_g - I_g^{(0)}$ is proportional to the absorption light intensity. The direction of the current corresponds to a reverse biased diode.

The excess carrier recombination under the forward bias results in photon emission from a diode. This is why forward biased diodes can be used as light emitting diodes (LED's) or even as lasers (see Section 11.4 and Fig.11.4.2 further).

10.2. SCHOTTKY BARRIER

The contact of two different materials also results in a potential barrier and in a nonlinear current-voltage characteristic.

The height of the potential barrier at the interface of two materials can be obtained from the general condition of equilibrium, which requires the equality of electrochemical potentials throughout the structure, as expressed by Eq.(10.1.5). To have a case substantially different from the $p-n$ junction considered above let us consider the contact of two metals. Each metal can be characterized by the electron work function W, which is the energy needed to remove an electron from the Fermi-level in the metal to infinity. The potential energy of electron at infinity can be taken to be zero. Figure 10.2.1a shows the energy diagram of two separated metals with the Fermi levels μ_1 and μ_2, and their work functions W_1 and W_2. We assume that $W_1 < W_2$. When these two metals are put in contact, electrons flow from the metal 1 in order to fill empty energy levels below μ_1 and above μ_2 in the metal 2. The process stops when the highest electron level in the metal 1 and the highest electron level in the metal 2 have the same energy. Metal 1 becomes positively charged and metal 2 is negatively charged. This means that at equilibrium the height of the interface potential barrier $eV_b = W_2 - W_1$ is defined by the difference of the work functions as it is shown in Fig.10.2.1b. The electrochemical potentials of the metals become equal

$$\mu_1 + e\phi_1 = \mu_2 + e\phi_2 ,$$

where $e\phi_1$ is the decrease of the chemical potential in the metal 1, and $e\phi_2$ is the increase of the chemical potential μ_2 in the metal 2.

As equilibrium is established, a potential difference between the metals occurs,

$$e(\phi_1 - \phi_2) = e\, V_b = W_2 - W_1 ,$$

which is defined by the difference of work functions. The work functions are measured experimentally for many metals. There is almost no effect of the surface potential barrier on the electric current through the contact of two metals. The width of the barrier in metals is smaller then the de Broglie wavelength of an electron because the concentration of electrons in the metal is high and the electron gas is

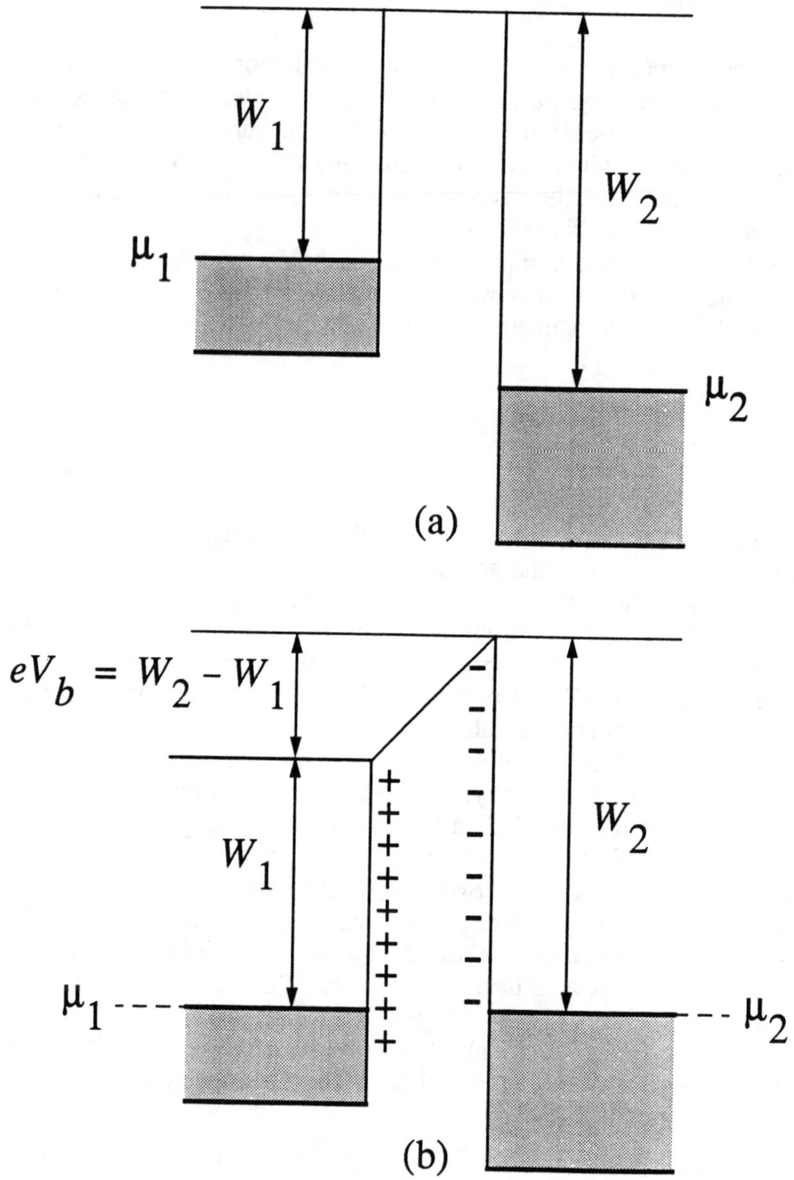

Figure 10.2.1 Energy diagram of two separated metals (a), and two metals in contact (b).

strongly degenerate. The quantum mechanical effect of tunneling through the potential barrier occurs, and the resistivity of the barrier is zero for all practical reasons.

The circumstances are different in case of a metal-semiconductor contact. The concentration of electrons in a semiconductor is lower than in metals. The width of the potential barrier is large, and there might be no tunneling of electrons through it. Let us consider a Schottky barrier which is formed by the contact of an n-type semiconductor with a metal. The energy band diagrams of metal and semiconductor

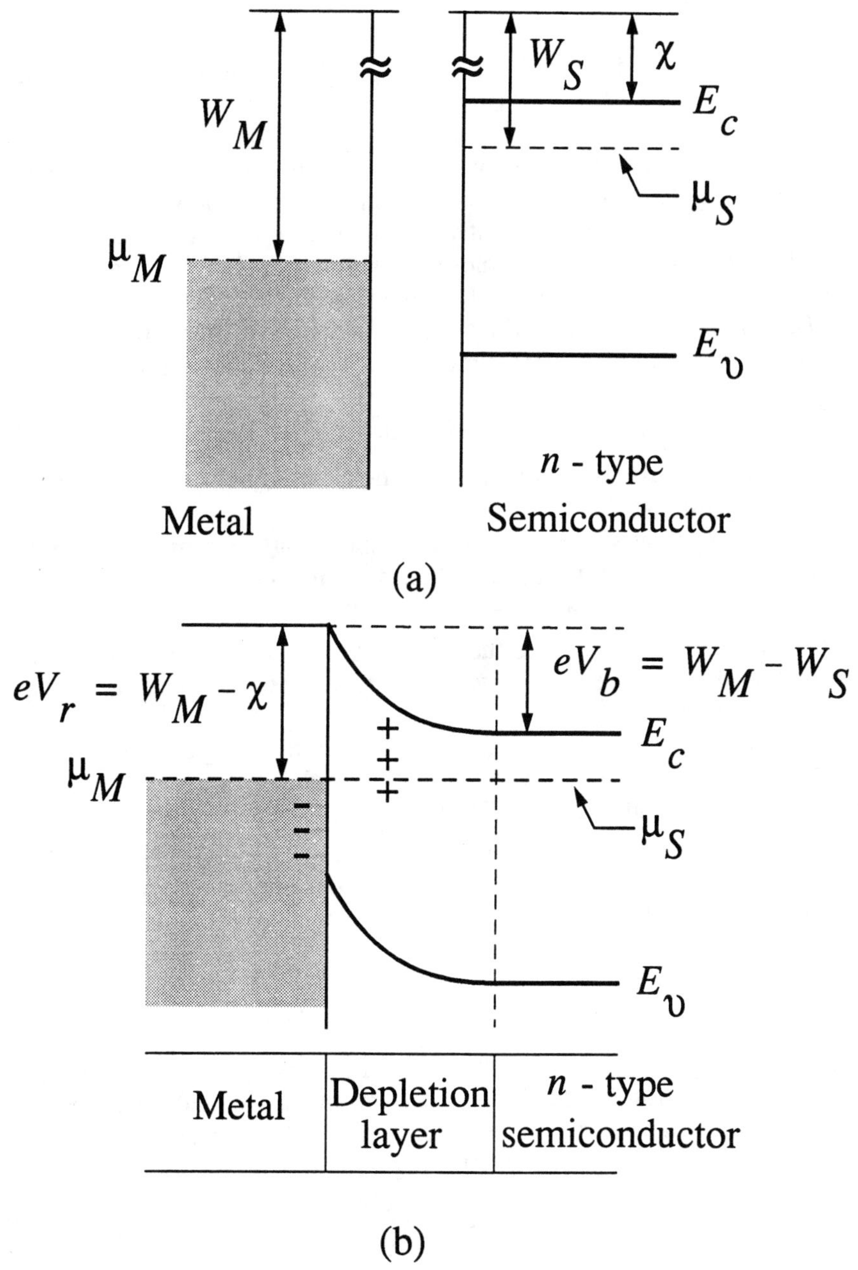

Figure 10.2.2 Energy diagram of a metal and a n- type semiconductor when they are separated (a), and when they are in contact (b).

before joining are shown in Fig. 10.2.2a. The work function of the metal is W_M and the work function of the semiconductor is W_S. In order to take an electron from the metal to the semiconductor one needs the energy $W_M - \chi$, where χ is the electron affinity, which is equal to the energy needed to remove an electron from the bottom of the conduction band to infinity, i.e. $W_S = \chi + (E_c - \mu_s)$, where μ_s is the Fermi level in the semiconductor: $E_c - \mu_s < 0$ for the degenerate case and $E_c - \mu_s > 0$ for the nondegenerate case.

When the semiconductor is put in contact with the metal, electrons go from the higher energy levels of the semiconductor to the lower energy levels of the metal. These electrons leave behind positively charged donors near the interface. The semiconductor surface layer becomes positively charged. This forms the depletion layer of the junction. The surface of the metal acquires negative charge.

The charge transfer from the semiconductor to the metal proceeds until the electrochemical potentials of the metal and semiconductor become equal. The condition of equilibrium is

$$\mu_M + e\phi_M = \mu_S + e\phi_S , \tag{10.2.1}$$

where ϕ_M is the electrostatic potential of the metal, and ϕ_S is the electrostatic potential of the semiconductor.

A built-in potential difference V_b between metal and semiconductor appears, which is defined by the difference of their work functions:

$$eV_b = e\,(\phi_M - \phi_S) = W_M - W_S \,.$$

The built-in potential V_b creates the upward band bending shown in Fig.10.2.2b. The band bending in the semiconductor creates a potential barrier for the electrons of the metal $eV_r = W_M - \chi$.

The electron concentration $n(x)$, and the electrostatic potential $\phi(x)$ are functions of the coordinate x. They vary from the value at the metal-semiconductor boundary to their bulk values. The electron concentration in the bulk of the semiconductor equals $n(x)_{|x\to\infty} = n_0 = N_D$. The quantities $n(x)$ and $\phi(x)$ are defined by the continuity equation (9.3.1) for the total current (9.1.10), together with the Poisson equation (10.1.9a). In the one-dimensional approximation the Poisson equation takes the form

$$\frac{d^2\phi(x)}{dx^2} = -\frac{e}{\varepsilon}(N_D - n(x)), \tag{10.2.2}$$

where N_D is the concentration of the ionized donors in the depletion layer. The electron concentration inside the potential barrier is defined by the Boltzmann distribution with the electron potential energy $U(x) = -e\phi(x)$. It is convenient to introduce the dimensionless potential energy

$$\tilde{V}(x) = -\frac{e\phi(x)}{k_B T}. \tag{10.2.3}$$

If we take the contact plane to be at $x = 0$ and assume that the potential is zero inside the semiconductor far from the barrier,

$$\phi(x)_{|x=0} = \phi(0) = -V_b, \quad \phi(x)_{|x=\infty} = 0 \,;$$

$$\text{or} \quad \tilde{V}(x)_{|x=0} = \tilde{V}_b = \frac{V_b e}{k_B T}, \quad \tilde{V}(x)_{|x=\infty} = 0, \tag{10.2.4}$$

the electron concentration is defined by the equation

$$n(x) = n_0 e^{\frac{e\phi(x)}{k_B T}} = n_0 e^{-\tilde{V}}. \qquad (10.2.5)$$

Substitution of Eqs.(10.2.4) and (10.2.5) transforms the Eq.(10.2.2) into the following form

$$\frac{d^2\tilde{V}(x)}{dx^2} = \frac{1}{r_s^2}\left(1 - e^{-\tilde{V}(x)}\right). \qquad (10.2.6)$$

The quantity

$$r_s = \left(\frac{\varepsilon k_B T}{n_0 e^2}\right)^{1/2} \qquad (10.2.7)$$

is the Debye screening radius defined in Section 7.2. When the band bending is small, $\left|\tilde{V}\right| \ll 1$, by expanding the exponent and keeping the first two terms of that expansion one obtains from Eq.(10.2.6) the equation satisfied by $\tilde{V}(x)$,

$$\frac{d^2\tilde{V}(x)}{dx^2} = \frac{1}{r_s^2}\tilde{V}(x). \qquad (10.2.8)$$

Taking into account the boundary conditions (10.2.4) one finds from Eq.(10.2.8)

$$\tilde{V}(x) = \tilde{V}_b e^{-\frac{x}{r_s}} \quad \text{and} \quad \phi(x) = -V_b e^{-\frac{x}{r_s}}. \qquad (10.2.9)$$

It follows from Eq.(10.2.8) that r_s is the length which corresponds to the decrease of $\tilde{V}(x)$ to $1/e$ of its value at $x = 0$. If $n \approx 10^{16} cm^{-3}$ and $\varepsilon = 16$, the Debye screening radius is equal to $r_s \approx 4\times10^{-6} cm$. The screening radius r_s is larger than the lattice parameter a. This justifies the use of the macroscopic Poisson equation.

In the opposite limiting case, when the band bending is large, $\left|\tilde{V}\right| \gg 1$, the second term on the right hand side of Eq.(10.2.6) is small and can be neglected. The resulting equation

$$\frac{d^2\tilde{V}(x)}{dx^2} = \frac{1}{r_s^2}, \quad \text{or} \quad \frac{d^2\phi(x)}{dx^2} = -\frac{e}{\varepsilon}N_D, \qquad (10.2.10)$$

coincides with the equation solved in Section 10.1 in the depletion layer approximation. The entire potential V_b decreases in a semiconductor which is doped by donors, so we can use all the results of Section 10.1 assuming that $N_A \to \infty$ (or $N_A >> N_D$). This leads to $d_p = 0$ and $d_n = d$ in Eq.(10.1.22).

The applied external voltage, V, can decrease or increase the equilibrium potential barrier V_b. In the depletion approximation the depleted layer near the barrier is an insulator. The entire external potential difference, as well as the built-in potential drops across the barrier. On the metal-semiconductor boundary, the electrostatic potential in the presence of the external bias voltage V equals

$$\phi(x)|_{x=0} = \phi(0) = -(V_b - V), \qquad (10.2.11)$$

where the sign is chosen in the same way as in section 10.1. If a positive potential is applied to the metal, the barrier for electrons from the semiconductor decreases. In this case we will obtain for the Schottky barrier the same type of rectifying current-voltage characteristic as described by Eqs.(10.1.26) and (10.1.31). To do this let us rewrite Eq.(9.1.10) for the electron current density in the presence of an electric field $\mathcal{E}_x$ and concentration gradient $dn(x)/dx$:

$$j_x = en(x)b_n\mathcal{E}_x + eD_n\frac{dn(x)}{dx}, \tag{10.2.12}$$

where both the electron mobility, b_n, and the electron diffusion coefficient, D_n, are assumed to be coordinate-independent. We consider the case when the mean free path of electrons l in the semiconductor is much shorter than the Debye radius

$$l \ll r_s\,. \tag{10.2.13}$$

Equation (10.2.13) means that inside the barrier an electron suffers many collisions, so that the use of Eq.(10.2.12) is justified. This is the so-called ***diffusion approximation***, in which diffusion contributes to the current across the junction in the same way as the drift of electrons in the electric field caused by the depletion region does. Using the Einstein relation between the diffusion coefficient and the mobility, Eq.(9.1.17), substituting the potential instead of the electric field $\mathcal{E}_x = -d\phi/dx$, and dividing Eq.(10.2.12) by eD_n, we obtain

$$\frac{dn(x)}{dx} - \frac{e}{k_BT}\frac{d\phi(x)}{dx}n(x) = \frac{j_x}{b_nk_BT}. \tag{10.2.14}$$

Equation (10.2.14) is a first order linear differential equation for the unknown function $n(x)$. The homogeneous part of this equation

$$\frac{dn(x)}{dx} - \frac{e}{k_BT}\frac{d\phi(x)}{dx}n(x) = 0. \tag{10.2.15}$$

has a simple exponential solution

$$n(x) = Ce^{\frac{e\phi(x)}{k_BT}} = Ce^{-\tilde{V}(x)}. \tag{10.2.16}$$

In accordance with the conventional method of solution of linear differential equations the substitution of Eq.(10.2.16) with $C = C(x)$ into Eq.(10.2.14) gives the solution of Eq.(10.2.14) in the form

$$n(x) = n_0e^{\frac{e\phi(x)}{k_BT}} - \frac{j_x}{b_nk_BT}\int_x^\infty e^{\frac{e}{k_BT}[\phi(x)-\phi(\xi)]}d\xi. \tag{10.2.17}$$

Here the integration constant has been chosen so that $n(x)$ satisfies the boundary condition

$$n_{x\to\infty} = n_0\,. \tag{10.2.18}$$

The solution (10.2.17) can be checked by direct substitution into Eq.(10.2.14).

Let us consider the concentration (10.2.17) at the semiconductor surface $x = 0$.

$$n(0) = n_0e^{-\frac{eV_b}{k_BT}}e^{\frac{eV_a}{k_BT}} - \frac{j_x}{b_nk_BT}\int_0^\infty e^{\frac{e[\phi(0)-\phi(\xi)]}{k_BT}}d\xi \tag{10.2.19}$$

We consider the case when each of the electron fluxes in Eq.(10.2.14) from metal to semiconductor and back are larger than the resulting flux j_x. The electron concentration at the junction $n(0)$ remains almost equal to its equilibrium value, which follows from Eq.(10.2.19) with $j_x = 0$ and $V = 0$ (compare with Eq.(10.2.5))

$$n(0) = n_0 e^{-\frac{eV_b}{k_B T}}. \tag{10.2.20}$$

Solving Eq.(10.2.19) for j_x and using Eq.(10.2.20), we obtain

$$j_x = n_0 b_n k_B T \left(\int_0^{\infty} e^{\frac{e[\phi(0)-\phi(\xi)]}{k_B T}} d\xi \right)^{-1} \left(e^{\frac{eV}{k_B T}} - 1 \right). \tag{10.2.21}$$

The integral in the denominator of Eq.(10.2.21) has a weak dependence on ϕ. The strongest dependence is in the numerator. This is why

$$j_x = j_{0x} \left(e^{\frac{eV}{k_B T}} - 1 \right), \quad \text{where} \quad j_{0x} = n_0 b_n k_B T \left(\int_0^{\infty} e^{\frac{e[\phi(0)-\phi(\xi)]}{k_B T}} d\xi \right)^{-1}. \tag{10.2.22}$$

If $V > 0$, the metal is a positive electrode and the semiconductor a negative electrode, see Fig.10.2.3a, and we speak about a ***forward bias***. It follows from Eq.(10.2.22) that $j_x > 0$. This results in an electron flux from the semiconductor to the metal, see Fig.10.2.3a. For a forward bias the contact potential, Eq.(10.2.11), is reduced from V_b to $(V_b - V)$, compare with Fig.10.2.2b. As a result, electrons in the semiconductor conduction band diffuse across the depletion region into the metal. The current flows in the opposite direction (from metal to semiconductor). This is a ***forward current***.

If $V < 0$, the metal is a negative electrode and the semiconductor is a positive electrode, see Fig.10.2.3b. The potential barrier increases, and the resulting electron flux flows from the metal into the semiconductor when a negative external voltage $V < 0$ is applied, compare with Fig.10.2.2.b. For $V < 0$, the electric current from the semiconductor to the metal is negligible. In both cases the flow of electrons from the metal to the semiconductor is retarded by the barrier $eV_b = (W_m - \chi)$ and j_{0x} is the reverse saturation current that depends on the barrier height. This barrier is not changed by the applied voltage.

We see that a Schottky barrier rectifies the electric current with a large current in the forward direction and a very small current in the opposite direction. The current voltage characteristic has the same shape as the one depicted in Fig.10.1.6. The Schottky barrier works as a rectifier. The large forward current is due to the injection of majority carriers from the semiconductor into the metal. The absence of minority carriers and a reduced storage delay time are advantages of a Schottky barrier diode. The high frequency behavior and the switching speed are better than in $p-n$ junctions.

Metal-semiconductor contacts are made by photolithographic technology. A

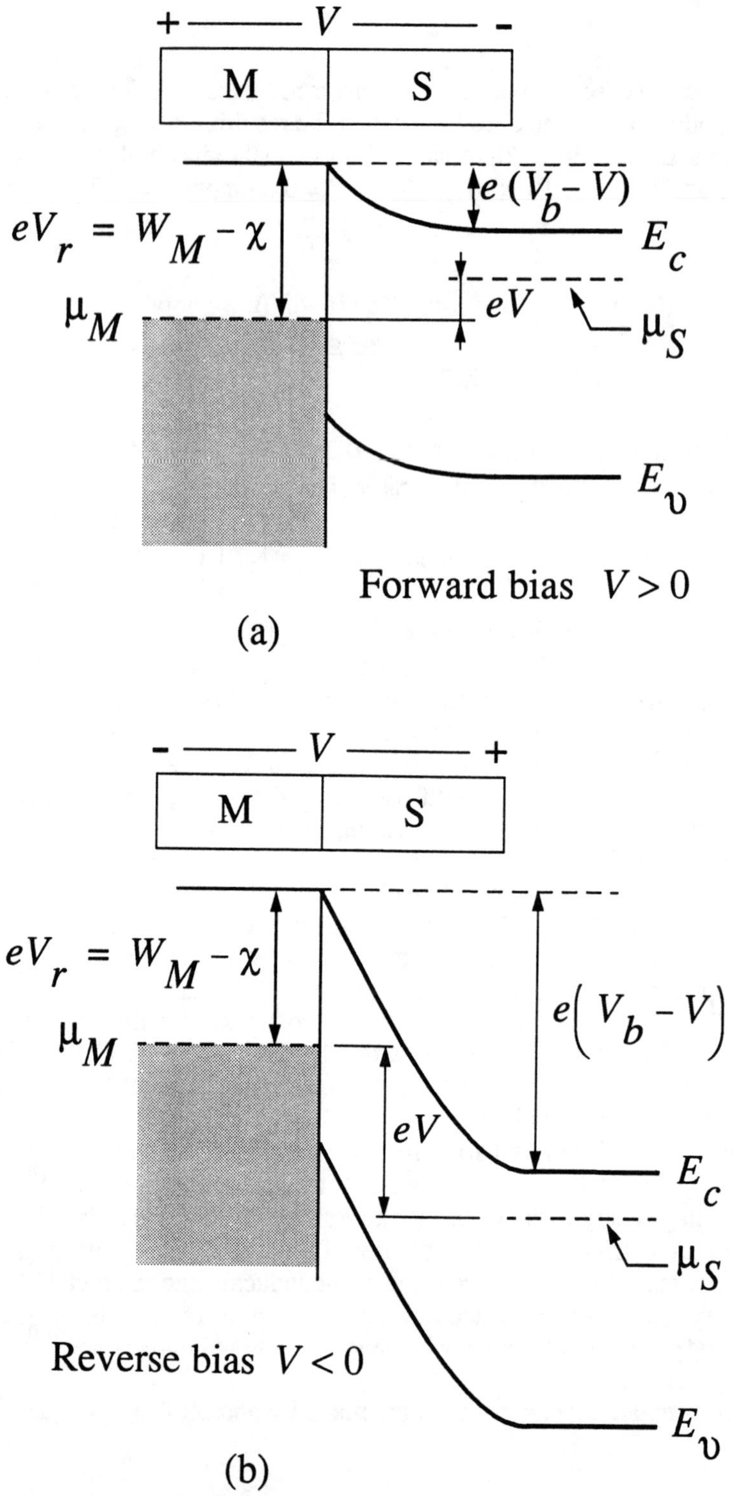

Figure 10.2.3 Forward (a), and reverse (b) biased Schottky barrier.

film of metal is deposited on a semiconductor surface, and with the help of photoresist the metal is partly removed from the surface outside the designated area.

Metal-semiconductor contacts are widely used in integrated circuits as ohmic semiconductor-metal contacts with linear current-voltage characteristics for both directions of the biasing voltage. A practical method of forming ohmic contacts consists of doping a semiconductor heavily in the contact region. It was shown in Section 7.4 that in a heavily doped material we have almost free electrons with a mean free path l longer than the screening radius r_s.

$$l > r_s \,. \tag{10.2.23}$$

There are no screening effects in the barrier region in this case. The depletion region is reduced to a very narrow layer which allows carriers to tunnel through it. For example, gold containing a small percentage of antimony (Sb) alloyed on n – type of Si is a very popular ohmic contact.

One disadvantage of Schottky contacts is the fact that a contact consists of two different materials. As a result, surface electronic states are possible, which can cause significant deviations from the properties of the ideal junction considered above.

10.3 BIPOLAR TRANSISTOR

One of the most important technical applications of semiconductor p-n junctions is their use in devices called ***transistors***. Transistors work as amplifiers of an alternating current or as switchers.

The first known device was a bipolar transistor. The adjective "bipolar" means that the device is designed on the basis of the phenomenon of injection when both electrons and holes participate in bipolar conductivity. A bipolar transistor consists of a semiconductor single crystal, say Si, doped with acceptors on one side, with donors in the central part of the crystal, and with acceptors on the other side of the crystal. The device has the form of two closely coupled p-n junctions shown in Fig.10.3.1a. The heavily doped p^+ region is called the ***emitter***, the narrow central region is referred to as the ***base***, and the p-region on the right with a low level of doping is called a ***collector***. Figure 10.3.1 shows the bipolar transistor in thermal equilibrium, with all three leads of emitter, base, and collector grounded. The entire description of a p-n junction given in Section 10.1 is applicable to the emitter-base and base-collector junctions. The base region is made narrow enough to make these two junctions dependent on each other. Figure 10.3.1b shows the charge density in the three doped regions in the depletion region approximation. The emitter is more heavily doped then the collector. The base doping is less than that of the emitter and is greater than that of the collector doping. The width of the depletion region on the base side of the emitter-base junction is larger than that on the emitter side. Similarly, the width of the depletion region of the base-collector junction is larger in the collector than in the base. The shaded areas on Fig.10.3.1b are the corresponding depletion regions.

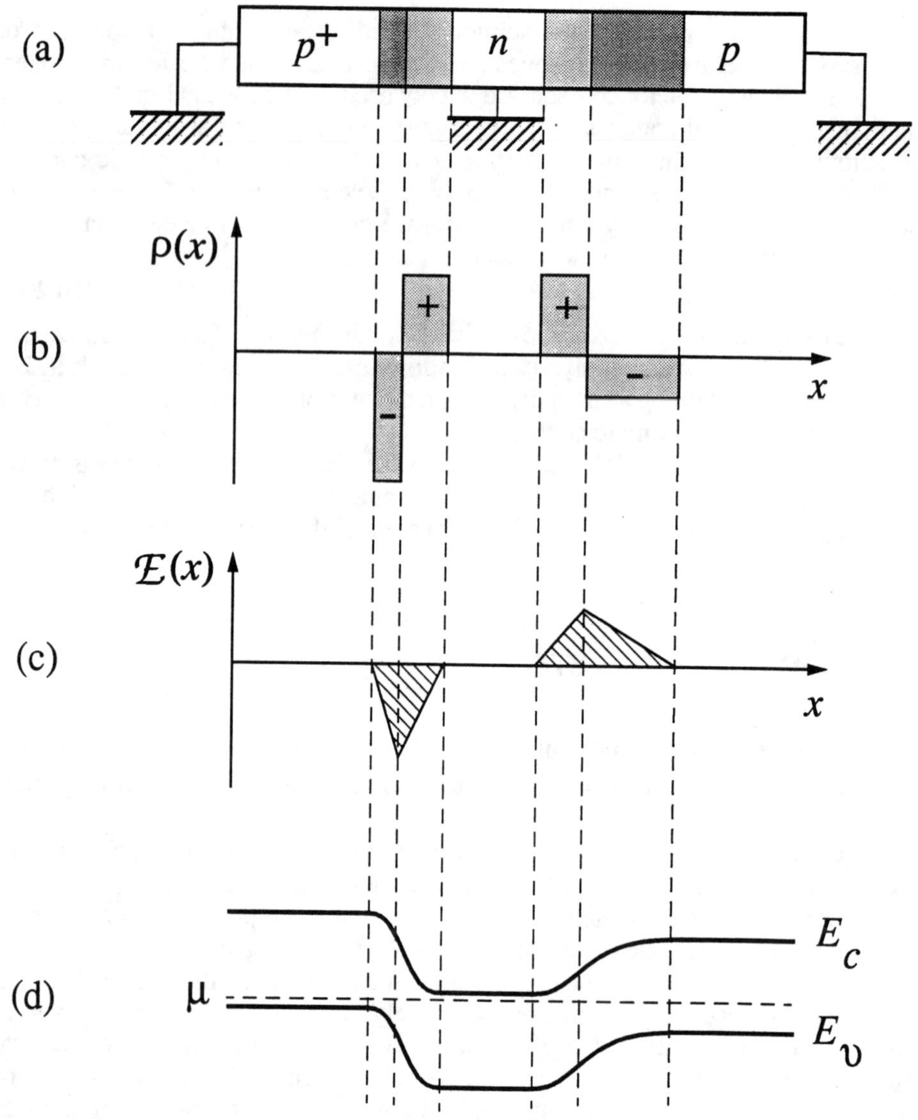

Figure 10.3.1 Schematics of a bipolar transistor in a depletion approximation (a), with its charge (b), electric field (c), and energy-band diagram (d) at equilibrium.

The electric field in the depletion layer, shown in Fig.10.3.1c, originates from the space charge of the ionized donors and acceptors. The built-in electric field prevents the motion of the majority carriers (e.g. electrons from the n-region into the p-region) through the depletion layers, while it is favorable to the motion of the minority carriers (e.g. electrons from the p-regions into the n-region) through the p-n junctions. The corresponding energy band diagram is given in Fig.10.3.1d. There is no current flow in the system in thermal equilibrium. The Fermi level (chemical potential, μ) has constant value along the structure.

The characteristic dimensions L of the active region of the device are considerably larger than the de Broglie wavelength λ of the current carriers

$$L \gg \lambda . \qquad (10.3.1)$$

Replacing the wavelength, λ, by the momentum, p , from the de Broglie relation (1.3.4b), we find that

$$pL \gg \hbar . \qquad (10.3.2)$$

Comparing Eq.(10.3.2) with the uncertainty principle, Eq.(1.3.7), we conclude that quantum effects can be completely neglected, and the principles of operation of a bipolar transistor are given in terms of the classical transport of current carriers.

A transistor is a three terminal device. There are three currents in the transistor called the emitter current $I^{(E)}$, the collector current $I^{(C)}$, and the base current $I^{(B)}$, which connect this device to the external circuit. These three currents satisfy the first Kirchoff law

$$I^{(E)} = I^{(B)} + I^{(C)} . \qquad (10.3.3)$$

Equation (10.3.3) indicates that only two of these three currents are independent. It is convenient to choose the input and the output currents as the independent ones. However, in different ways of connecting of a bipolar transistor to the external circuit, different currents act as the input and output currents. Since a bipolar transistor is a current controlled device, the important parameter characterizing it is the ratio of the input current to the output current. This ratio is called the current gain

$$\alpha = \frac{I_{output}}{I_{input}} . \qquad (10.3.4)$$

We begin with connection when the base lead is common to the emitter-base and to the base-collector circuits, as depicted in Fig.10.3.2a. This configuration is referred to as the ***common base*** configuration. One can see from Fig.10.3.2a that in the common base configuration the emitter current $I^{(E)}$ is the input current, and $I^{(C)}$ is the output current. The emitter-base junction is biased forward, and holes are injected from the p^+ region into the base causing a current $I_p^{(E)}$. These holes are extrinsic minority carriers in the base region.

The injection of electrons from the n-base into the emitter region results in the current $I_n^{(E)}$. The total emitter current is

$$I^{(E)} = I_p^{(E)} + I_n^{(E)} . \qquad (10.3.5)$$

The doping of the emitter is much heavier than that of the base, and

$$I_n^{(E)} \ll I_p^{(E)} . \qquad (10.3.5)$$

Hence one can neglect the electron component of the current with respect to the hole current and find

$$I^{(E)} \approx I_p^{(E)} . \qquad (10.3.6)$$

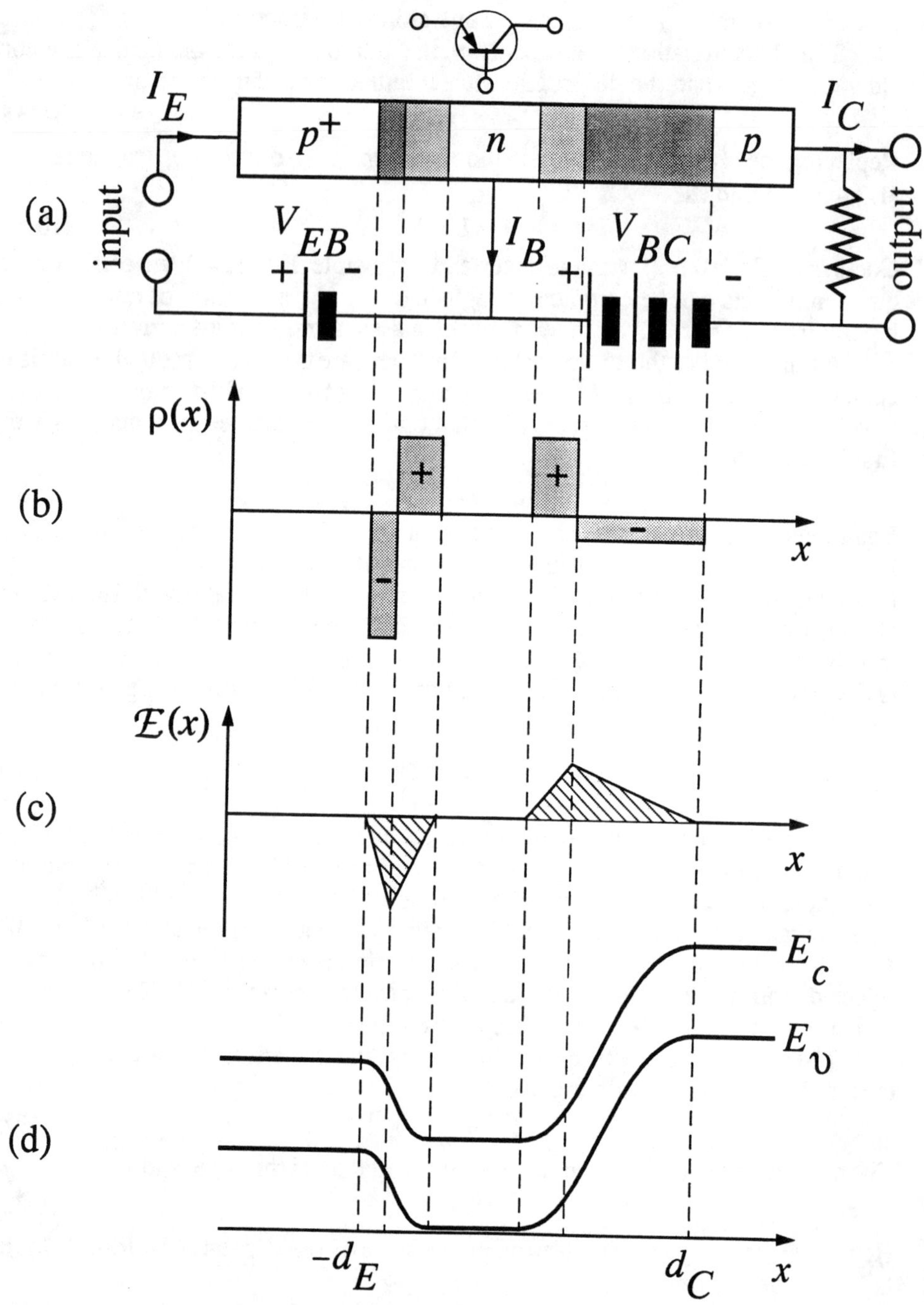

Figure 10.3.2 Schematics of a bipolar transistor in a circuit with a common base configuration in the depletion approximation (a), with its charge (b), electric field (c), and energy-band diagram (d) at equilibrium.

The condition that the crystal be neutral requires that the extrinsic minority carrier concentration decrease due to recombination. The characteristic time of this process is of the order of the electron-hole recombination time at equilibrium. However, the width of the base W_B is made to be sufficiently smaller than the diffusion length of the minority holes, L_D:

$$W_B < L_D \, . \tag{10.3.7}$$

Under this condition holes can drift through the entire base into the depletion region of the base-collector junction. The emitter-base junction is biased forward and holes are injected from the p^+ region into the base. As a result, most of the injected holes are able to reach the base-collector junction. Typically W_B is of the order of 0.1 μm in Si. As a result, more than 90% of the current through the emitter passes into the collector.

The base-collector junction is under a large reverse bias (the potential of the collector is negative with respect to that of the base), and there is large electric field in the base-collector junction that extracts holes into the collector region. These holes contribute a current $I_p^{(C)}$ to the total collector current $I^{(C)}$,

$$I^{(C)} = I_p^{(C)} + I_n^{(C)} \, , \tag{10.3.8}$$

where $I_n^{(C)}$ is the electron current from the collector to the base. The concentration of holes in the base is much higher than the equilibrium concentration. Therefore, $I_n^{(C)} << I_p^{(C)}$. The collector current reduces to

$$I^{(C)} \approx I_p^{(C)} \approx I_p^{(E)} \approx I^{(E)} \, . \tag{10.3.9}$$

A change of the emitter current $I^{(E)}$ results in the same change of the collector current $I^{(C)}$. This means that the behavior of a semiconductor bipolar transistor is similar to that of a vacuum triode, where the cathode represents for the emitter, the anode - the collector, and the control grid - the base. This similarity allows using the symbol of the triode shown at the top of Fig.10.3.2 for the semiconductor transistor.

It should be pointed out that the recombination of carriers results in a very small base current $I^{(B)}$ which maintains the neutrality of the base region. There is also an extra collector current $I_n^{(C)}$ which correspond to the "thermal" current of the reverse biased p-n junction. Both currents are small and so far were neglected. In the case of the common-base configuration the current gain α from Eq.(10.3.4) has the value

$$\alpha_0 \approx \frac{I^{(C)}}{I^{(E)}} \approx 1 \, . \tag{10.3.10}$$

This means that there is no gain in current for the common-base configuration of the bipolar transistor. However, the voltage V_{BC} across the reverse biased base-collector junction is substantially larger than the voltage V_{EB} across the forward biased emitter-base junction. This means that the power output of the collector circuit is much larger than that of the emitter circuit. A bipolar transistor with the common-base configuration acts as a ***power amplifier.***

There are two other possible configurations of a bipolar transistor in an external circuit. In the ***common emitter*** configuration the input current is the base current $I^{(B)}$ and the output current is the collector current $I^{(C)}$. It follows from the Kirchoff law (10.3.3) that the base current $I^{(B)}$ is

$$I^{(B)} = I^{(E)} - I^{(C)} .$$

The common emitter configuration is shown in Fig.10.3.3. The current gain coefficient from Eq.(10.3.4) is

$$\gamma_0 = \frac{I^{(C)}}{I^{(B)}} = \frac{I^{(C)}}{I^{(E)} - I^{(C)}} \approx \frac{\alpha_0}{1 - \alpha_0} \gg 1 . \qquad (10.3.11)$$

Since $I^{(B)} << I^{(C)}, I^{(E)}$, the current gain in Eq.(10.3.11) can be much larger than unity. A small variation of the input current $I^{(B)}$ results in a large variation of the output current $I^{(C)}$. The bipolar transistor in the common emitter configuration acts as a ***current amplifier***.

The ***common collector*** configuration is shown in Fig.10.3.4. This type of configuration also gives rise to current amplification.

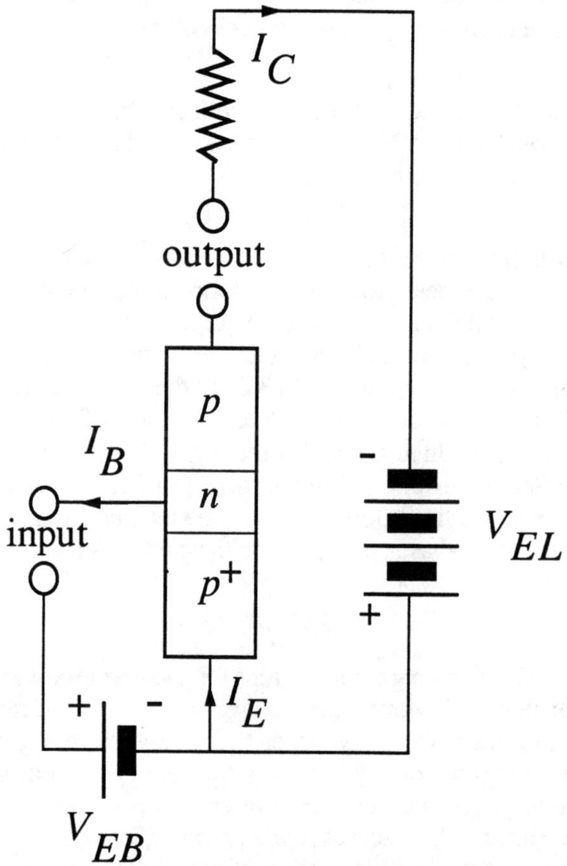

Figure 10.3.3 Bipolar transistor in a circuit with a common emitter configuration.

Bipolar semiconductor transistors were invented in late 1950's. They revolutionarized the electronics industry and influenced strongly our everyday life. The creators of the first bipolar transistor, Shockley, Brattain, and Bardeen were awarded the Nobel Prize in Physics in 1956.

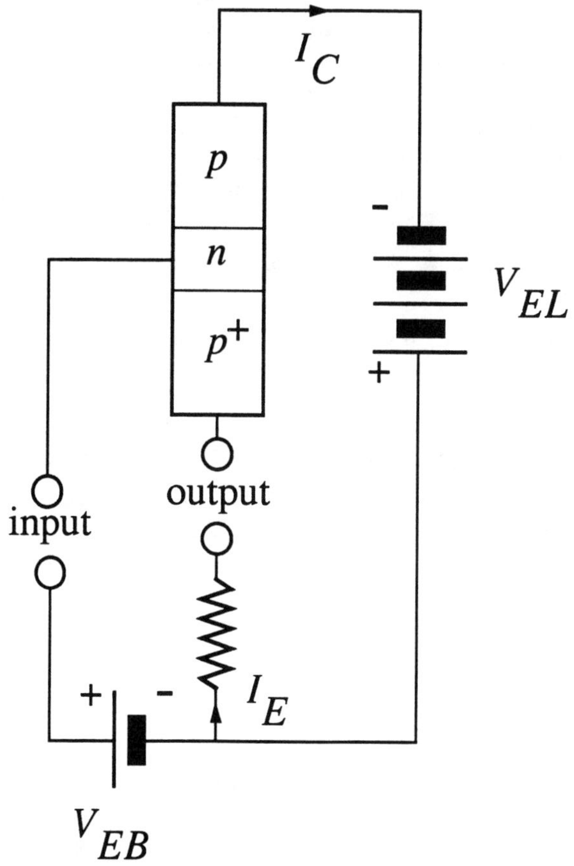

Figure 10.3.4 Bipolar transistor in a circuit with a common collector configuration.

10.4 FIELD EFFECT TRANSISTOR

A field effect transistor (FET) is a three terminal semiconductor device in which electric current through two of its terminals is controlled by the third terminal. The FET consists of a conducting channel (n- or p-type) with two ohmic contacts, one acting as a ***source*** of the carriers and the other as the ***drain*** of carriers. The fabrication of an n-channel FET begins with a lightly n-doped semiconductor material, say Si, which is called the ***substrate***. p^+-type heavily doped region is formed within the substrate to create a p^+-n junctions. A sketch of an n-channel device is given in Fig.10.4.1. The third electrode, the ***gate***, forms a rectifying p^+-n junction with the conducting channel. Since the p^+-region is heavily doped, the depletion region is almost entirely in n-region. The depletion region is shaded in Fig.10.4.1. The right and left gates are tied together. FET requires two external bias voltages for proper operation. One of them is the gate voltage V_{GS} and the second is the drain voltage V_{DS}. The source is grounded, and V_{GS} and V_{DS} are measured with respect to the source. Under the operating conditions the gate is reverse biased $(V_G \leq 0)$, and the drain is forward biased $(V_D \geq 0)$ as shown in Fig.10.4.1. When there is no gate voltage $V_{GS} = 0$ and the small drain voltage is applied, electrons flow from the source to drain (the current flow is in the opposite direction) through the conducting n-channel. The channel acts as a resistor, and the drain current I_D is proportional to the drain voltage V_{DS},

$$I_D = \frac{V_{DS}}{R}. \tag{10.4.1}$$

The resistance R of the channel is defined by its geometry

$$R = \rho \frac{L}{A}, \tag{10.4.2}$$

where ρ is the resistivity of the channel material, L is the channel length, and A is the area of the cross-section of the channel which is defined by the width of the depletion layer W. The current in Eq.(10.4.1) varies linearly with V_{DS}.

The second voltage V_{GS} is applied between the gate and the source. It is used

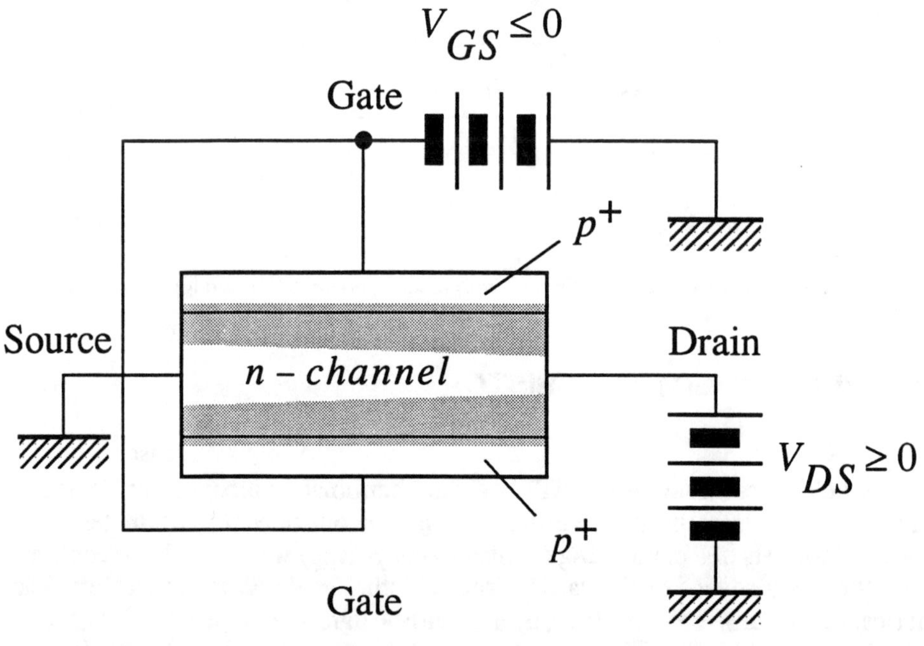

Figure 10.4.1 Schematics of n-channel field effect transistor in the circuit.

to control the amount of the current flowing through the channel. When a small reverse bias V_{GS} is applied between the p^+ region and the n-channel, this effectively reverse biases the two p^+-n junctions formed between the p^+-type gate and the n-type channel. As V_{DS} increases, the width of the depletion region increases proportionally to $\sqrt{V_{DS}+V_b}$, according to Eq.(10.1.28). The average cross-sectional area, A, of the channel is reduced. The resistance from Eq.(10.4.2) increases, and the current I_D from Eq.(10.4.1) decreases.

When V_{GS} increases to a sufficiently high value, the depletion region increases to a point where the channel is entirely depleted, see Fig.10.4.2. I_D decreases almost to zero. The value of V_{GS} required to deplete the entire conducting region is called the gate to the source ***pinch-off voltage.***

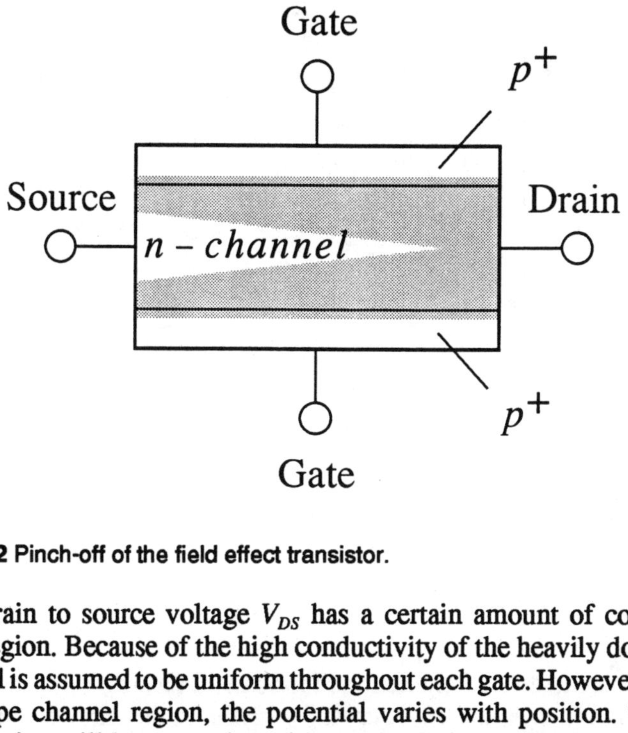

Figure 10.4.2 Pinch-off of the field effect transistor.

The drain to source voltage V_{DS} has a certain amount of control over the depletion region. Because of the high conductivity of the heavily doped p^+ region, the potential is assumed to be uniform throughout each gate. However, in the lightly doped n-type channel region, the potential varies with position. Therefore, the depletion region will be somewhat wider at the drain and as shown in Fig.10.4.1. This is because V_{DS} adds to V_{GS} so that the voltage across the drain end of the p^+-n junction is higher than the voltage across the source end. As V_{DS} increases the drain current also increases. At the same time the increase of the positive potential in the n region is equivalent to the increase of the reverse bias between the n and p^+ regions in the vicinity of the drain. A continued increase in V_{DS} results in a pinch-off in the vicinity of the drain, as shown in Fig.10.4.2, and I_D levels off. The n-type channel becomes so depleted of majority carriers that it will not allow I_D to increase. At this drain voltage the source and the drain are pinched-off or completely separated by the depletion region. A large drain current, called the saturation current, can flow across the depletion region. The dependence of I_D on V_{DS} is shown in Fig.10.4.3. The effect is very similar to the situation caused by injecting carriers into a reverse biased depletion region such as the collector-base depletion region of a bipolar transistor.

Similarly to bipolar transistors, FET find many applications in electric circuits. FET can amplify electronic signals. But a FET differs from bipolar transistors in two aspects. First, the current in a FET involves only the majority carriers. In a bipolar transistor the minority carrier current operates the device. Second, a FET is a voltage controlled device, rather than current controlled, as is the bipolar transistor.

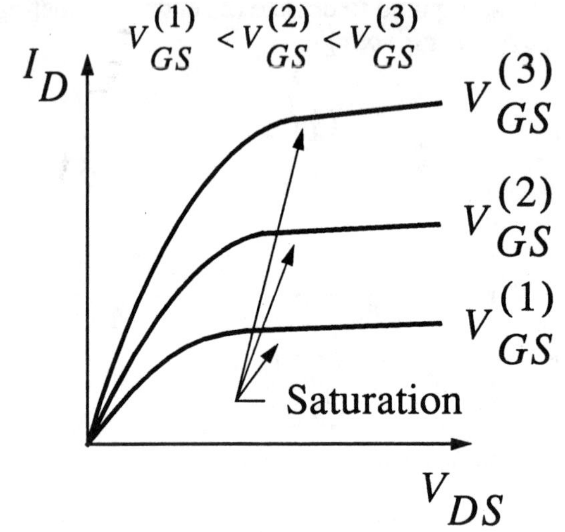

Figure 10.4.3 Current-voltage characteristic of field effect transistor for different gate to source voltages.

10.5 HETEROJUNCTIONS

We have considered so far junctions fabricated in a single semiconductor by appropriate doping (p-n junction). A crystal with the same lattice constant, energy gap, electron affinity, and refractive index is doped differently on the left and the right sides of the junction. Such junctions are called homojunctions.

The other important class of junctions is the junction of two different semiconductors with different band gaps and different electron affinities. These junctions are called ***heterojunctions***. Heterojunctions are fabricated in the process of the growth of one semiconductor single crystal on the substrate of another semiconductor. Perfect growth of this type is possible when the lattice parameters of the growing crystal and substrate are close to each other. Examples of perfect heterojunctions are: Ge-GaAs, GaAs-$Ga_xAl_{1-x}As$, GaAs-$GaAs_xP_{1-x}$, CdTe-CdSe. Only lattice matched heterojunctions are discussed below.

If both semiconductors of a heterojunction have the same type of conductivity, the junction is called an isotype heterojunction *(n-n^+, p-p^+)*. When the conductivity types are different, the junction is called an anisotype heterojunction *(p^+-n or p-n^+)*.

The most important feature of a heterojunction is the difference of the energy gaps, which results in band discontinuities or offsets between the two conduction bands as well as between the two valence bands. These discontinuities act as potential steps which can be used together with an external bias to control the flow and distribution of current carriers in the heterostructure.

We consider an example of anisotype heterojunctions consisting of a wide bandgap n-type semiconductor and a narrow bandgap p-type semiconductor. Figure 10.5.1a shows the energy band profile for the separated materials. The subscripts "1" and "2" refer to the wide and narrow bandgap materials, respectively. The edges of the conduction and valence bands, and the bandgaps are shown in Fig.10.5.1a for both materials. The positions of the conduction bands with respect to the vacuum level are given by the electron affinities χ_1 and χ_2. In our discussion of metal-semiconductor junctions, we used the work function of the two materials in order to find the built-in potential. The work function is the energy necessary to remove an electron from the Fermi-level to the vacuum level. The Fermi-level in a semiconductor depends on the impurity concentration. Therefore the work function of a semiconductor also varies with the impurity concentration. For this reason it is more convenient to use electron affinities, which are believed to be constant characteristics of semiconductors, than work functions. The electron affinities are obtained from photoelectric emission measurements for many semiconductors.

It is seen from Fig.10.5.1a that the discontinuity of the conduction band at the interface is defined by the difference of electron affinities

$$\Delta E_c = \chi_1 - \chi_2 \, . \tag{10.5.1}$$

Using values of the bandgaps E_{g1} and E_{g2} one finds the discontinuity of the valence band

$$\Delta E_v = E_{g2} - E_{g1} - \Delta E_c \, . \tag{10.5.2}$$

When the semiconductors are put into contact, the equilibrium condition requires the Fermi-levels to align. Therefore, band bending occurs. The built-in potential and the depletion region on each side of the junction can be found by solving the Poisson equation with proper boundary conditions. The resulting energy band diagram is shown qualitatively in Fig.10.5.1b. This diagram demonstrates that the barriers for electrons, eV_n, and for holes, eV_p are different. This behavior is used to alter the relative injection of electrons and holes through heterojunction.

It is necessary to note that the energy band diagram depicted in Fig.10.5.1b corresponds to a widely used heterojunction which consists of heavily doped n-type wide bandgap $Al_xGa_{1-x}As$ grown on the lightly doped narrow bandgap GaAs. Due to the discontinuity of the conduction band electrons flow from the n^+-AlGaAs, and they become trapped in the triangular potential well in the GaAs conduction band. The electron conductivity in the triangular potential well corresponds to a very high mobilities because electrons come from the doping impurities of the wide bandgap AlGaAs. The thickness of this channel is about 100 Å. The thickness of the triangular well is smaller than the de Broglie wavelength of an electron in the semiconductor.

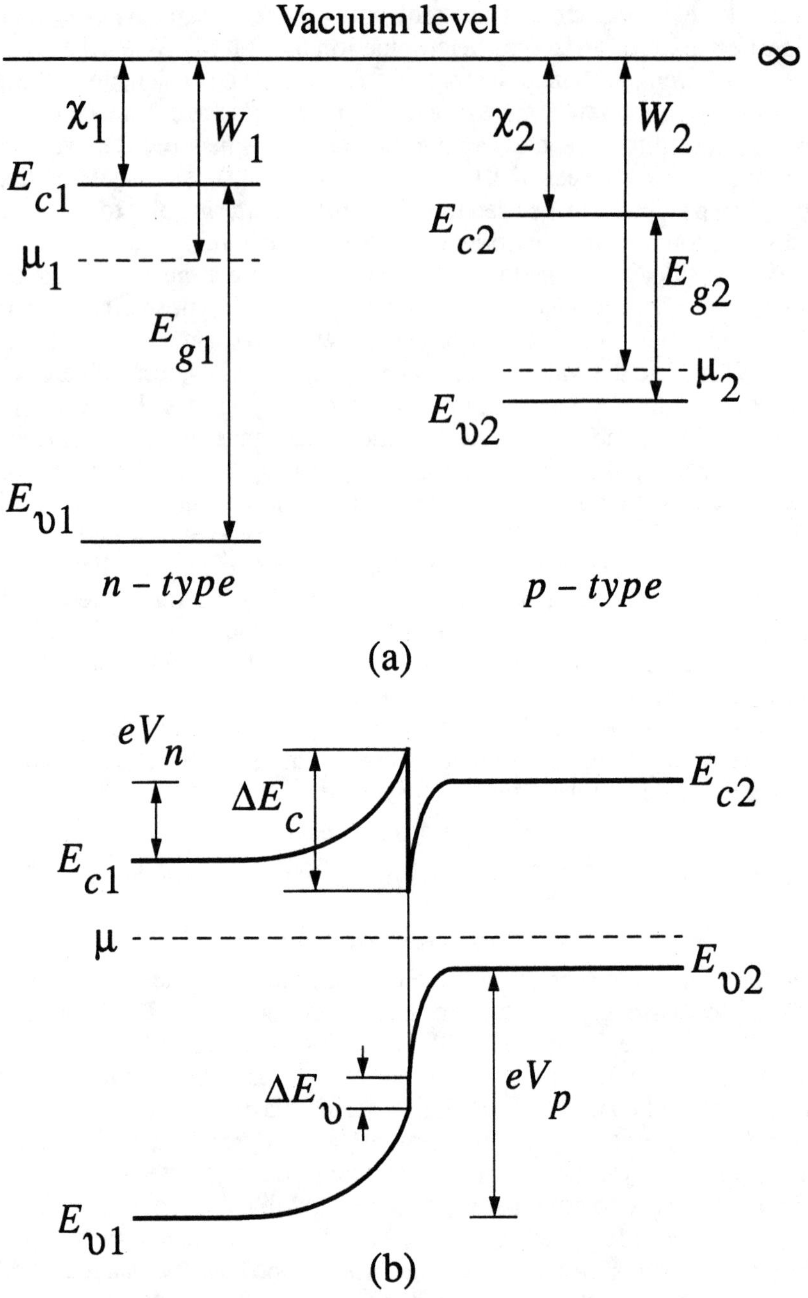

Figure 10.5.1. Energy-band diagram of a wide bandgap n-type and a narrow p-type semiconductors (AlGaAs/GaAs) when they are separated (a), and when they are in contact (b).

Therefore, the electron energy is quantized. This means that the electron gas in the triangular well is practically two-dimensional. Many properties of a two-dimensional electron gas have been verified for just this triangular potential well.

Heterojunctions provide some improvement in the performance of electronic and optoelectronic devices. The potential barriers produced by the band discontinuities are used in the new types of bipolar and field effect transistors and in semiconductor lasers.

BIBLIOGRAPHY

M. Shur *Physics of Semiconductor Devices* ed. N. Holonyak, Jr. (Prentice Hall Series in Solid State Physical Electronics, 1990).

K. Hess *Advanced Theory of Semiconductor Devices* ed. N. Holonyak, Jr. (Prentice Hall Series in Solid State Physical Electronics, 1990).

CHAPTER **11**

Optical Properties of Semiconductors

11.1 ABSORPTION COEFFICIENT

Light absorption by the electronic interband transitions is at present a well developed method of studying the parameters of semiconductor band structures. Since transitions occur between valence and conduction band across the energy gap, it is possible to take into account only these two bands to obtain the main features of the light absorption.

The quasi-classical description of the interaction of an electron with an electromagnetic wave is used below. The quantum-mechanical description of an electron presented in Chapter 4 is combined with a classical description of the electromagnetic field. The electric field intensity $\vec{\mathcal{E}}$ of a transverse monochromatic electromagnetic wave with frequency ω and wave vector $\vec{k}$ can be taken in the form

$$\vec{\mathcal{E}} = -\frac{\partial \vec{A}}{\partial t} \tag{11.1.1}$$

where

$$\vec{A}=\vec{A}_0 e^{i\vec{k}\cdot\vec{r}-i\omega t} \tag{11.1.2}$$

is the vector potential of the transverse electromagnetic wave. It will be shown below (see Eq.11.1.10) that $\vec{k}=0$ for the process we are interested in.

It is known from classical mechanics that in the presence of an external electromagnetic wave defined by the vector potential $\vec{A}$, the electron momentum operator $\hat{\vec{p}}_e=-i\hbar\nabla$ is replaced by the operator

$$\hat{\vec{p}}_e \rightarrow \left(\hat{\vec{p}}_e+e\vec{A}\right) \tag{11.1.3}$$

and the electron kinetic energy operator becomes

$$\frac{\hat{\vec{p}}_e}{2m_0}\rightarrow\frac{\left(\hat{\vec{p}}_e+e\vec{A}\right)^2}{2m_0}=\frac{\hat{\vec{p}}_e^{\,2}}{2m_0}+\frac{e}{m_0}\left(\vec{A}\cdot\hat{\vec{p}}_e\right)+\frac{e^2}{2m_0}A^2 . \tag{11.1.4}$$

The second term on the right hand side of Eq.(11.14) depends on both the electron momentum operator $\hat{\vec{p}}_e$ and on the vector potential $\vec{A}$ of the electromagnetic wave. Therefore, this term represents the interaction of the electron with the electromagnetic wave.

The probability $W_{v\rightarrow c}(\vec{p},\vec{p}')$ of an electronic transition from the valence to the conduction band can be found in the lowest approximation of time dependent perturbation theory from Section 1.16:

$$W_{i\rightarrow f}=\frac{2\pi}{\hbar}\sum_f |\langle i\,|\hat{H}^{(1)}|\,f\rangle|^2\,\delta(E_f-E_i-\hbar\omega) . \tag{11.1.5}$$

Here the interaction operator is

$$\hat{H}^{(1)}=\frac{e}{m_0}e^{i\vec{k}\cdot\vec{r}}e^{-i\omega t}\left(\vec{A}_0\cdot\hat{\vec{p}}_e\right)=-i\frac{e\hbar}{m_0}e^{i\vec{k}\cdot\vec{r}}e^{-i\omega t}(\vec{A}_0\cdot\nabla) , \tag{11.1.6}$$

$E_i=E_v$ and $E_f=E_c$ are the energies of the electron in the valence band before the transition and in the conduction band after the transition, respectively. It should be noted that the interaction $\hat{H}^{(1)}$ depends on the mass of the free electron m_0. There is no way to introduce an effective mass here, since it is a high frequency interaction.

In the initial state , there is a photon in the crystal, and there is neither an electron in the conduction band nor a hole in the valence band. Calculations are considerably simplified when the wave function of the initial state is taken in the simple form

$$\psi_i(\vec{r}_e,\vec{r}_h)=\delta(\vec{r}_e-\vec{r}_h) . \tag{11.1.7}$$

Here $\vec{r}_e$ and $\vec{r}_h$ are the coordinates of the electron and the hole, and $\delta(\vec{r}_e - \vec{r}_h)$ is the δ-function of the type defined by Eq.(1.16.19), which has a sharp maximum at $\vec{r}_e = \vec{r}_h$ and vanishes otherwise. This wave function corresponds to the electron and the hole being very close to each other in the valence band.

In the final state of the absorption process, the photon is absorbed and an electron appears in the conduction band and a hole in the valence band, as is shown in Fig. 4.7.1. In the case of a noninteracting electron and hole, the wave function of the electron-hole pair is a simple product of the electron Bloch wave function and the hole Bloch wave function:

$$\psi_f(\vec{r}_e,\vec{r}_h) = e^{\frac{i}{\hbar}\vec{p}_e\cdot\vec{r}_e} u_{c\vec{p}_e}(\vec{r}_e) e^{\frac{i}{\hbar}\vec{p}_h\cdot\vec{r}_h} u_{v\vec{p}_h}(\vec{r}_h) . \tag{11.1.8}$$

The quasimomentum conservation law requires the momentum of the light $\hbar\vec{k}$ to equal the sum of the electron and the hole momenta,

$$\hbar\vec{k} = \vec{p}_e + \vec{p}_h . \tag{11.1.9}$$

It has been already shown in Section 4.2 that the light momentum $\hbar\vec{k}$ is considerably smaller than the electron and hole momenta

$$\hbar\vec{k} \ll \vec{p}_e, \vec{p}_h . \tag{11.1.10}$$

Neglecting $\hbar\vec{k}$ in Eq.(11.1.9), one obtains

$$\vec{p}_e = -\vec{p}_h = \vec{p} . \tag{11.1.11}$$

Equation (11.1.11) means that the optical transition corresponds to a constant value of $\vec{p}$. On the band diagram, an optical transition is vertical, and $\vec{k}$ can be taken zero in Eq.(11.1.6). Taking into the account Eq.(11.1.11), the wave function of the final state given by Eq.(11.1.8) takes the form

$$\psi_f(\vec{r}_e,\vec{r}_h) = e^{\frac{i}{\hbar}\vec{p}\cdot(\vec{r}_e-\vec{r}_h)} u_{c\vec{p}}(\vec{r}_e) u_{v\vec{p}}(\vec{r}_h) . \tag{11.1.12}$$

The matrix element of the interaction Hamiltonian $\hat{H}^{(1)}$ equals

$$< i\,|\,\hat{H}^{(1)}\,|\,f > = \frac{e}{m_0}\int \psi_i^*(\vec{r}_e,\vec{r}_h)\left(\vec{A}_0\cdot\hat{\vec{p}}\right)\psi_f(\vec{r}_e,\vec{r}_h)d\vec{r}_e d\vec{r}_h .$$

The substitution of the wave functions given by Eqs.(11.1.7) and (11.1.12) results in

$$< i\,|\,\hat{H}^{(1)}\,|\,f > = \frac{e}{m_0}\int \delta(\vec{r}_e-\vec{r}_h)\left(\vec{A}_0\cdot\hat{\vec{p}}\right)e^{\frac{i}{\hbar}\vec{p}\cdot(\vec{r}_e-\vec{r}_h)} u_{v\vec{p}}(\vec{r}_h) u_{c\vec{p}}(\vec{r}_e)d\vec{r}_e d\vec{r}_h . \tag{11.1.13}$$

Using the rule of integration with a δ-function, Eq.(1.16.20), one can evaluate the integral over $\vec{r}_h$ with the result that

$$< i\,|\,\hat{H}^{(1)}\,|\,f > = \frac{e}{m_0}(\vec{A}_0\cdot\vec{p}_{vc}(\vec{p})) , \tag{11.1.14}$$

where the quasimomentum matrix element

$$\vec{p}_{vc}(\vec{p}) = \int u_{c\vec{p}}(\vec{r}_e)\hat{\vec{p}} u_{v\vec{p}}(\vec{r}_e)d\vec{r}_e \tag{11.1.15}$$

defines the transition probability.

We see from Eq.(11.1.14) that interband optical transitions occur at any fixed point $\vec{p}$ of the Brillouin zone. This is why they are called ***direct*** or ***vertical*** transitions.

The calculation of the matrix element Eq.(11.1.14) requires, in principle, the knowledge of the electron wave functions and the energy band structure. This leads to tedious numerical calculations. To proceed in a simple way, one can expand the matrix element $\vec{p}_{vc}(\vec{p})$ in powers of $\vec{p}-\vec{p}_0$ in the vicinity of an arbitrary point $\vec{p}_0$ in the first Brillouin zone,

$$[p_{vc}(\vec{p})]_\alpha = [p_{vc}(\vec{p}_0)]_\alpha + C_{\alpha\beta}(p_\beta - p_{0\beta}) , \qquad (11.1.16)$$

where $\alpha, \beta = x, y, z$ and $C_{\alpha\beta}$ is a dimensionless second rank tensor (coefficient of expansion). This expansion is valid when $|\vec{p}-\vec{p}_0| \ll |\vec{p}_0|$. Depending on the symmetry of the wave functions at $\vec{p}_0$, the matrix element $\vec{p}_{vc}(\vec{p}_0)$ can vanish or be nonzero. A direct symmetry analysis reveals to conclude which elements are equal to zero. Some examples are given below in Section 11.2.

We consider the case when the first term in the expansion (11.1.16) does not vanish. This is called an ***allowed transition***. Substituting the first term of Eq.(11.1.16) into the matrix element (11.1.14), and the result into Eq.(11.1.6) one obtains for the transition rate

$$W_{v\to c}(\vec{p}) = \frac{2\pi}{\hbar}\frac{e^2}{m_0^2}(\vec{A}_0 \cdot \vec{p}_{vc}(\vec{p}_0))^2 \delta(E_c - E_v - \hbar\omega) . \qquad (11.1.17)$$

The case of a direct gap semiconductor with electron energy minima at the Γ-point of the Brillouin zone ($\vec{p}_0 = 0$) is of special importance for applications of semiconductors. In order to find the average energy $Q_{v\to c}$ absorbed per unit time and per unit volume, the probability Eq.(11.1.17) should be multiplied by the light energy $\hbar\omega$ absorbed in the process of transition. In addition, an average over the initial electron distribution $f_v^{(0)}(\vec{p})$ in the valence band and all possible empty final states, $[1 - f_c^{(0)}(\vec{p})]$ in the conduction band should be taken,

$$Q_{v\to c} = \frac{2\pi\omega e^2}{V\, m_0^2}(\vec{A}_0 \cdot \vec{p}_{vc}(\vec{p}_0))^2 \sum_{\vec{p}} f_v^{(0)}(\vec{p})[1 - f_c^{(0)}(\vec{p})]\delta(E_v(\vec{p}) - E_c(\vec{p}) - \hbar\omega) , \qquad (11.1.18a)$$

where

$$\vec{p}_{vc}(p_0) = \int u_{vp_0}(\vec{r}_e)(-i\hbar\nabla)u_{cp_0}(\vec{r}_e)d\vec{r}_e . \qquad (11.1.18b)$$

We can calculate the same quantity $Q_{v\to c}$ using the rules of classical electromagnetic field theory. The average density of energy absorbed by an electron from an electromagnetic wave, Q_{abs}, is equal to the dot-product of the current density $\vec{j} = \sigma\vec{\mathcal{E}}$ by the electric field intensity $\vec{\mathcal{E}}$

$$Q_{abs} = <\vec{j}\cdot\vec{\mathcal{E}}> = \sigma <\vec{\mathcal{E}}^2> . \qquad (11.1.19)$$

The symbol < ... > here means an average over the period of the electromagnetic wave.

In an absorbing semiconductor medium with low losses, the propagation of an electromagnetic wave is given in terms of the complex wave vector (propagation constant)

$$k = i\alpha + \beta , \qquad (11.1.20)$$

where α is the attenuation constant responsible for absorption in the material and β is the propagation constant. It is known from the physics of electromagnetism that in the case of low loss semiconductors the attenuation constant α is connected with the conductivity by the relation

$$\alpha=\frac{\sigma}{2\sqrt{Re\varepsilon}}, \tag{11.1.21}$$

where $Re\varepsilon$ is the real part of the dielectric permitivity of the crystal. The density of absorbed electromagnetic energy depends on the amplitude of the electromagnetic wave propagating in the x-direction

$$\vec{\mathcal{E}}=\omega\vec{A}=\omega\vec{A}_0e^{-\alpha x}e^{i\beta x-\omega t}. \tag{11.1.22}$$

The real exponential here represents the attenuation of the electromagnetic wave in an absorbing medium. Substituting $\vec{j}$ from Ohm's law and $\vec{\mathcal{E}}$ from Eq.(11.1.22), we obtain

$$Q_{abs}=\sigma\omega<A^2>=\frac{1}{2}\sigma\omega^2A_0^2e^{-2\alpha x}. \tag{11.1.23}$$

The quantity

$$\gamma=2\alpha \tag{11.1.24}$$

has the dimension of an inverse length and is the reciprocal of the distance over which the energy density decreases to $1/e$ of its initial value. It is called the ***linear absorption coefficient***. Substituting Eqs. (11.1.24) and (11.1.21) into Eq. (11.1.23) one obtains

$$Q_{abs}=\frac{1}{2}\gamma\sqrt{Re\varepsilon}\,\omega^2\vec{A}_0^2e^{-\gamma x}. \tag{11.1.25}$$

Comparing Eq.(11.1.25) with Eq.(11.1.18a) and taking into account that Eq. (11.1.18a) was obtained for the electric field Eqs.(11.1.1) and (11.1.2) with a constant amplitude, i.e. for $\gamma\to 0$ when $e^{-\gamma x}\approx 1$, one can find the absorption coefficient γ in the form

$$\gamma=\frac{4\pi}{V}\frac{e^2}{m_0^2\omega\sqrt{Re\varepsilon}}\frac{1}{A_0^2}\sum_{\vec{p}}|\vec{A}_0\cdot\vec{p}_{vc}|^2f_v^{(0)}[1-f_c^{(0)}]\delta(E_c(\vec{p})-E_v(\vec{p})-\hbar\omega).$$

Using the rule (2.3.24) and taking into account that each state is occupied by two electrons we replace the sum over $\vec{p}$ by an integral over $\vec{p}$ and find

$$\gamma=\frac{4\pi}{V}\frac{e^2}{m_0^2\omega\sqrt{Re\varepsilon}}\frac{2}{(2\pi\hbar)^3}\frac{1}{A_0^2}\int|\vec{A}_0\cdot\vec{p}_{vc}|^2f_v^{(0)}[1-f_c^{(0)}]\delta(E_c(\vec{p})-E_v(\vec{p})-\hbar\omega)d\vec{p}.$$

In nondegenerate semiconductors, the valence band is almost completely occupied by electrons and the conduction band is empty so that the factor $f_v^{(0)}[1-f_c^{(0)}]\approx 1$. If the incident light is unpolarized, we can substitute for $|\vec{A}_0\cdot\vec{p}_{vc}|^2$ its value averaged over the solid angle, $A_0^2|\vec{p}_{vc}|^2/3$. The absorption coefficient takes the form

$$\gamma = \frac{4\pi e^2}{3m_0^2\omega\sqrt{Re\,\varepsilon}} |\vec{p}_{vc}|^2 \frac{2}{(2\pi\hbar)^3}\int \delta(E_c - E_v - \hbar\omega)d\vec{p}\,. \tag{11.1.26}$$

Comparing Eq.(11.1.26) with electron density of states (6.2.16), we see that the quantity

$$\rho_{opt} = \frac{2}{(2\pi\hbar)^2}\int \delta(E_c(\vec{p}) - E_v(\vec{p}) - \hbar\omega)d\vec{p} \tag{11.1.27}$$

has the dimensions and the meaning of a density of states. The only difference is that the density of states here is defined for both bands $E_c(\vec{p})$- $E_v(\vec{p})$, instead of being defined separately for the conduction, $E_c(\vec{p})$, and valence, $E_v(\vec{p})$, bands as in Eqs.(6.2.15), (6.2.16). ρ_{opt} is called the ***optical density of states***.

It is seen from Eq.(11.1.27) that ρ_{opt} is nonzero only for frequencies

$$\omega > \omega_{\min} = \frac{E_g}{\hbar}\,. \tag{11.1.28}$$

This means that interband absorption in a direct gap semiconductor has a low frequency "red" threshold. The semiconductor is transparent only for $\omega < E/\hbar$.

For the experimental study of semiconductor band structures, the frequency dependence of the absorption coefficient near the threshold frequency $\omega_{\min}$ is of special value. If the energy-quasimomentum dependences in both bands are parabolic and isotropic, as in Eqs.(4.1.7) and (4.7.1), ρ_{opt} is equal to

$$\rho_{opt} = \frac{2}{(2\pi\hbar)^3}\int \delta\left(E_c + \frac{\vec{p}^2}{2m_n} - E_v + \frac{\vec{p}^2}{2m_p} - \hbar\omega\right)d\vec{p}\,. \tag{11.1.29}$$

Carrying out the integration in spherical coordinates, where $d\vec{p} = p^2 dp \sin\theta d\theta d\phi$, we first evaluate the angular part of the integral and find

$$\rho_{opt}(\hbar\omega) = \frac{2}{(2\pi\hbar)^3} 4\pi \int_0^\infty p^2 \delta\left(E_g + \frac{\vec{p}^2}{2m_r} - \hbar\omega\right)dp\,, \tag{11.1.30}$$

where

$$m_r = \frac{m_n m_p}{m_n + m_p} \tag{11.1.31}$$

is called the ***reduced mass*** of the electron and the hole. The integral in Eq.(11.1.30) is evaluated using the rule of integration with the δ-function given by Eq.(1.16.20),

$$\rho_{opt}(\hbar\omega) = \frac{1}{\pi^2\hbar^3}[2m_r^3(\hbar\omega - E_g)]^{1/2}\,. \tag{11.1.32}$$

We see that in the simple case of parabolic energy-momentum dependences, the frequency dependence of the absorption coefficient takes the form

$$\gamma(\omega) = \frac{4}{3}\frac{e^2(2m_r^3\hbar)^{1/2}}{\pi m_0^2}\frac{(\omega - \omega_{\min})^{1/2}}{\omega}\,. \tag{11.1.33a}$$

The frequency dependence of the absorption coefficient has been displayed before, in Section 4.2. This dependence allows the direct determination of E_g.

In case of a forbidden optical transition when $\vec{p}_{vc} = 0$, the second term in the expansion Eq.(11.1.16) must be substituted into the absorption coefficient Eq.(11.1.18a). This results in the following frequency dependence of the absorption coefficient γ

$$\gamma \sim \frac{(\omega - \omega_m)^{3/2}}{\omega}. \qquad (11.1.33b)$$

Along with the direct band-to-band absorption considered above, there are some other mechanisms of light absorption, such as:

a. Intraband absorption by free carriers. This mechanism of absorption has no threshold. The absorption coefficient is proportional to the concentration of free carriers, and is several orders of magnitude smaller than γ for intraband absorption, Eq. (11.1.33).

b. Absorption by carriers bound by impurities or some other defects. This results in an ionization of impurities, and the transitions of electrons into the conduction band or holes into the valence band.

c. Light absorption in indirect band gap materials such as many valley semiconductors. The absorption process in this case is accompanied by simultaneous absorption or emission of phonons. The energy dependence of the absorption coefficient for such a process has been shown in Fig.4.2.4.

11.2 SELECTION RULES

The absorption coefficient given by Eq.(11.1.26) is proportional to the squared modulus of the matrix element $\vec{p}_{vc}(\vec{p}_0)$ of the dipole electronic transition (11.1.18b) from the valence band into the conduction band. Since the Bloch modulating wave functions $u_{v\vec{p}_0}(\vec{r})$ and $u_{c\vec{p}_0}(\vec{r})$ are very often not known, we are not able to calculate $p_{vc}(\vec{p}_0)$. Application of symmetry arguments allows us to find out whether this matrix element vanishes or not or, in other words, whether the transition is allowed or forbidden. Symmetry-based predictions that certain optical transitions are allowed or forbidden are called ***selection rules.***

We illustrate the derivation and application of selection rules for two important semiconductors: Si and GaAs. The symmetry analysis is similar to the one applied in considering the nonvanishing matrix elements in the effective mass approximation for degenerate energy bands in Section 4.3. The momentum operator $\hat{\vec{p}}$ in Eq.(11.1.18) transforms under the symmetry operations S of the crystal in the same way as the radius-vector $\vec{r}$. This means that if the coordinate component transforms as $x \rightarrow y$, the transformation of the corresponding momentum component is

$$p_x = -i\hbar\frac{\partial}{\partial x} \rightarrow -i\hbar\frac{\partial}{\partial y} = p_y.$$

Silicon is an indirect band gap semiconductor of O_h cubic symmetry, whose band structure shown in Fig.4.1.2. The minimum of the electron energy occurs at the X point of the Brillouin zone. The extremum of the valence band is at the Γ point. The momentum conservation law expressed by Eq.(11.1.12) requires the optical transition to be vertical in the band diagram shown in Fig.4.2.2. There is another minimum of the electron energy at the Γ point of the Brillouin zone. Optical absorption of Si results from the vertical electronic transition from the valence band with symmetry Γ'_{25} into this high conduction band minimum with symmetry Γ'_2, see Fig.4.3.1. Possible wave functions of a cubic crystal at the Γ point are listed in Tables 3.4.2 and 3.4.3.

We consider the x – component of $\vec{p}_{vc}$ and calculate the matrix element using possible wave function at Γ point

$$(p_{vc})_x = \int [yz]\left(-i\hbar\frac{\partial}{\partial x}\right)[yz(y^2 - z^2)]d\vec{r}. \qquad (11.2.1)$$

The integrand in Eq.(11.2.1) transforms under the 48 symmetry operations of the silicon crystal as the product $\left[yz\frac{\partial}{\partial x}yz(y^2-z^2)\right]$. It is an odd function of x. Therefore, symmetry operation of inversion I, which exists in the symmetry group of the silicon crystal, leads to $x \rightarrow -x$, and

$$I\left[yz\frac{\partial}{\partial x}yz(y^2-z^2)\right] = -\left[yz\frac{\partial}{\partial x}yz(y^2-z^2)\right]. \qquad (11.2.2)$$

Equation (11.2.2) means that the matrix element $(p_{vc})_x$ vanishes. Applying this symmetry operation to the other components of $\vec{p}_{vc}$, one finds that they vanish as well, so that the direct optical transition in silicon is ***forbidden.***

The second example is GaAs. It is a direct band gap semiconductor with tetraheadral symmetry T_d. The extrema of the conduction and valence bands are located at the Γ point of the Brillouin zone. It is seen from the band structure of GaAs shown in Fig.4.3.1 that the three wave functions of the lower valence band have Γ_{15} symmetry. The wave function of the upper conduction band has Γ_1 symmetry. Wave functions of these types are given in Table 4.3.2.

We analyze again the contribution $(p_{vc})_x$ to $\vec{p}_{vc}$, which has the form

$$(p_{vc})_x = \int [x]\left(-i\hbar\frac{\partial}{\partial x}\right)[1]d\vec{r}. \qquad (11.2.3)$$

Under the operations of T_d symmetry given in Table 3.4.1, the integrand in Eq.(11.2.3) transforms as the product $\left[x\frac{\partial}{\partial x}\right]$. It is an even function of x which does not change the sign under operations of symmetry from the Table 4.3.2. This means that the matrix element p_{vc} does not vanish for GaAs, and the dipole optical transition is ***allowed.***

11.3 EXCITON ABSORPTION

The calculation of the frequency dependence of the absorption coefficient carried out above is based on the single-electron band structure discussed in Chapter 4, which considers electrons and holes as noninteracting charged particles. However, it is found experimentally that the structure of the optical absorption spectrum near the threshold differs considerably from the result obtained earlier and depicted in Fig.4.2.3. When the light energy $\hbar\omega$ approaches E_g, the Coulomb interaction between an electron and a hole becomes important at the energies which satisfy the condition

$$|\hbar\omega - E_g| \le \frac{m_r e^4}{2\varepsilon^2\hbar^2}. \qquad (11.3.1)$$

The right hand side of this inequality contains the Bohr energy calculated with the reduced mass, m_r, of the electron and the hole defined in Eq.(11.1.31).

We consider now the process of light absorption taking into account the Coulomb interaction of the electron and the hole. The simple model of a semiconductor material with nondegenerate isotropic energy bands is used in this consideration.

The probability of an interband optical transition is still defined by Eq.(11.1.5), and the wave function of the initial state is given by Eq.(11.1.7). The wave vector of the electromagnetic wave can be taken equal to zero, Eq. (11.1.10). Considering the Coulomb interaction of the electron and the hole in the final state as a small perturbation to the previously used approximation of a noninteracting electron and hole, we can take the wave function of the final state Ψ_f as a linear combination of the wave functions given by Eq.(11.1.12):

$$\Psi_f = \sum_{\vec{p}} C_{\vec{p}} e^{\frac{i}{\hbar}\vec{p}\cdot(\vec{r}_e - \vec{r}_h)} u_{c\vec{p}}(r_e) u_{v\vec{p}}(\vec{r}_h). \qquad (11.3.2)$$

In our case of a weak electron-hole interaction, only the functions belonging to the top of the valence band and to the bottom of the conduction band contribute to the sum in Eq.(11.3.2). This means that the momentum $\vec{p}$ transferred in the process of absorption is small

$$p \ll p_{\max} = \frac{\pi}{a}. \qquad (11.3.3)$$

Therefore, keeping small $\vec{p}$ in the exponential, one can evaluate the functions $u_{cp}(\vec{r}_e)$ and $u_{vp}(\vec{r}_h)$ at the Γ point ($\vec{p} = 0$) of the Brillouin zone. The wave function Ψ_f then takes the form

$$\Psi_f = u_{c0}(\vec{r}_e)u_{v0}(\vec{r}_h)\cdot \sum_{\vec{p}} C_p e^{\frac{i}{\hbar}\vec{p}\cdot(\vec{r}_e - \vec{r}_h)} = u_{c0}(\vec{r}_e)u_{v0}(\vec{r}_h)\Phi(\vec{\rho}), \qquad (11.3.4)$$

where Φ depends only on the relative coordinate of the electron and the hole $\vec{\rho} = \vec{r}_e - \vec{r}_h$,

$$\Phi(\vec{\rho}) = \sum_{\vec{p}} C_p e^{\frac{i}{\hbar}\vec{p}\cdot\vec{\rho}} \qquad (11.3.5)$$

The effective mass approximation gives $\Phi(\vec{\rho})$ as a solution of the Schrödinger equation for the relative motion of the electron and the hole, which is similar to the Schrödinger equation for hydrogen atom discussed in Section 1.13

$$\left(-\frac{\hbar^2}{2m_r}\nabla_\rho^2-\frac{e^2}{\varepsilon\rho}\right)\Phi(\vec{\rho})=E\Phi(\vec{\rho}), \tag{11.3.6}$$

where the mass of the electron m_0 is replaced by the reduced mass m_r. The eigenvalues E_n and the wavefunctions Φ_n are already known from the results of Section 1.13. In a bound state of the electron and hole, the energy equals

$$E_n=-\frac{m_re^4}{2\hbar^2\varepsilon^2}\frac{1}{n^2}=-E_{ex}\frac{1}{n^2}, \quad E_{ex}=\frac{m_re^4}{2\hbar^2\varepsilon^2}. \tag{11.3.7}$$

Here n is the principal quantum number. Comparing E_{ex} with the Bohr energy, Eq. (1.13.33) we obtain

$$E_{ex}=13.5\left(\frac{m_r}{m_0}\frac{1}{\varepsilon^2}\right)\quad eV. \tag{11.3.8}$$

For example, it equals 4 meV in GaAs and 3 meV in InSb. This hydrogen-like energy is a hundred to a thousand times smaller than the energy of the hydrogen atom.

The wave function of the relative electron-hole motion corresponding to the discrete level of the hydrogen atom with $n = 1$ ($1s$-orbital) is given by Eq.(1.13.29),

$$\Phi_1(\vec{\rho})=\frac{1}{\sqrt{\pi a_0^2}}e^{-\frac{\rho}{a_0}}, \tag{11.3.9}$$

where the Bohr radius a_0 is

$$a_0=\frac{\hbar^2\varepsilon}{m_re^2}=0.53\left(\frac{m_0}{m_r}\varepsilon\right)\quad \mathring{A}. \tag{11.3.10}$$

The radius of this hydrogen-like state is two order of magnitudes larger than the Bohr radius of the hydrogen atom. It follows from Eq.(11.3.7) that optical transitions occur from the valence band to the set of hydrogen-like levels below the bottom of the conduction band, see Fig. 11.3.1. The bottom of the conduction band corresponds to the ionization energy of an exciton.

The motion of an exciton as a particle is not represented in the hydrogen-like equation (11.3.6). In fact, this motion exists in the crystal. When it is taken into account, each discrete hydrogen-like level transforms into an energy band. The momentum conservation law requires that only excitons with zero total momentum will be created as a result of light absorption. Substituting wave functions of the initial, Eq. (11.1.7), and final, Eq. (11.3.4), states into Eq. (11.1.5), one can calculate the matrix element

$$<i\,|\hat{H}^{(1)}|\,f>=\frac{e}{m_0}\sum_{\vec{p}}C_{\vec{p}}\int\delta(\vec{r}_e-\vec{r}_h)\left(\vec{A}_0\cdot\hat{\vec{p}}_e\right)u_{c0}(\vec{r}_e)u_{v0}(\vec{r}_h)e^{\frac{i}{\hbar}\vec{p}(\vec{r}_e-\vec{r}_h)}d\vec{r}_e d\vec{r}_h\,.$$

Using the property of the δ-function given by Eq.(1.16.19) one can evaluate the integral over $d\vec{r}_h$ and find

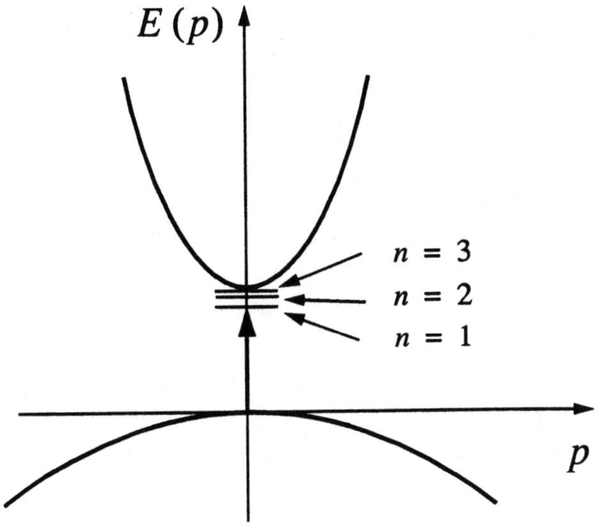

Figure 11.3.1 Hydrogen-like levels of exciton in a direct bandgap semiconductor.

$$< i \,|\, \hat{H}^{(1)} \,|\, f > = \frac{e}{m_0} \sum_{\vec{p}} C_{\vec{p}} \int u_{c0}(\vec{r}_e)\left(\vec{A}_0 \cdot \hat{\vec{p}}_e\right) u_{v0}(\vec{r}_e) d\vec{r}_e \,. \qquad (11.3.11)$$

The quantity

$$\vec{p}_{ve}(0) = \int u_{c0}(\vec{r}_e) \hat{\vec{p}}_e u_{v0}(\vec{r}_e) d\vec{r}_e \qquad (11.3.12)$$

corresponds to Eq.(11.1.18) at $\vec{p}_0 = 0$. It defines whether the transition from the valence to the conduction band is allowed by the symmetry of the modulating Bloch wave functions or not, see Section 11.2.

Taking into account that according to Eq. (11.3.5)

$$\sum_p C_p = \Phi(0)$$

one can find the matrix element of the transition $V_{i \to f}$,

$$< i \,|\, \hat{H}^{(1)} \,|\, f > = \Phi(0) p_{ve}(0) \frac{e}{m_0} (\vec{A}_0 \cdot \vec{p}_{ve}(0)) \,.$$

The transition rate Eq.(11.1.17) equals

$$W_{v \to c}(0) = \frac{2\pi\, e^2}{\hbar\, m_0^2} \,|\, \vec{A}_0 \vec{p}_{vc}(0) \,|^2 \sum_n \,|\, \Phi_n(0) \,|^2\, \delta(E_c + E_n - E_v - \hbar\omega) \,. \qquad (11.3.13)$$

We start with an electronic transition from the valence band to the discrete hydrogen-like exciton levels. This is why the initial energy is taken to be equal to E_v while the final energy is the negative of the energy of the n –th state of an exciton, and $\Phi_n(0)$ is its wave function evaluated at $\rho = 0$. The sum over the final states remains in (11.3.13) as a sum over the exciton's states.

In the simple case of two bands of spherical symmetry with effective masses m_n and m_p, $\Phi(0)$ is nonzero only for s hydrogen-like states corresponding to $l=0$, and

$$|\Phi_n(0)|^2 = \frac{1}{\pi a_0^3 n^3}. \tag{11.3.14}$$

The wave functions of states with $l \neq 0$ ($p-, d-$ ..... states) are odd functions of the coordinates, and they vanish at $\rho = 0$. Therefore, a series of lines that correspond to the energies

$$\hbar\omega_n = E_g - E_{ex}\frac{1}{n^2}, \tag{11.3.15}$$

appears below the absorption edge, $\hbar\omega_{\min} = E_g$. These lines are the allowed transitions from the valence band into the s-states of the hydrogen-like exciton.

It follows from Eqs.(11.3.13) and (11.3.14) that the intensities of these lines are defined by the factor

$$|\Phi_n(0)|^2 \sim \frac{1}{n^3}.$$

In the hydrogen-like spectrum given by Eq.(11.3.7), the separation of levels ΔE_n between the levels n and $n+1$ decreases with increasing n,

$$\Delta E_n \approx \frac{1}{(n+1)^2} - \frac{1}{n^2} \approx \frac{1}{n^3}.$$

The absorption coefficient γ is defined by the density of discrete states per unit interval of energies, that is by the ratio

$$\gamma \sim \frac{|\Phi_n(0)|^2}{\Delta E_n} = Constant\ . \tag{11.3.16}$$

We see that the absorption coefficient γ does not depend on n.

The probability of transition from the valence band into the continuous spectrum of the conduction band is defined by the hydrogen atom wave functions in the continuous spectrum. These functions are known from quantum mechanics. The squared modulus of these functions evaluated at $\rho = 0$ has a quite simple form,

$$|\Phi(0)|^2 = \frac{2\pi\xi}{1-e^{-2\pi\xi}}, \tag{11.3.17}$$

where

$$\xi = \left(\frac{E_{ex}}{\hbar\omega - E_g}\right)^{1/2}. \tag{11.3.18}$$

In our case of attraction between the electron and hole, $\xi > 0$. Substituting Eq. (11.3.17) into Eq. (11.3.13), and summing over all possible $\vec{p}$ of the continuous spectrum instead of n, one obtains the frequency dependence of the absorption coefficient

$$\gamma \sim \frac{\xi}{1-e^{-2\pi\xi}}\sqrt{\hbar\omega - E_g}\,. \tag{11.3.19}$$

Here the square root originates from the frequency dependence of the density of final states in the continuous spectrum defined by Eq.(11.1.32). Near the absorption edge, the quantity $\xi \gg 1$ and Eq. (11.3.19) leads to

$$\gamma = Constant\ . \tag{11.3.20}$$

Comparing Eq. (11.3.20) with Eq. (11.3.16), we may conclude that the limiting values of the absorption coefficient as the absorption edge is approached from the side of the discrete spectrum and from the side of the continuous spectrum are the same.

Far from the absorption edge, $\hbar\omega - E_g >> E_{ex}$ and $\xi \ll 1$, the frequency dependence of α reduces to one that follows from the electron density of states

$$\gamma \approx \sqrt{\hbar\omega - E_g}\ .$$

The schematic frequency dependence of γ in a wide region of frequencies is shown in Fig. 11.3.2. The hydrogen-like model gives the exciton series to within experimental errors as a set of exciton lines with $n = 1,2$, and so on. A real exciton spectrum is shown in Fig.11.3.3 for pure GaAs. Absorption by exciton increases as the temperature decreases. Levels for different n are not resolved in this experiment.

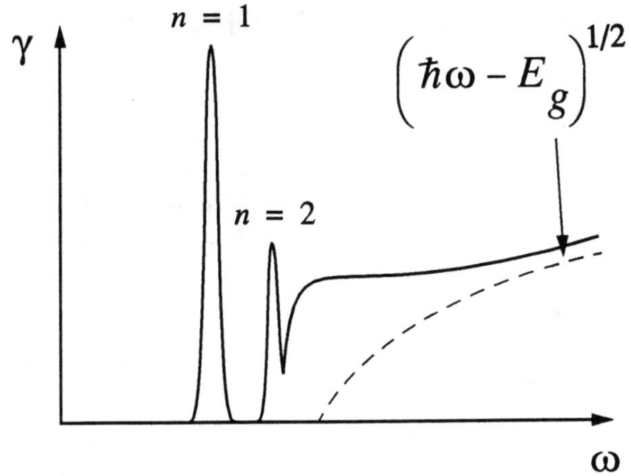

Figure 11.3.2 Schematics of frequency dependence of absorption coefficient, γ.

The Coulomb interaction considered above increases the tendency for two particles to recombine and to emit a photon. An electron not coupled with a hole is more likely to encounter a nonradiative recombination center. It will return to the valence band through that center, and an energy equal to E_g will be converted into heating the crystal instead of being emitted as light. In quantum wells and superlattices narrow electron and hole confining layers enhance the probability of the radiative recombination of electrons and holes.

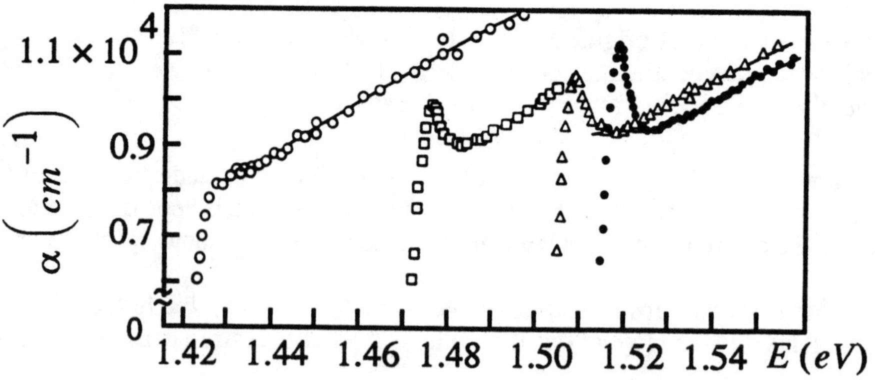

Figure 11.3.3 Experimental absorption spectra of pure GaAs at different temperatures. Spectrum of exciton absorption is well pronounced at low temperatures (o - $T = 294$ K, □ - $T = 186$ K, △ - $T = 90$ K, • - $T = 21$ K). (After M.D. Sturge "Optical absorption of Gallium Arsenide between 0.6 and 2.75 eV", Phys. Rev., v.127, N3, p.771, 1962).

11.4 STIMULATED EMISSION OF RADIATION

Along with processes of light absorption there are processes of light emission accompanied by the transition of an electron from a higher energy state in the conduction band E_2 into a lower energy state E_1 in the valence band, see Fig.11.4.1. There are two different types of emission processes: spontaneous emission and stimulated emission. The rate of spontaneous emission, $Q^{sp}_{2\to1}$, is equal to the product of the probability that the initial state E_2 is occupied by an electron, and the probability that the final state E_1 is empty or rather, that it is occupied by a hole:

$$Q^{sp}_{2\to1} = A_{21} f_2 (1 - f_1)\,, \tag{11.4.1}$$

where the proportionality coefficient A_{21} gives the quantum mechanical probability that the process occurs. The occupation probabilities are given by the Fermi-Dirac distributions

$$f_1 = \frac{1}{e^{\frac{E_1 - \mu_1}{k_B T}} + 1} \quad \text{and} \quad f_2 = \frac{1}{e^{\frac{E_2 - \mu_2}{k_B T}} + 1}\,, \tag{11.4.2}$$

where μ_1 and μ_2 are the chemical potentials (Fermi levels) of the electrons and holes, respectively. Equation (11.4.1) shows that at any finite temperature T there is a certain number of electrons in any excited state E_2. These electrons can undergo a ***spontaneous*** transition from the upper state E_2 into the lower state E_1. The energy conservation law requires that the process be accompanied by the emission of a photon with frequency

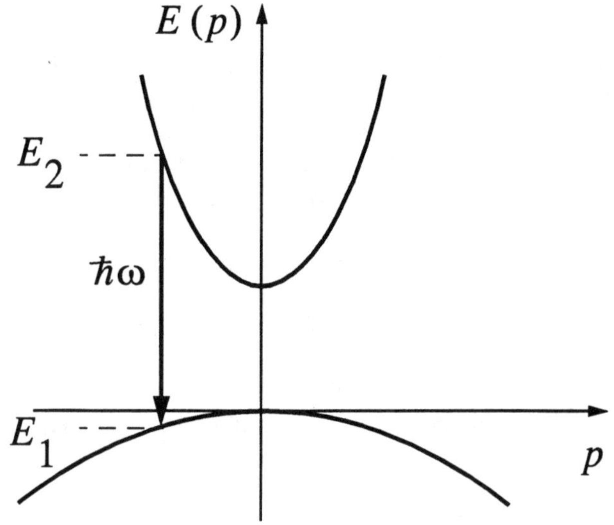

Figure 11.4.1 Electron transition from the upper level E_2 to a lower empty state E_1.

$$\omega_{21} = \frac{E_2 - E_1}{\hbar}. \tag{11.4.3}$$

The mean life-time of an electron in the upper state E_2 is of the order of the life-time of an electron in an atom, which is $10^{-8}s$.

The interaction of an electron in the upper state E_2 with an external field results in the ***stimulated emission*** of photons. Stimulated emission occurs when the crystal is in an external electromagnetic field with spectral density $u(\omega_{21})$ defined in Section 1.1. The rate of stimulated emission is proportional to the occupation probability $f_2(1 - f_1)$, and to the spectral density of radiation $u(\omega_{21})$,

$$Q^{st}_{2\to 1} = B_{21} f_2 (1 - f_1) u(\omega_{21}) . \tag{11.4.4}$$

The proportionality coefficient B_{21} gives the quantum mechanical probability of the transition. It depends on the wave functions in the states E_2 and E_1. The main difference between the rates given by (11.4.1) and (11.4.4) is in the factor $u(\omega_{21})$ for stimulated emission. Due to this factor all emitted photons have the same frequency, polarization, and phase as the external electromagnetic wave. This means that radiation emitted in the process of stimulated emission is ***coherent***.

The process of stimulated emission has an analog in the classical mechanics of an oscillator. If an external periodic mechanical force with frequency Ω is applied to a system (oscillator, pendulum) with natural frequency ω_0, the system obeys the external force and finally oscillates with the frequency of the external force Ω, i.e. the process of oscillation is stimulated by the external force.

The rate of photon absorption, $Q^{abs}_{1\to 2}$, which has been discussed in Section 11.1, can be represented in a similar way,

$$Q^{abs}_{1\to 2} = B_{12} f_1 (1 - f_2) u(\omega_{12}) , \tag{11.4.5}$$

where $B_{12}u(\omega_{12})$ is the probability of transition from the valence to the conduction band given by Eq.(11.1.17). Let us transform Eq.(11.1.17) to obtain an expression for B_{12}. Absorption occurs in the presence of external radiation with spectral density $u(\omega_{12})$ which is defined by the squared amplitude of the external field $\mathcal{E}_0^2$. The energy density of the electromagnetic field averaged over a period of its oscillations is

$$<W> = \frac{\varepsilon\mathcal{E}_0^2}{8}, \tag{11.4.6}$$

where $\varepsilon\mathcal{E}_0^2 = \mu B^2$ for a plane electromagnetic wave. $<W>$ can be expressed in terms of the spectral density of the radiation $u(\omega)$,

$$<W> = \hbar\omega u(\omega) . \tag{11.4.7}$$

Combining Eqs.(11.4.6) and (11.4.7), and using Eq.(11.1.1) one finds

$$\mathcal{E}_0^2 = \frac{8}{\varepsilon}\hbar\omega u(\omega) = \omega^2 A_0^2 . \tag{11.4.8}$$

Using the same procedure of replacing $|\vec{A}_0 \cdot \vec{p}_{vc})|^2$ by its value averaged over a solid angle, $A_0^2 p_{vc}^2/3$, which led to Eq.(11.1.26), and substituting A_0^2 given by Eq.(11.4.8) into Eq.(11.1.17), we obtain that $W_{v\to c}$ is proportional to $u(\omega)$ with the coefficient of proportionality B_{12} equal to:

$$B_{12} = \frac{16\pi}{3}\frac{e^2}{m_0^2\omega_{12}\varepsilon}p_{vc}^2\delta(E_2 - E_1 - \hbar\omega) . \tag{11.4.9}$$

All three processes of spontaneous emission, stimulated emission, and absorption occur together. In equilibrium the upward transition rate (11.4.5) should be balanced by the downward transition rates of spontaneous and stimulated emissions:

$$B_{12}f_1(1-f_2)u(\omega_{21}) = A_{21}f_2(1-f_1) + B_{21}f_2(1-f_1)u(\omega_{21}) . \tag{11.4.10a}$$

The chemical potentials μ_1 and μ_2 in the distribution functions f_1 and f_2 should be equal due to the equilibrium between electrons and holes,

$$\mu_1 = \mu_2 . \tag{11.4.10b}$$

The equilibrium condition Eq.(11.4.10a) was obtained by Einstein in connection with the problem of blackbody radiation. Einstein introduced for the first time the concept of stimulated emission given by Eq.(11.4.4) which appears to be a very important contribution to the equilibrium condition (11.4.10a). Proportionality coefficients A_{12}, B_{21} and B_{12} are called Einstein coefficients.

Solving Eq.(11.4.10a) for the spectral density of radiation $u(\omega_{21})$, one finds

$$u(\omega_{21}) = \frac{A_{21}f_2(1-f_1)}{B_{12}f_1(1-f_2) - B_{21}f_2(1-f_1)} = \frac{A_{21}/B_{21}}{\frac{B_{12}f_1(1-f_2)}{B_{21}f_2(1-f_1)} - 1} .$$

Using the explicit forms of f_1 and $1 - f_2$, the radiation spectral density can be written in the form

$$u(\omega_{21})=\frac{A_{21}/B_{21}}{\frac{B_{12}}{B_{21}}e^{\frac{\hbar\omega_{21}}{kT}}-1}. \qquad (11.4.11)$$

The radiation spectral density $u(\omega_{21})$ in thermal equilibrium is given by the Planck blackbody radiation density

$$u(\omega,T)=\frac{8\pi n_r h\nu^3}{c^3}\frac{1}{e^{\frac{\hbar\omega}{k_BT}}-1}, \qquad (11.4.12)$$

where $\nu=\omega/2\pi$ and n_r is the refractive index. Equating the temperature dependent and temperature independent terms in Eqs.(11.4.11) and (11.4.12) results in

$$B_{12}=B_{21}, \qquad (11.4.13)$$

$$\frac{A_{21}}{B_{21}}=\frac{8\pi n_r h\nu_{21}^3}{c^3}. \qquad (11.4.14)$$

These are the Einstein relations which show that the three radiative processes are related to each other, and that the knowledge of one of them gives the other two at equilibrium. Equation (11.4.13) means that the coefficient B_{21} for stimulated emission and the coefficient B_{12} for absorption are equal. This is an obvious result of the quantum mechanical relation between transition probabilities $W_{v\to c}(\vec{p})=W_{c\to v}(\vec{p})$. The ratio A_{21}/B_{21} (11.4.14) depends on the separation of the levels ν_{21}. The larger is the separation, the higher is the probability of spontaneous emission with respect to stimulated emission.

The ratio of the rates for stimulated and spontaneous emission is very small in equilibrium

$$\frac{Q_{2\to1}^{st}}{Q_{2\to1}^{sp}}=\frac{B_{21}}{A_{21}}u(\omega_{21})=\frac{1}{e^{\frac{\hbar\omega_{21}}{kT}}-1}\ll 1, \qquad (11.4.15)$$

because $\hbar\omega_{21}>>kT$. To enhance the stimulated emission, a very strong electromagnetic field with a high spectral density $u(\omega_{21})$ should be applied to the system.

The absorption coefficient that is measured experimentally is defined, in fact, by the difference of the absorbed and emitted energy. In obtaining γ in Section 11.1 (Eq.(11.1.26)), we took into account absorption only. Substracting stimulated emission, taking into account Eq.(11.4.13), and using the relation

$$f_1(1-f_2)-f_2(1-f_1)=f_1-f_2,$$

one obtains the final equation for the absorption coefficient from Eq.(11.1.25),

$$\gamma=\frac{4\pi e^2}{3m_0^2\omega}\frac{2}{(2\pi\hbar)^3}\int|\vec{p}_{vc}|^2(f_1-f_2)\delta(E_2-E_1-\hbar\omega)d\vec{p}. \qquad (11.4.16)$$

The only difference between Eqs.(11.4.16) and (11.1.26) is the factor f_1-f_2, which takes into account the difference in the populations of the states with electron energies E_1 and E_2. At equilibrium (Eq.(11.4.10b)) the difference of the Fermi-Dirac distributions is positive

$$f_1^{(0)} - f_2^{(0)} \geq 0\,, \tag{11.4.17}$$

because the lower state E_1 is more highly populated than the upper state E_2. The absorption coefficient is therefore positive

$$\gamma > 0\,. \tag{11.4.18}$$

A positive value of the absorption coefficient means that an electromagnetic wave travelling in a medium is absorbed and its amplitude is decaying.

Energy flow from an external source drives the system out of equilibrium. If the electrons are in their own equilibrium state characterized by a quasi-Fermi level μ_1, and the holes are in their own equilibrium state with quasi-Fermi level μ_2, we can use Fermi-Dirac functions f_1 and f_2 with their own quasi-Fermi levels $\mu_1 \neq \mu_2$. Under this condition the population of the upper level E_2 can be larger than the population of the lower level E_1. One can say that the ***population inversion*** occurs. Then

$$(f_1 - f_2) < 0\,, \tag{11.4.19}$$

and the stimulated emission exceeds the absorption. The net absorption coefficient becomes negative

$$\gamma < 0\,. \tag{11.4.20}$$

A negative value of the absorption coefficient means that an electromagnetic wave travelling in a medium is amplified. The system emits radiation. The absolute value of the negative absorption coefficient is called the ***gain coefficient***

$$g(\omega) = -\gamma(\omega) > 0\,. \tag{11.4.21}$$

Under the conditions of stimulated emission an external electromagnetic wave interacts with electrons in their inversely populated states causing emission of ***identical photons.***

Substituting f_1 and f_2 from Eq.(11.4.2) into Eq.(11.4.19) results in the following condition for population inversion:

$$e^{\frac{\mu_2 - \mu_1}{k_B T}} > e^{\frac{E_2 - E_1}{k_B T}}\,, \tag{11.4.22}$$

which leads to

$$\mu_2 - \mu_1 > E_2 - E_1\,. \tag{11.4.23}$$

The condition for population inversion (11.4.23) can be achieved in a forward biased p-n junction diode if both sides of the junction are heavily doped to produce degenerate electrons in the conduction band and the degenerate holes in the valence band. The energy band diagram of a *p-n* junction with degenerate electrons and holes under equilibrium conditions is shown in Fig.11.4.2a. The equilibrium Fermi level μ is inside the conduction band in the *n*-region and it is in the valence band in the *p*-region.

Under a forward bias the band diagram takes the form shown in Fig.11.4.2b. Nonequilibrium holes are injected into the *n*-region and nonequilibrium electrons into the *p*-region. These nonequilibrium concentrations decrease in the bulk over a distance of the order of their diffusion lengths. The equilibrium Fermi level μ splits

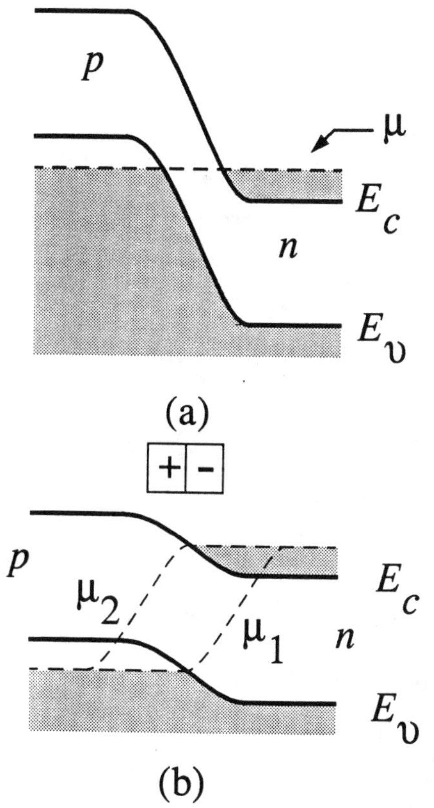

Figure 11.4.2 Heavily doped $p-n$ junction in equilibrium (a), and under forward bias (b).

into quasi-Fermi levels μ_1 and μ_2, which have a coordinate dependence shown in Fig.11.4.2b by dashed lines.

Under a forward bias a very thin region with population inversion appears in the vicinity of the junction: the high concentration of electrons in the conduction band is directly above the high concentration of holes in the valence band. It follows from the energy conservation law Eq.(11.4.3) and the amplification condition given by Eq.(11.4.23) that the energy of the emitted photon should satisfy the condition

$$\hbar\omega < \mu_2 - \mu_1 . \tag{11.4.24}$$

On the other hand, the photon energy $\hbar\omega$ should be larger than the minimum separation of the energy levels $(E_2 - E_1)_{\min} = E_g$

$$\hbar\omega > E_g . \tag{11.4.25}$$

Combining (11.4.24) and (11.4.25) one obtains the conditions imposed on the energy of the emitted photons

$$\mu_2 - \mu_1 > \hbar\omega \geq E_g . \tag{11.4.26}$$

Condition (11.4.26) is illustrated on the energy band diagram in Fig.11.4.3. The energy levels in the valence band with $E_1 \geq \mu_1$ are empty, and the energy levels in the conduction band with $E_2 \leq \mu_1$ are occupied by electrons. Therefore, photons within the interval defined by (11.4.26) are not able to excite an electron from the valence to the conduction band. These photons are therefore not absorbed. At the

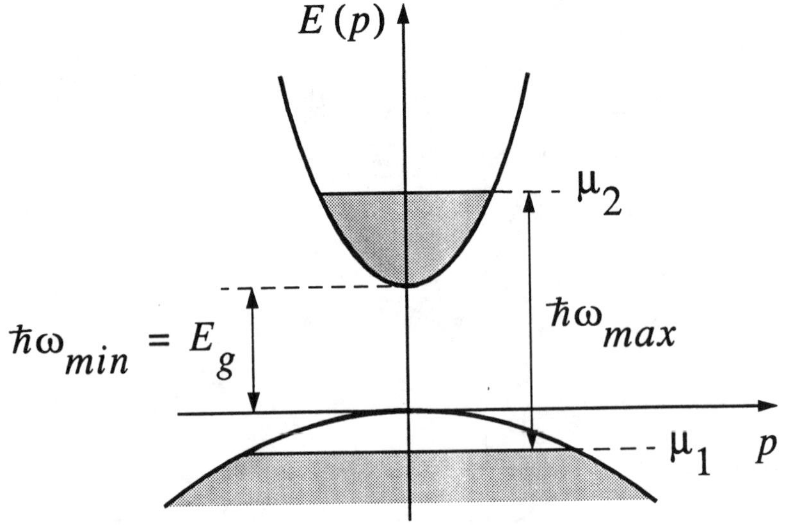

Figure 11.4.3 Region of inversion population in semiconductor.

same time, radiative transitions from the conduction band to the valence band are allowed. Thus, emission of photons occurs in the energy interval defined by Eq.(11.4.26), i.e. in the vicinity of the *p-n* junction there is an ***active*** layer which amplifies radiation in the frequency interval given by (11.4.26).

In order to achieve amplification the injection current should reach a certain value, I_{th}, required to produce population inversion. This is the so-called ***threshold current***, which is easy to estimate because it is defined by the radiative recombination rate R and by the quantum efficiency η. The quantum efficiency η is the ratio of the number of photons emitted, N_f, to the number of electrons, N_e, injected into the active region. In the ideal case $\eta = 1$. The number of electrons, N_e, injected from the external circuit into the p-n junction per second is equal to the current I divided by the magnitude of the electron charge e. For a semiconductor active layer of thickness d and a cross-section S the number of emitted photons per unit volume per second is equal to the radiative recombination rate R which leads to $N_f = RSd$. Taking this into account we obtain the following equation for the quantum efficiency:

$$\eta = \frac{N_f}{N_e} = \frac{eRSd}{I}.$$

The threshold current density

$$J_{th}^{(1)} = \frac{I}{S} = \frac{eRSd}{\eta S} = eR\frac{d}{\eta} \tag{11.4.27}$$

is proportional to the recombination rate and to the thickness of the active layer d. The smaller d is, the lower is the threshold current density $J_{th}^{(1)}$ which leads to amplification. A decrease of the quantum efficiency η leads to an increase of the threshold current.

High concentrations of electrons and holes in the active region can be achieved using different techniques. Besides injection into the p-n junction as described above, they can also be obtained through illumination by an intense light source, or by high energy electron bombardment.

11.5 PRINCIPLES OF SEMICONDUCTOR LASERS

The effect of stimulated emission in the active medium with population inversion allows one to build sources of coherent radiation called lasers. The word "laser" stands for "Light Amplification by Stimulated Emission of Radiation". In order to obtain a device with a continuous emission of radiation, feedback with the emitted radiation should be established in the device. For this purpose an active medium is placed between two parallel, partially-transparent, mirrors in order to form a Fabry-Perot type optical resonator, see Fig. 11.5.1. The condition for radiative emission can be obtained by considering the successive plane wave reflections between the parallel mirrors. Since the mirrors are made partially-transparent, a portion of the radiation comes out of the optical cavity. A portion of the radiation is lost in the optical resonator due to scattering and absorption in the laser medium, and due to imperfect reflection.

The total energy stored in the resonator will increase with time. This causes the amplification coefficient to decrease as a result of gain saturation. The oscillation level keeps increasing until the saturated gain per pass just equals the losses. At this point the net gain per pass is unity, and no further increase of radiation intensity is possible. This means that steady state oscillation occurs. The device becomes a source of coherent monochromatic radiation.

We now consider the propagation of the incident electromagnetic wave $\mathcal{E}_i = \mathcal{E}_0 e^{-i(\omega t - \vec{k}\vec{r})}$ inside the optical resonator shown in Fig. 11.5.1. The transmission and

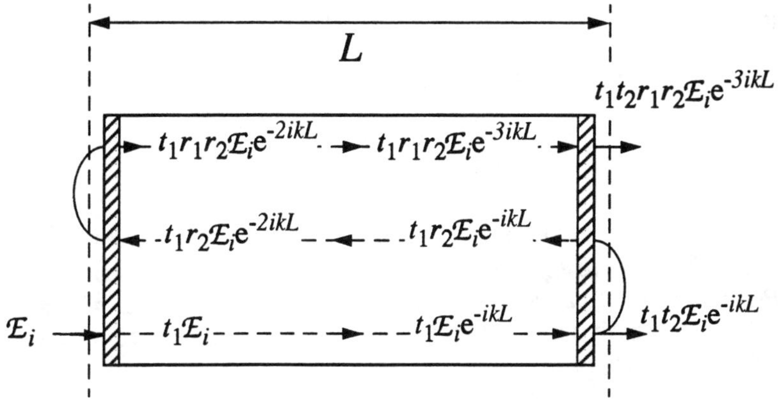

Figure 11.5.1 Fabri-Perot resonator of the length L (After A. Yariv "Optical Electronics", Sounders College Publishing, 1991).

reflection coefficients at the left mirror are t_1 and r_1. The transmission and reflection coefficients at the right mirror are t_2 and r_2. If the wave $\mathcal{E}_i$ is incident on the left mirror from outside, then inside the left boundary its amplitude becomes $t_1\mathcal{E}_i$. The propagation factor which corresponds to a single transit inside the cavity is e^{ikL}, where L is the length of the cavity. After a travelling to the right mirror the amplitude of the wave $t_1\mathcal{E}_i$ becomes $t_1\mathcal{E}_i e^{ikL}$. A portion of the wave is reflected back by the partially-transparent right mirror. Its amplitude equals $t_1 r_2 \mathcal{E}_i e^{ikL}$. The other portion is transmitted by the right mirror, its amplitude becomes $t_1 t_2 \mathcal{E}_i e^{ikL}$, and it contributes to the output energy. The reflected portion of amplitude $t_1 r_2 \mathcal{E}_i e^{ikL}$ propagates to the left mirror where it is transformed by reflection from it into a wave with amplitude $t_1 r_1 r_2 \mathcal{E}_i e^{i2kL}$. Travelling to the right boundary it is again partly transmitted, and its amplitude becomes $t_1 r_2 r_1 r_2 \mathcal{E}_i e^{i3kL}$. The portion which is reflected travels back to the left mirror, and so on.

Addition of the transmitted fields at the output give the total outgoing wave amplitude $\mathcal{E}_t$

$$\mathcal{E}_t = t_1 t_2 \mathcal{E}_i e^{ikL}[1 + r_1 r_2 e^{i2kL} + r_1^2 r_2^2 e^{i4kL} + \ldots] .$$

Summing this geometric series, one obtains

$$\mathcal{E}_t = \mathcal{E}_i \left[\frac{t_1 t_2 e^{ikL}}{1 - r_1 r_2 e^{i2kL}} \right] . \tag{11.5.1}$$

The complex propagation constant, k, contains both real and imaginary parts $k = \beta + i\alpha$, as was defined by Eq.(11.1.20). Here β is the wave vector of the travelling wave, and α is the attenuation constant, which is equal to half of the absorption coefficient $\gamma(\omega)$, Eq.(11.1.24). In an active medium the negative of the absorption coefficient defines the gain: $g = -\gamma(\omega)$, Eq.(11.4.21). There are some other absorption processes in the medium due to scattering or due to absorption by the crystal. Therefore, α for a medium with losses will be defined by the difference between the absorption coefficient of the medium γ_a and the gain $g(\omega)$. Taking into account Eq.(11.1.24) we have now

$$\alpha = \frac{1}{2}[\gamma_a(\omega) - g(\omega)] . \tag{11.5.2}$$

Substituting $k = \beta + i\alpha$, with α given by Eq.(11.5.2), into Eq.(11.5.1) one obtains

$$\mathcal{E}_t = \mathcal{E}_i \left[\frac{t_1 t_2 e^{i\beta L} e^{-\frac{1}{2}(\gamma_a - g)L}}{1 - r_1 r_2 e^{i2\beta L} e^{-(\gamma_a - g)L}} \right] . \tag{11.5.3}$$

When $g > \gamma_a$ the denominator in Eq.(11.5.3) can be very small, and the amplitude of the transmitted wave $\mathcal{E}_t$ can be larger than that of the incident wave $\mathcal{E}_i$. The Fabry-Perot cavity with an active medium works as an amplifier with power gain and $|\mathcal{E}_t| > |\mathcal{E}_i|$. In a passive Fabry-Perot cavity there is no power gain, and $|\mathcal{E}_t| < |\mathcal{E}_i|$.

The denominator in Eq.(11.5.3) is zero when

$$r_1 r_2 e^{i2\beta L} e^{-(\gamma_a - g)L} = 1 . \tag{11.5.4}$$

The ratio $\mathcal{E}_t/\mathcal{E}_i$ then becomes infinite. This means that one has a nonzero transmitted wave $\mathcal{E}_t$ with no incident wave ($\mathcal{E}_i = 0$): the system displays ***self-sustained oscillations***. Physically, the condition (11.5.4) corresponds to the case in which a wave makes a complete round trip inside the resonator and returns to the starting plane with the ***same*** amplitude and with the ***same phase*** except for some integral multiple of 2π.

We can separate the oscillation condition (11.5.4) into an amplitude condition

$$r_1 r_2 e^{-(\gamma_a - g)L} = 1 \qquad (11.5.5a)$$

and a phase condition

$$2\beta L = 2\pi m, \quad m = 1,2,3,\ldots. \qquad (11.5.5b)$$

Equation (11.5.5a) defines the threshold ***gain constant*** g_t,

$$g_t = \gamma_a - \frac{1}{L}\ln r_1 r_2\,, \qquad (11.5.6a)$$

while Eq.(11.5.5b) defines the resonant wave length

$$\lambda = \frac{2L}{m}\,. \qquad (11.5.6b)$$

The conventional relation $\beta = 2\pi/\lambda$ was used to transform Eq.(11.5.5b) into Eq.(11.5.6b). Equation (11.5.6a) means that the oscillations become self-sustained when the gain g_t becomes equal to the sum of the losses due to absorption in the medium, γ_a, and the losses due to imperfect reflection from the mirrors. Note that the term with $\ln r_1 r_2 \le 0$ because $r_1 r_2 \le 1$.

The phase condition (11.5.6b) in a long resonator is satisfied for an infinite set of wavelengths (frequencies), which correspond to the different values of the integer m. If the gain condition is satisfied at one of these frequencies, the resonator will oscillate at that particular frequency.

There is the third source of losses which comes from the fact that electromagnetic radiation is not entirely confined in the active layer of thickness d. A portion of the electromagnetic field outside the active layer does not contribute to gain. The ratio of the energy of the propagating mode within the active layer to the total energy of the propagating mode is called the ***confinement coefficient*** Γ. The gain coefficient from Eq.(11.5.6a) is reduced to the quantity

$$\Gamma g(\omega) = \gamma_a(\omega) - \frac{1}{L}\ln r_1 r_2\,. \qquad (11.5.7)$$

In order to achieve maximum confinement in the active region a configuration of semiconductor layers with different refractive indices n_r is used. It provides well-defined waveguiding. A suitable structure for a semiconductor waveguide is the symmetric three-layer slab depicted in Fig.11.5.2, whose refractive indices satisfy the condition

$$n_{r_2} > n_{r_1}\,.$$

If the angles Θ_{21} and Θ_{23} measured in radians, exceed the critical angle for total internal reflection at the (1,2) and (2,3) interfaces,

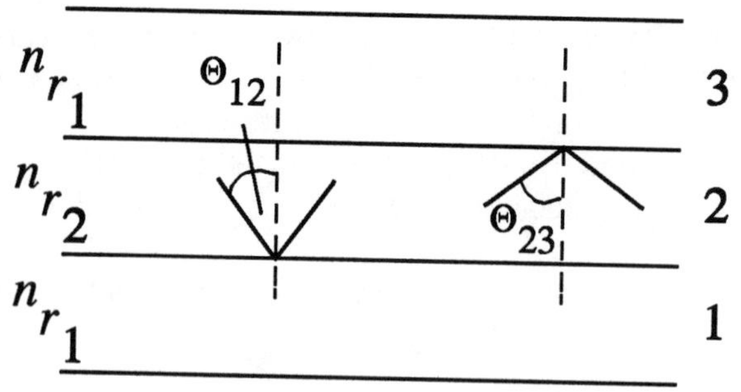

Figure 11.5.2 Optical waveguiding in the semiconductor layer with larger refractive index n_{r_2} than cladding layers.

$$\Theta_{12}, \Theta_{23} > \Theta_c = \sin^{-1}\left(\frac{n_{r_1}}{n_{r_2}}\right),$$

the flow of electromagnetic radiation is well guided in a direction parallel to the layer interfaces.

Because the gain coefficient $g(\omega)$ is proportional to the current injected into the $p-n$ junction $Bg(\omega) = J_{injected}$, we could define an additional contribution to the threshold current density

$$J_{th}^{(2)} = B\frac{1}{\Gamma}\left[\gamma_a(\omega) - \frac{1}{L}\ln r_1 r_2\right], \tag{11.5.8}$$

where B is a proportionality coefficient. Combining Eqs.(11.5.8) and (11.4.27) one obtains the total current density at threshold

$$J_{th} = A\frac{d}{\eta} + B\frac{1}{\Gamma}\left[\gamma_a(\omega) - \frac{1}{L}\ln r_1 r_2\right]. \tag{11.5.9}$$

At the present time the basic device which uses light confinement and operates at $J > J_{th}$, Eq.(11.5.9), is the double heterojunction (DH) laser. The band diagram of a heterojunction has been discussed in Section 10.5. The energy gaps of the neighboring crystals are different. This difference is distributed as a conduction band discontinuity ΔE_c and a valence band discontinuity ΔE_v. The double heterojunction used in lasers is a structure in which a layer of a semiconductor with a relatively narrow energy gap is sandwiched between two layers of wider energy gap semiconductors. It is intended that radiative recombination occurs in the narrow energy gap material where the injected carriers are confined. An example of heterojunction is a thin p-GaAs layer between p^+-$Al_xGa_{1-x}As$ and n^+-$Al_xGa_{1-x}As$ layers as depicted in Fig.11.5.3. Under a forward bias electrons are injected from n^+-$Al_xGa_{1-x}As$ into p-GaAs. Population inversion is easily achieved in p-GaAs, since the conduction band discontinuity ΔE_c prevents electrons from diffusing into the p^+-$Al_xGa_{1-x}As$ region, while valence band discontinuity ΔE_v prevents leaking of holes from the p –layer to the n^+-layer. The double heterostructure confines electrons in

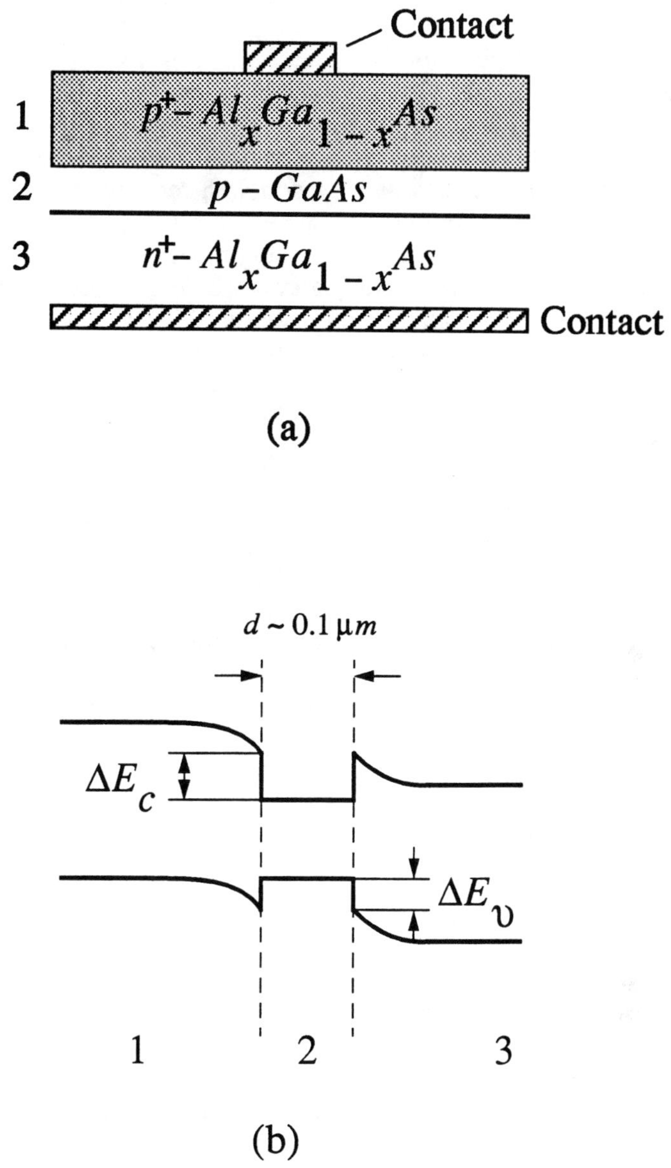

Figure 11.5.3 An example of heterojunction laser (a) with a narrow bandgap semiconductor GaAs embedded into a wide bandgap $Al_xGa_{1-x}As$ p-n junction. (b) Energy band diagram under the forward bias.

the active region of p-GaAs. The threshold current density can be considerably reduced due to reduction of the thickness of the active layer. In a $Al_xGa_{1-x}As$/GaAs double heterojunction d is of the order of $0.1\mu m$ or even less.

The double heterojunction structure allows increasing the electromagnetic field confinement inside the p-GaAs region. The refractive index of the center narrow gap material (*p*-GaAs) is larger than that of the other two cladding layers $Al_xGa_{1-x}As$. As a result, the light wave is also confined in the *p*-GaAs region. A small fraction of the propagating mode outside the active layer is in the wider energy gap semiconductor, and is not absorbed.

The optical cavity is formed by cleaving semiconductor perpendicular to the plane of diffusion in p-n junctions. The double heterojunction $GaAs/Al_xGa_{1-x}As$ lasers which provide both well-defined dielectric waveguide and a narrow active region were the first injection lasers permitting room temperature cw emission of radiation. The emitted power is of the order of 10 Watt. Since the thickness of an active layer is small as $10^{-4} cm$, the power density in DH lasers is high: 10^2 - 10^3 kW/cm^2. The magnitude of the threshold current density is of the order of 10^3 A/cm^2 at 300 K.

The principal materials for semiconductor injection lasers are direct gap compounds of III and V group elements in the Periodic Table. GaAs is of special interest since it has a very efficient radiative recombination rate. Recently researches paid special attention to other III-V materials and, in particular, to GaN, which lases in a visible frequency range. Semiconductor compounds of II and VI as well as of IV and VI group elements of the Periodic Table are also used for stimulated emission of radiation.

For indirect gap semiconductors (Ge, Si), free carrier absorption increases faster than gain with increasing intensity of excitation. This makes stimulated emission from them very unlikely. Only direct gap semiconductors have been used as active layers in lasers so far.

11.6 NANOSTRUCTURE LASERS

It is quite possible that further developments of semiconductor lasers will be based on the use of semiconductor nanostructures. A quantum well laser is similar in many aspects to the conventional double heterojunction laser except for the thickness of the active layer. In double heterojunction lasers the thickness of the layer which confines electrons and holes is about 1000 $\AA$ whereas in quantum well lasers it is about 50 - 100 $\AA$. This behavior of quantum well lasers has nothing to do with quantum confinement of carriers in the well. Moreover, some improvement in quantum well lasers originates from the small physical size of the active region and from the step-like behavior of the density of states. We will not discuss here consequences and advantages of the quantum confinement of carriers in the well. Reduction of the width of the active region by a factor of 10 results, according to the first term of Eq.(11.5.9), in a factor of 10 decrease of the threshold current density which creates the population inversion. On the other hand the small physical size results in a very weak confinement of the optical modes. The optical confinement factor Γ scales down with the size of the well. As a result there is no substantial overall advantage in single quantum well lasers. The problem of weak optical confinement Γ in a single quantum well is circumvented by using the multiple quantum wells in the optical region of the laser. Assuming that injected electrons are equally distributed between all wells, the optical confinement factor for the multiple quantum well laser is simply multiplied by the number of wells.

Attempts to reach further reduction in the width of the active region below 100 Å have brought no reduction of the threshold current.

In the studies of the quantum well lasers the negative role of the valence band states was recovered. To improve the situation it was proposed to reduce the heavy hole mass using strained structures. Quantum wells with a built-in strains allow valence band engineering. Tailoring of the valence band structure is achieved by growing high quality strained structures in which two quantum well layers are composed of a material with a lattice constant that differs significantly from that of the barrier material. The lattice mismatch is accommodated by a tetragonal distortion of the quantum well layer, leading to built-in uniaxial strain. The strain removes the degeneracy of the light and heavy holes states at the center of the Brillouin zone (at the Γ-point) as is shown, for example, in Fig. 11.6.1. This built-in strain eliminates some loss mechanisms of Auger recombination and intravalence band absorption.

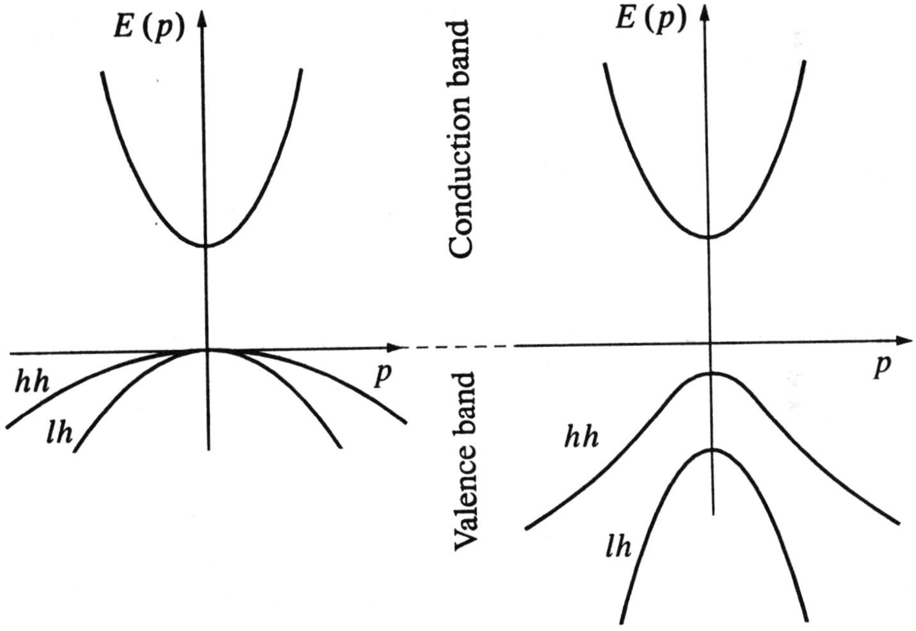

Figure 11.6.1 Schematic diagram of a direct bandgap semiconductor, such as GaAs, showing an approximately parabolic conduction band and a degenerate valence band when it is unstrained (a), and under biaxial compression (b). Under the strain energy of both light and heavy hole valence bands increase relatively to conduction band edge. There is also separation of bands, which removes their degeneracy because of the different effective masses in the valence band.

A reduced density of states at the top of the valence band appears, leading to a lower threshold current density.

The first quantum well lasers consisted of a single 200 $\AA$ quantum well and had a threshold current of about 3 kA/cm^2 at 300 K. In a multiple quantum well laser the threshold is as low as 250 A/cm^2 with four 120 $\AA$ quantum wells.

Further attempts to decrease the threshold current could come from shrinking the lateral dimensions of the laser. This would lead to quantum wire and quantum dot lasers.

The first attempts to fabricate quantum wires and quantum dots consisted of etching the wires and dots from quantum well heterostructures. But the damaged wire and dot interfaces caused by etching result in a very high threshold current. In addition, laser operation occurs at low temperatures only. It became clear that postgrowing processes should be avoided. Quantum wire and quantum dot structures are now obtained by specially arranged growth procedures. These low-dimensional semiconductor heterostructures may form the basis for future optoelectronic devices to be used in computer-optical interconnections or optical computing. It is expected that lasers with thresholds in the tens of microamperes or even microampere range will become a reality in the very near future.

BIBLIOGRAPHY

A. Yariv *Optical Electronics* (Sounders College Publishing, 1991).

Appendix

APPLICATIONS OF TENSORS AND MATRICES IN SOLID STATE PHYSICS

One of the most impressive attributes of crystals is their anisotropy. Solids have properties of different types. Some properties as the energy or the temperature are represented by the scalar quantities. A scalar has only one characteristic, its magnitude. This magnitude remains constant under any transformation of a coordinate system.

Some other properties, such as an electric field intensity $\vec{\mathcal{E}}$ or an electric current density $\vec{j}$, are vectors. Vectors are characterized by their magnitude and direction. Vectors can also be expressed in terms of their components:

$$\vec{A} = (A_x, A_y, A_z).$$

When a symmetry operation S is applied to a vector, its components change their magnitudes. An example of a vector transformation has been discussed in Chapter 3. The transformation of the radius-vector $\vec{r}$ from the old to the new primed coordinate system and back were defined by Eqs.(3.3.4) and (3.3.5):

$$\vec{r} = S\vec{r}', \quad \vec{r}' = S^{-1}\vec{r}. \qquad (A.1)$$

It is shown in vector algebra that the transformation of any vector from one coordinate system to another one is given by an equation similar to (A.1):

$$\vec{A} = S\vec{A}', \quad \vec{A}' = S^{-1}\vec{A}. \qquad (A.2)$$

Properties of solids are not exhausted by scalars and vectors. For example, considering the electric current density, we have Ohm's law

$$\vec{j} = \sigma\vec{\mathcal{E}}, \qquad (A.3)$$

where $\vec{j}$ and $\vec{\mathcal{E}}$ are vectors. The proportionality coefficient σ is not a scalar quantity for a crystal. If an electric field $\vec{\mathcal{E}} = (\mathcal{E}_x, 0, 0)$ is applied to an anisotropic crystal, the effective current density can have all three components:

$$j_x = \sigma_{xx}\mathcal{E}_x, \quad j_y = \sigma_{yx}\mathcal{E}_x, j_z = \sigma_{zx}\mathcal{E}_x. \qquad (A.4)$$

Experimental data confirm the relations (A.4).

If the electric field intensity has all three components $\vec{\mathcal{E}} = (\mathcal{E}_x, \mathcal{E}_y, \mathcal{E}_z)$, one can obtain the more general relations:

$$j_x = \sigma_{xx}\mathcal{E}_x + \sigma_{xy}\mathcal{E}_y + \sigma_{xz}\mathcal{E}_z$$
$$j_y = \sigma_{yx}\mathcal{E}_x + \sigma_{yy}\mathcal{E}_y + \sigma_{yz}\mathcal{E}_z \quad (A.5)$$
$$j_z = \sigma_{zx}\mathcal{E}_x + \sigma_{zy}\mathcal{E}_y + \sigma_{yz}\mathcal{E}_z \ .$$

One can see that in case of an anisotropic crystal there is the quantity

$$\sigma = \begin{pmatrix} \sigma_{xx} & \sigma_{xy} & \sigma_{xz} \\ \sigma_{yx} & \sigma_{yy} & \sigma_{yz} \\ \sigma_{zx} & \sigma_{zy} & \sigma_{zz} \end{pmatrix} \quad (A.6)$$

consisting of nine components $\sigma_{\alpha\beta}$, $\alpha, \beta = x, y, z$. Quantities of this type are called ***second rank tensors***. Equation (A.5) can be written in the matrix form

$$\begin{pmatrix} j_x \\ j_y \\ j_z \end{pmatrix} = \begin{pmatrix} \sigma_{xx} & \sigma_{xy} & \sigma_{xz} \\ \sigma_{yx} & \sigma_{yy} & \sigma_{yz} \\ \sigma_{zx} & \sigma_{zy} & \sigma_{zz} \end{pmatrix} \begin{pmatrix} \mathcal{E}_x \\ \mathcal{E}_y \\ \mathcal{E}_z \end{pmatrix} . \quad (A.7)$$

The order of the tensor is defined by the number of subscripts it possesses. For example, the quantity $\lambda_{\alpha\beta\gamma}$, where $\alpha, \beta, \gamma = x, y, z$ is called a third rank tensor. The quantity $\lambda_{\alpha\beta\gamma\delta}$, where $\alpha, \beta, \gamma, \delta = x, y, z$ is called a fourth rank tensor, and so on.

Under a transformation from one coordinate system to another, the components of a tensor change their values. In order to find the law of transformation for second rank tensor, we continue using the example of Eq.(A.3).

We wish to find a way of transforming of Eq.(A.3) from an old coordinate system to a new, primed, coordinate system. Let S be the matrix of a rotation or reflection of a body. The current density $\vec{j}$ transforms under S according to Eq.(A.2),

$$\vec{j} = S\vec{j}' , \quad (A.8a)$$

and

$$\vec{j}' = S^{-1}\vec{j} . \quad (A.8b)$$

There is similar transformation for $\vec{\mathcal{E}}$

$$\vec{\mathcal{E}} = S\vec{\mathcal{E}}' \quad \text{and} \quad \vec{\mathcal{E}}' = S^{-1}\vec{\mathcal{E}} . \quad (A.9)$$

Substitution of $\vec{j}$ from Eq.(A.8) and $\vec{\mathcal{E}}$ from Eq.(A.9) results in

$$S\vec{j}' = \sigma S\vec{\mathcal{E}}' , \quad (A.10)$$

where $S\sigma$ is the product of two matrices. Multiplaying Eq.(A.10) by S^{-1} leads to

$$\vec{j}' = S^{-1}\sigma S\vec{\mathcal{E}}' ,$$

where the second rank tensor

$$\sigma' = S^{-1}\sigma S \quad (A.11)$$

represents the conductivity in the new primed coordinate system. Equation (A.11) gives the transformation rule of a second rank tensor from the old coordinate system to the new coordinate system. Equation (A.11) holds for any second rank tensor.

Problems

CHAPTER 1

1.1 Verify that the operator of momentum $\vec{p} = -i\hbar\nabla$ is Hermitian. Give the physical meaning of $\vec{p}$ being Hermitian.

1.2 Sketch qualitatively the electron probability density $|\phi|^2$ for the following potential barriers.

a)

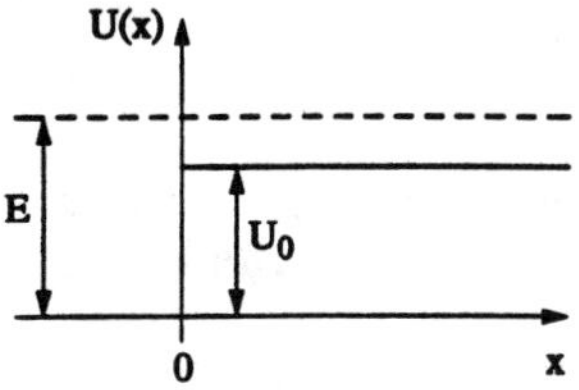

$$U(x) = \begin{cases} 0 & \text{for } x \le 0 \\ U_0 & \text{for } x \ge 0 \end{cases}$$

$$E > U_0$$

b)

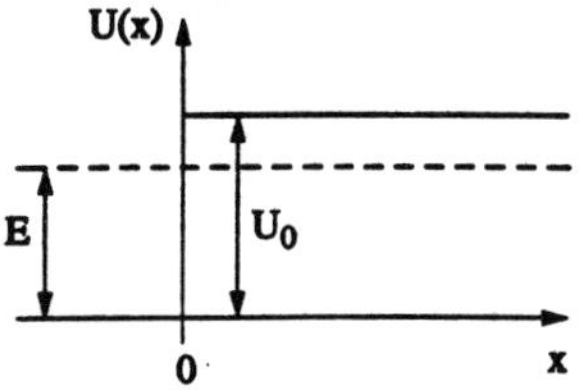

$$U(x) = \begin{cases} 0 & \text{for } x \le 0 \\ U_0 & \text{for } x \ge 0 \end{cases}$$

$$E < U_0$$

c)

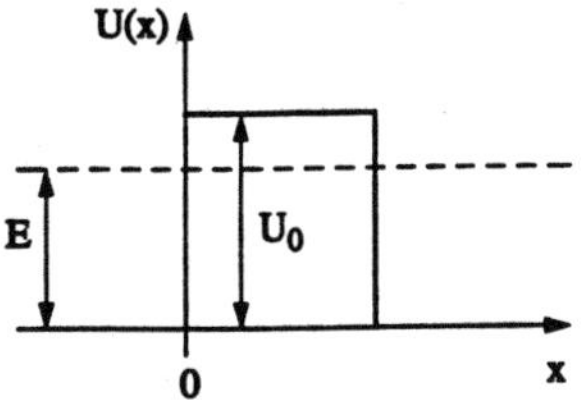

$$U(x) = \begin{cases} 0 & \text{for } x \le 0 \\ U_0 & \text{for } 0 \le x \le a \\ 0 & \text{for } x \ge a \end{cases}$$

$$E < U_0$$

d)

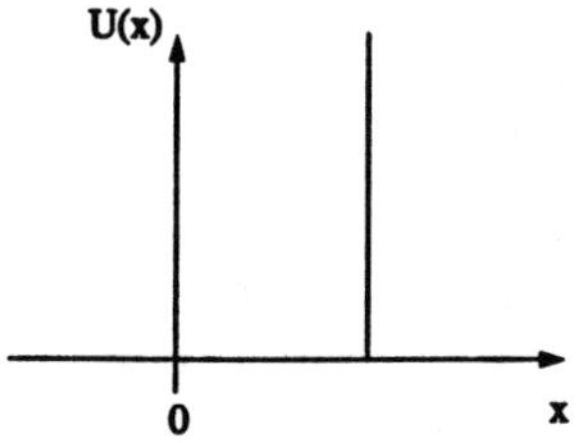

$$U(x) = \begin{cases} +\infty & \text{for } x \le 0 \\ 0 & \text{for } 0 \le x \le a \\ +\infty & \text{for } x \ge a \end{cases}$$

1.3 A 5 eV beam of electrons is directed towards a potential barrier of height 30 eV. The width of the barrier is 10 $\AA$. What percent of the incident flux goes through the barrier per second?

1.4 Compute the transmission coefficient of two similar potential barriers of width a separated by the distance b. Note that resonance can occur for some energy of incident flux of particles.

1.5 What is the maximum number of electrons for the d-state of the hydrogen atom?

1.6 The operator of electron angular momentum is $\hat{L} = \left(\hat{\vec{r}} \times \hat{\vec{p}}\right)$, $\hat{\vec{r}}$ and $\hat{\vec{p}}$ being operators of coordinate and momentum of an electron. Calculate the commutator $\left[\hat{\vec{L}}_x, \hat{\vec{L}}_y\right]$ and give a qualitative interpretation of the result.

CHAPTER 2

2.1 Describe the electronic configurations for Boron (B), Phosphorous (P), Neon (Ne), Germanium (Ge), and Antimonide (As). Use the Periodic Table of elements.

2.2 Describe the Bravais lattices and sketch the 1st and the 2nd Brillouin zones for the 2D crystal lattices shown in Figure.

a)

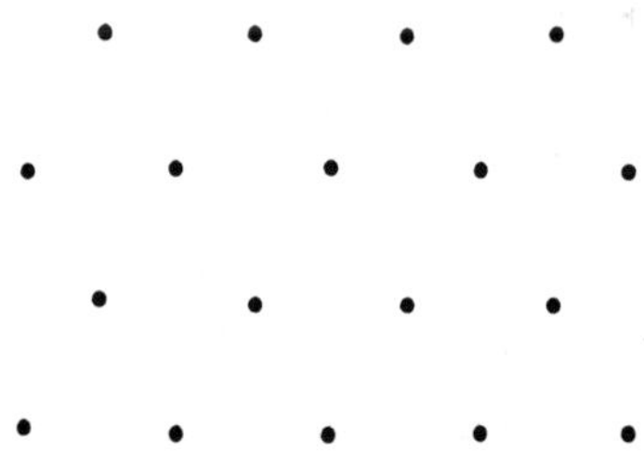

b)

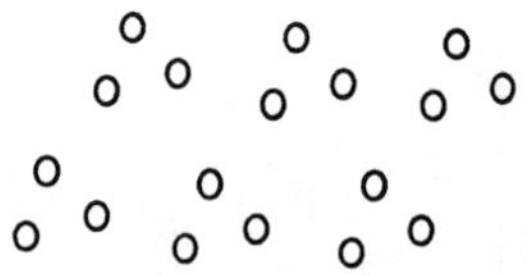

2.3 Does the lattice in the Figure have a basis? Sketch the primitive cell and primitive vectors. Calculate the area of the primitive cell.

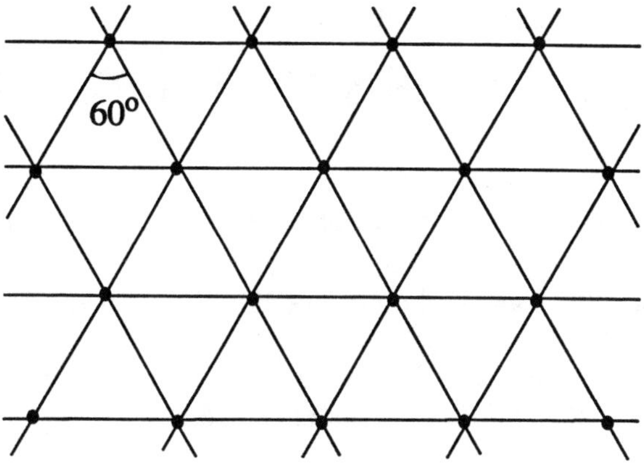

2.4 Plot the Fermi-Dirac function *f(E)* as a function of the energy *E* at four different temperatures: 0 *K*, 4.2 *K* (liquid helium temperature), 77 *K* (liquid nitrogen temperature), and 300 *K* (room temperature). The chemical potential of electron is $\mu = 0.0067 eV$. Find the criterion for electron gas degeneracy.

2.5 Consider an ideal electron gas with density *n*. Sketch the chemical potential μ as a function of temperature, *T*, for $T \ll E_F$ through $T \gg E_F$.

2.6 Calculate the number of allowed electronic states in the 1st Brillouin zone of a 2D square lattice with period, *a*. The linear dimensions of the sample are L_x, L_y.

2.7 Consider a 2D square lattice with crystal lattice potential

$$U(x, y) = U_0\left(1 + \cos\left(\frac{2\pi}{a}x\right)\right)\cos\left(\frac{2\pi}{a}y\right).$$

Calculate the energy gap at the points $(x = \pi/a, y = \pi/a)$ and $(x = 0, y = \pi/a)$ in the 1st Brillouin zone. Find the effective mass of the electron at these points.

2.8 The energy of an electron in a superlattice has the form

$$E(p_z) = E_0 - A\cos\left(\frac{p_z}{\hbar}d\right).$$

Find the velocity of the electron $v_z(t)$ when the external effective field $\vec{\mathcal{E}} = (0, 0, \mathcal{E})$ is applied to the superlattice. The initial condition is $v_z(t = 0) = 0$.

CHAPTER 3

3.1 Find the matrix for reflection in the zy-plane.

3.2 Find the matrix for rotation through the angle α with consequent reflection in the plane σ_h.

3.3 Find the matrix for the rotation designated C_2 with consequent reflection in σ_v

3.4 Sketch the first and the 2nd Brillouin zones for a two-dimensional square lattice.

3.5 Sketch the nesting surfaces of constant energy for a two-dimensional square lattice using reduced zone scheme and extended zone scheme.

3.6 List all operations of symmetry for a:
 a) rectangle,
 b) bilateral triangle.

3.7 List the matrices for 8 different symmetry operations of the square.

3.8 List 5 possible Bravais lattices for a two-dimensional crystal.
Hint: Bravais net can have only axes of two-, three-, four-, and six-fold symmetry. There are altogether 5 Bravais lattices in two dimensions.

3.9 Prove that the molecule shown in Figure has S_4 symmetry.

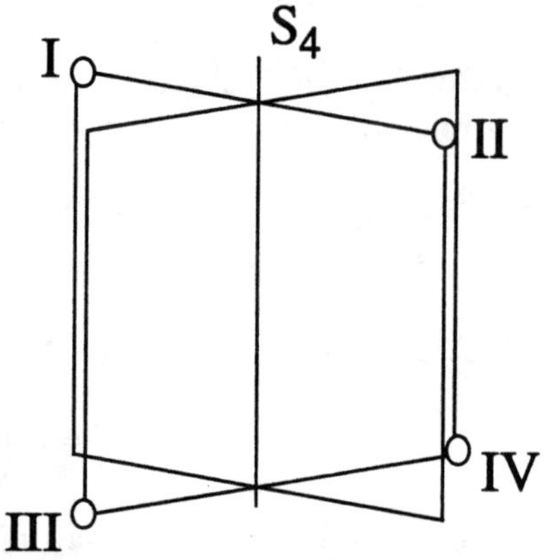

CHAPTER 4

4.1 Why there are 6 valleys in the electron energy spectrum of Si and 8 valleys in the electron energy spectrum of Ge?

4.2 The minimum of energy occurs at the point $\vec{p}_0$ of the Brillouin zone. The energy has the form

$$E(\vec{p}) = \frac{(p_z - p_{z0})^2}{2m_l} + \frac{(p_x - p_{x0})^2 + (p_y - p_{y0})^2}{2m_t},$$

where $m_l = 0.98\, m_0$ and $m_t = 0.19\, m_0$ are longitudinal and transverse masses. What is the ratio of major to minor axes for the ellipsoidal surface of constant energy?

4.3 A family of the anisotropic surfaces of constant energy has the form

$$E(\vec{p}) = \frac{1}{2m_0}\left(Ap^2 - B\frac{p_x^2 p_y^2 + p_y^2 p_z^2 + p_z^2 p_x^2}{p^2}\right),$$

where $A,\ B > 0$. Find the elements of the effective mass second rank tensor $m_{\alpha\beta}^{-1}, \alpha, \beta = x, y, z$.

4.4 In the geometry of cyclotron resonance, calculate the effective mass m_c for $B =$ 1.5 *Tesla* and microwave field frequency of 35 *GHz*.

4.5 Derive the Fermi energy of an electron gas confined to a one-dimensional square well.

4.6 Find the direction of rotation of a hole in an external magnetic field B.

4.7 Calculate the electron wave function $\psi(x)$ and the corresponding energy $E < U_0$ for a semiconductor structure with potential energy

$$U(x) = \begin{Bmatrix} +\infty & \text{for} & x < 0 \\ 0 & \text{for} & 0 < x < a, \\ U_0 & \text{for} & x > a \end{Bmatrix}$$

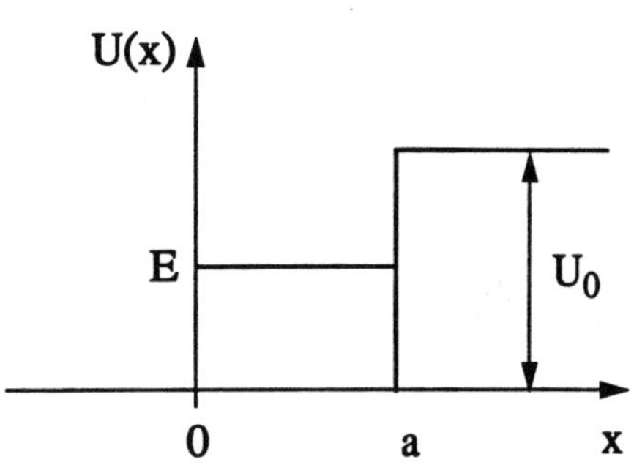

(see Figure).
Hint: The boundary condition for $\psi(x)$ at $x = 0$ is $\psi(0) = 0$.

CHAPTER 5

5.1 Consider a one-dimensional crystal lattice with two atoms of equal masses M in the basis. Atoms are connected by springs of alternating strength Ψ_1 and Ψ_2, see Figure.

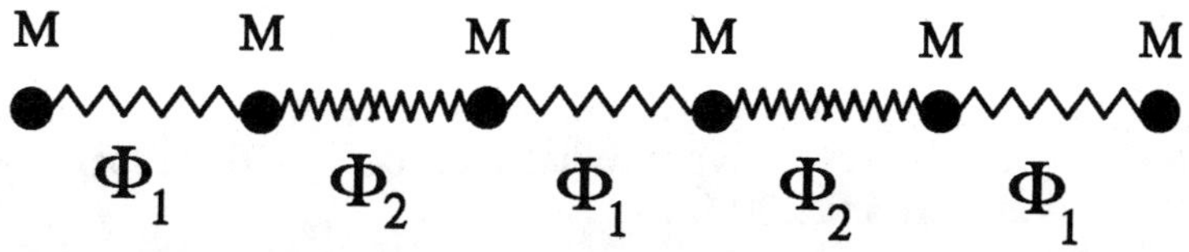

a) Derive the equations of motion considering only nearest neighbor interactions.
b) Find frequencies of vibrations $\omega_j^2(q)$ and the velocity of sound s.
c) Show that the group velocity $\frac{\partial\omega(\vec{q})}{\partial\vec{q}}$ vanishes at the Brillouin zone boundaries.
d) Show that the phase difference between the amplitudes is zero for acoustic modes and equals π for optical modes.

5.2 Find the number of acoustic modes for the crystal lattices shown below:

a) 1D - lattice,

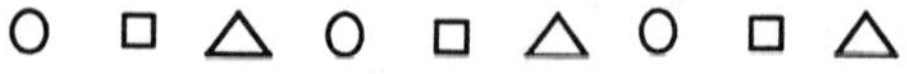

b) 2D - lattice,

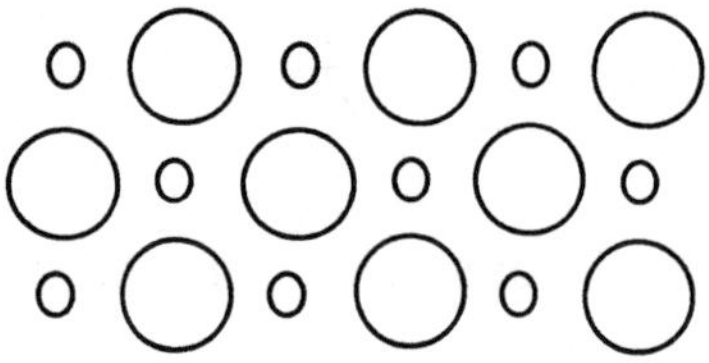

c) 2D - lattice,

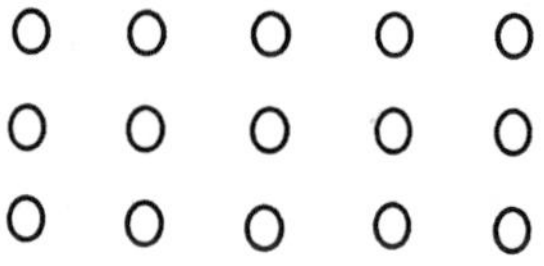

d) 3D -lattice (NaCl).

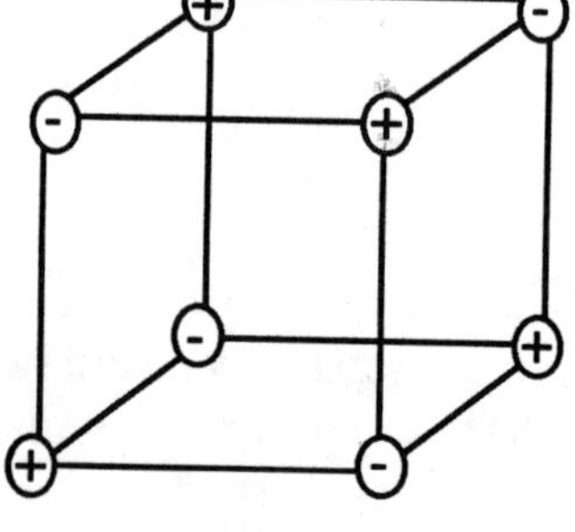

5.3 Optical phonon energies of GaAs and AlAs at the center of the Brillouin zone are 36 meV and 50 meV, respectively. What is the occupation probability for these optical phonons at 77 K and 300 K ?

CHAPTER 6

6.1 Given $n_i^{Ge} = 1.9 \times 10^{13} cm^{-3}$ in Ge and $n_i^{Si} = 0.9 \times 10^{10} cm^{-3}$ in Si at 300 K, find the intrinsic carrier concentrations of Ge and Si at 400 K.

6.2 Which elements of the Periodic Table can be used as impurities in Si in order to get an n-type semiconductor?

6.3 There are $N_D = 10^{13}$ donors per cubic centimeter in a semiconductor material. The ionization energy of the donors is $E_d = 1\ meV$. The effective mass of the carriers is 0.01 m_0. Find the concentration of electrons at 30 K and 300 K.

6.4 A crystal of n-type GaAs is doped $10^{17}\ cm^{-3}$. The donor ionization energy is 0.007 eV. Calculate the concentration of free electrons in the wide temperature interval from 10 K up to 500 K. Find the temperature dependence of the chemical potential (Fermi level).

6.5 Assume that the electron gas degeneracy in semiconductor is defined by the condition

$$\mu - E_c > 3k_B T.$$

Here μ is the chemical potential and E_c is the bottom of the conduction band. Calculate the electron concentration at which the electron gas in GaAs is degenerate at 300 K.

CHAPTER 7

7.1 Show that the Fourier transform of the screened Coulomb potential $U(r) = \frac{e^2}{r} e^{-\lambda r}$ is equal to $\frac{4\pi e^2}{q^2 + \lambda^2}$.

7.2 Derive the screening radius of the charged impurity atoms in a nondegenerate electron gas.

7.3 Calculate the electron plasma frequency for GaAs doped with $10^{17}\ cm^{-3}$ donors at 300 K.

7.4 Sketch a plasma dispersion curve $\omega_p(\overline{q})$ versus q.

7.5 Find the concentration of electrons at which the condition of heavy doping holds for GaAs.

7.6 Explain what a quasiparticle is. Give some examples.

CHAPTER 8

8.1 Given the effective mass of a hole, $m^* = 0.5\, m_0$, and the hole mobility, $b = 480$ cm^2/Vs, evaluate the hole momentum relaxation time $< \tau >$.

8.2 Find the energy interval near the Fermi surface in which the occupation number $< f_e^{(0)}(\overline{p}) >$ averaged over collisions is significantly affected by the external electric field $\vec{\mathcal{E}}$.

8.3 Calculate the electrical conductivity, σ, of intrinsic Ge at room temperature. The energy gap $E_g = 0.7\ eV$, and the electron and hole mobilities are $\mu_e = 0.38 m^2/Vs$, and $\mu_h = 0.18 m^2/Vs$, respectivelly.

8.4 Prove that the surface electronic conductivity (conductivity of 2D electrons) is anisotropic for the surface [110], see Figure.

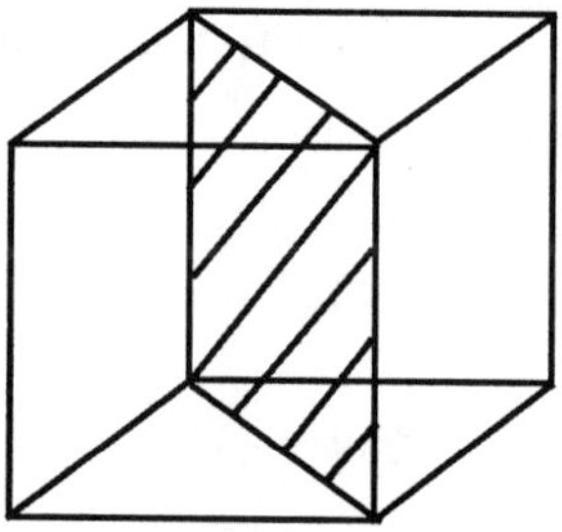

[110]- surface

8.5 Find independent components of the conductivity tensor of second rank, $\sigma_{\alpha\beta}$, where $\alpha, \beta = x, y, z$ for a crystal with three orthogonal C_4 axes (rotations through 90° about x-, y-, z - axes).

8.6 The dielectric permittivity tensor of the second rank $\varepsilon_{\alpha\beta}$ connects the displacement vector $\vec{D}$ with the electric field $\vec{\mathcal{E}}$:

$$D_\alpha = \sum_\beta \varepsilon_{\alpha\beta} \mathcal{E}_\beta .$$

Find the independent components of $\varepsilon_{\alpha\beta}$ for a crystal with mirror reflection plane normal to y - direction.

CHAPTER 9

9.1 Given a profile of the electron concentration $n(x)$, as shown in Figure below. Calculate the electric current density $\vec{j}$, if the electron mobility is equal to $b = 1000$ cm/Vs at $300K$.

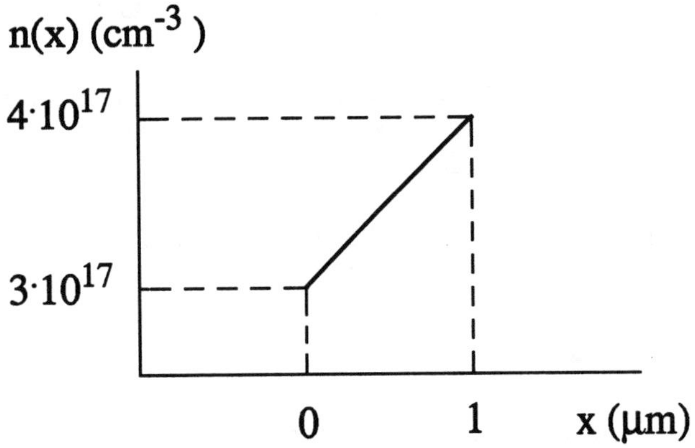

9.2 Assume that holes are injected into the semiinfinite n-type material ($x \geq 0$). The excess hole concentration at $x = 0$ is $\Delta p(0) = 10^{14}$ cm^{-3}. Find the nonequilibrium hole concentration at $x = 0.4$ cm, if the relaxation time $\tau(p) = 10^{-3}$ s and the hole diffusion coefficient equals $D_p = 40$ $cm^2 s^{-1}$.

9.3 Find the distribution of excess holes along the x-axis in a long n-Ge sample under stationary conditions in the external electric field $\mathcal{E} = 5V/m$. The hole concentration at $x = 0$ is $p_0 = 10^{18}$ cm^{-3} at $300K$, the diffusion length $L_p = 0.09$ cm and the hole mobility is $b_p = 1900$ cm^2/Vs.

9.4 A sample of n-GaAs has the electron concentration $n = 10^{17}$ cm^{-3}. It is illuminated with light of energy $\hbar\omega > E_g$, and a concentration of nonequilibrium holes is created. Calculate the time t which is necessary for 90 % of the nonequilibrium holes to recombine if the light is turned off at $t = 0$. Coefficient $\kappa = 10^{-10}$ cm^3/s.

CHAPTER 10

10.1 A current of 1.66 mA flows through a forward biased p-n junction with voltage of 0.15 V. Find the current in the reverse biased diode.

10.2 Evaluate the width of the depletion layer, d, for the abrupt p-n junction made of Si with concentration of acceptors and donors equal to $N_A = 10^{17}\ cm^{-3}$ and $N_D = 10^{18}\ cm^{-3}$. The energy gap $E_g = 1.10\ eV$ at room temperature 300 K. To simplify the problem you can neglect μ_n and μ_p with respect to E_g.

10.3 In the n-channel of a silicon field effect transistor (JFET) with width 0.3 μm the concentration of donors is $N_D = 10^{16}\ cm^{-3}$. The channel conductance is $G_0 = 0.14(Ohm)^{-1}$.
a) What is the highest voltage that can be applied to the drain while the drain operates in the pinch-off regime? It is assumed that the source and gate are grounded.
b) What current flows in the drain at this voltage?

10.4 An abrupt p-n junction with the cross section area $S = 2 \times 10^{-3}\ cm^2$ has a concentration of acceptors, $N_A = 4 \times 10^{18}\ cm^{-3}$, on the left side and a concentration of donors, $N_D = 10^{16}\ cm^{-3}$, on the right side. Calculate the built-in potential V_b and the maximum electric field $\vec{\mathcal{E}}_b$ at equilibrium.

10.5 Sketch the band diagrams of the perfect heterojunction $Al_{0.3}Ga_{0.7}/GaAs$ for the following structures:
a) p^+ $Al_{0.3}Ga_{0.7}As$ - n^+ GaAs,
b) n^+ $Al_{0.3}Ga_{0.7}As$ - intrinsic GaAs,
c) n^+ $Al_{0.3}Ga_{0.7}As$ - p GaAs.
Energy gaps of GaAs and $Al_{0.3}Ga_{0.7}As$ are 1.4 eV and 1.85 eV, respectively. Electron affinities are 3.56 eV for GaAs and 4 eV for $Al_{0.3}Ga_{0.7}As$.

10.6 Find the reflection coefficient for a potential step at a heterojunction. The effective mass on the left side of the junction is m_1^* and on the right side of the junction it is m_2^*.
Hint: The boundary condition for the first derivative of the electron wave function satisfies the condition of the electron flux continuity and has the form

$$\frac{1}{m_1^*}\psi_1' = \frac{1}{m_2^*}\psi_2'$$

CHAPTER 11

11.1 Show that the absorption of a photon is a vertical process.

11.2 Is the optical absorption edge in semiconductor the same thing as the band gap?

11.3 Compute the binding energy of an exciton in Ge taking the isotropic effective mass of electron $m_n = 0.22\, m_0$ and the hole mass $m_p = 0.34\, m_0$. The dielectric constant of Ge is $\varepsilon_0 = 16.0$.

11.4 Estimate the electric field in Volts per centimeter required to ionize the exciton in a semiconductor.

11.5 The energy gap of GaAs is $E_g = 1.4\, eV$. What is the maximum wavelength of the light which can be absorbed by a GaAs monocrystal?

List of Symbols

a	Lattice constant, width of a potential barrier, width of a potential well
$\vec{a}_i$	Primitive translation vector
a_0	Radius of the first Bohr orbit, radius of the electron orbit of the impurity atom
$\vec{B}$	Magnetic flux density
b	Mobility
b_a	Ambipolar mobility
$D(E)$	Density of states
D_a	Coefficient of ambipolar diffusion
$D_{c,v}(E)$	Density of states of electrons (c), and holes (v)
$D_{n,p}$	Diffusion coefficient for electrons (n), and holes (p)
d	Width of depletion layer
$d_{n,p}$	Width of depletion layer in n-material (n), and p-material (p)
E	Energy
$\mathcal{E}$	Electric field intensity
$E_{A,D}$	Acceptor (A), and donor (D) ionization energy
E_{ex}	Exciton ionization energy
E_F	Fermi energy

e	Charge of the electron
$e' = e/\sqrt{4\pi\varepsilon_0}$	Quantity used to simplify some equations
$\vec{F}$	Force
$f(E)$	Fermi-Dirac distribution function
$f^{(1)}(\vec{p})$	Small nonequilibrium perturbation to the Fermi-Dirac distribution function
$G_{n,p}$	Rate of generation of electrons (n), and holes (p)
G_{th}	Rate of thermal generation of electron-hole pairs
g_t	Gain constant
h	Planck constant
$\hbar = h/2\pi$	Reduced Planck constant
$I^{(B)}$	Base current
$I^{(C)}$	Collector current
$I^{(E)}$	Emitter current
J	Total angular momentum
$\vec{j}$	Current density
K	Compensation ratio
$\vec{K}_i$	Reciprocal lattice vector
$\vec{k}$	Wave vector
L	Diffusion length
$\vec{L}$	Angular momentum
l	Orbital quantum number, mean free path of electrons
$l_{d,u}$	Downstream (d), and upstream (u) diffusion length
M_{ls}	Mass of atom s in the l-th primitive cell
M_z	Magnetic dipole moment
m	Magnetic quantum number, effective mass
m_{eff}, m_d	Density of states effective mass
m_0	Mass of the free electron

$(1/m)_{\alpha\beta}$	Anisotropic reciprocal effective mass tensor
N	Number of states
$N_{A,D}$	Concentration of acceptors (A), and donors (D)
N_A^-	Concentration of ionized acceptors
$N_{c,v}$	Effective density of states in the conduction (c), and valence (v) bands
N_D^+	Concentration of ionized donors
n	Quantum number, principal quantum number, electron concentration
n_i	Equilibrium concentration of electrons or holes in an intrinsic semiconductor
$n_{n0,p0}$	Equilibrium concentration of electrons in n-material ($n0$), and in p-material ($p0$)
$\vec{p}$	Momentum, quasimomentum
$\vec{p}_F$	Fermi momentum
$\vec{p}_h$	Quasimomentum of the hole
$p_{n0,p0}$	Equilibrium concentration of holes in n-material ($n0$), and in p-material ($p0$)
R	Reflection coefficient
r	Distance
$\bar{r}$	Average separation of impurity atoms
r_s	Debye screening radius
S	Matrix of the coefficients of a transformation
T	Transmission coefficient
U	Potential energy
$U(\omega,T)$	Spectral radiancy
$\vec{u}$	Drift velocity
$u(\vec{r})$	Modulation Bloch function
$\vec{u}_{ls}$	Displacement of atom s in the l-th primitive cell
V	Volume, external voltage

$\tilde{V}(x)$	Dimensionless potential energy
V_b	Built-in potential drop across the depletion region
V_{DS}	Drain voltage
V_f	Forward bias
V_{GS}	Gate voltage
V_n	Voltage drop in *n*-region
V_p	Voltage drop in *p*-region
V_r	Reverse bias
v	Velocity
v_a	Ambipolar drift velocity
v_g	Group velocity
v_h	Velocity of the hole
v_{th}	Thermal velocity
W	Electron work function
$W(\vec{p},\vec{p}')$	Probability of scattering
$\vec{w}$	Acceleration
α	Current gain, attenuation constant
β	Propagation constant
Γ	Width of the energy level, optical confinement coefficient
γ	Gyromagnetic ratio, linear absorption coefficient, coefficient of the threshold current, central point of the Brillouin zone
$\Delta = \nabla^2$	Laplacian operator
$\delta(x)$	Dirac δ-function
$\delta_{nn'}$	Kronecker delta
$\varepsilon(\omega)$	Dielectric permittivity
$\varepsilon(\vec{q},\omega)$	Lindhardt dielectric function
$\varepsilon(0)$	Low frequency dielectric permittivity
ε_0	Permittivity of vacuum

ε_∞	High frequency dielectric permittivity
η_i	Internal quantum efficiency
λ	Wavelength
$\mu(T)$	Chemical potential
ν	Natural frequency
ρ	Charge density, resistivity
$\vec{\rho}$	Relative coordinate of the electron and hole
ρ_{opt}	Optical density of states
τ	Relaxation time
Φ	Potential energy
$\phi(\vec{r},t)$	Time dependent wave function
$\phi(x)$	Probability wave amplitude
χ	Electron affinity
$\psi(\vec{r})$	Wave function of the stationary state
ω	Angular frequency
ω_c	Cyclotron frequency
ω_p	Plasma oscillation frequency
$<\ldots>$	Average

Index